Automation and Human Performance: Theory and Applications

HUMAN FACTORS IN TRANSPORTATION

A Series of Volumes Edited by
Barry H. Kantowitz

Automation and Human Performance: Theory and Applications

Edited by

Raja Parasuraman
Catholic University of America

Mustapha Mouloua
University of Central Florida, Orlando

LEA
1996

Lawrence Erlbaum Associates, Publishers
Mahwah, New Jersey

Lawrence Erlbaum Associates, Inc., Publishers
10 Industrial Avenue
Mahwah, New Jersey 07430

Library of Congress Cataloging-in-Publication Data

Automation and human performance : theory and applications / Raja
    Parasuraman, Mustapha Mouloua, editors.
        p.   cm. -- (Series on human factors in transportation)
    Includes bibliographical references and indexes.
    ISBN 0-8058-1616-X (cloth : alk. paper)
    1. Man-machine systems. 2. Automation--Human factors.
I. Parasuraman, Raja. II. Mouloua, Mustapha. III. Series.
TA167.A96 1996
620.8'2--dc20   96-10760
                                                        CIP

Books published by Lawrence Erlbaum Associates are printed
on acid-free paper, and their bindings are chosen
for strength and durability.

Printed in the United States of America

10 9 8 7 6 5 4 3 2 1

Contents

Series Foreword

Barry H. Kantowitz
Battelle Human Factors Transportation Center
Seattle, Washington

The domain of transportation is important for both practical and theoretical reasons. All of us are users of transportation systems as operators, passengers, and consumers. From a scientific viewpoint, the transportation domain offers an opportunity to create and test sophisticated models of human behavior and cognition. This series covers both practical and theoretical aspects of human factors in transportation, with an emphasis on their interaction.

The series is intended as a forum for researchers and engineers interested in how people function within transportation systems. All modes of transportation are relevant, and all human factors and ergonomic efforts that have explicit implications for transportation systems fall within the series purview. Analytic efforts are important to link theory and data. The level of analysis can be as small as one person, or international in scope. Empirical data can be from a broad range of methodologies, including laboratory research, simulator studies, test tracks, operational tests, field work, design reviews, or surveys. This broad scope is intended to maximize the utility of the series for readers with diverse backgrounds.

I expect the series to be useful for professionals in the disciplines of human factors, ergonomics, transportation engineering, experimental psychology, cognitive science, sociology, and safety engineering. It is intended to appeal to the transportation specialist in industry, government, or academia, as well as the researcher in need of a testbed for new ideas about the interface between people and complex systems.

The present book is especially appropriate to launch this series because of the outstanding job of integrating theoretical and practical aspects of transportation human factors performed by the editors. I have always believed that the best practical tool is a good theory. The editors have honored this maxim by starting in Part I with four helpful theoretical perspectives. Part II explores human performance in the context of theory while Part III emphasizes practical applications. Part IV concludes with interesting speculations about the future that meld the theoretical and the practical. Forthcoming books in the series will continue this blend of practical and theoretical perspectives.

Foreword

Earl L. Wiener
University of Miami

"The question is no longer whether one or another function can be automated, but, rather, whether it should be."
(Wiener & Curry, 1980)

In 1979 Renwick Curry and I, at the request of NASA, embarked on a project to determine the influence of cockpit automation on flight safety. We found what we later called "promises and problems," good news and bad news. One of the questions we asked was whether cockpit automation had reached or passed a point of optimality, and was possibly doing more harm than good. Now, 15 years later, we still do not have an answer to that question.

The miracle of automated flight that the industry has witnessed over the last two decades was enabled by a device so tiny that you could hold hundreds of them in the palm of your hand. The development of the microprocessor chip ushered in a new generation of computer-based automatic devices in transport aircraft, and now in air traffic management. Automation has brought an era of fuel-efficient and potentially airspace-efficient flight. But certain incidents and accidents in recent years have led many to question the designers' claims for safety and even workload reduction in automatic flight. This is not due to shortcomings in the equipment itself, but to problems at the human interface. The equipment is highly reliable—as pilots like to say, it works "as advertised."

Still the problems persist. In early 1995, following a string of incidents and disastrous crashes of the most modern airliners, *Aviation Week and Space Technology* was moved to run a two-part series on human-computer interaction in the modern cockpit (Hughes and Dornheim, 1995). Emblazoned on the cover of the first issue was one of the questions that Curry and I had asked 15 years earlier, "Automated cockpits: who's in charge?"

Designers and operators of the modern equipment may have underestimated the changes brought by automation in the style of flying, the increase in mental workload, and the demands on the crews to monitor the equipment and the status of flight. Terms such as "situational awareness" suddenly appeared, and rapidly became part of the lexicon of human factors. The popular but elusive term "complacency" was seen as part of the automation problem. Authors began to demand "human-centered" automation. It seems ironic that after a half-century of human factors engineering, it was necessary to call for human-centered (rather than technology-centered) design.

In many ways the situation in aviation today resembles that which was first encountered in the World War II era: highly capable machines outstripping the ability of the human operators to manage them. The problem was attacked from every angle by the emerging field of human factors engineering, and was brought under control. By the time the jet airliners were introduced in the early '60s, cockpit human engineering was ready for the challenge.

With the microprocessor revolution upon us, we again find ourselves in the same position: an extremely sophisticated technology, which in many applications has not exploited its full potential, due to problems at the human interface. I once stated that when the history of cockpit design is written, the present age will be called the "era of clumsy automation."

As this volume indicates, the problem is ubiquitous. Aviation does not have exclusive rights to human factors problems, though it has generally been the case that other high-risk industries look in that direction for technological guidance. The papers assembled here paint a panorama of technology that has raced ahead of our understanding. They offer not despair, but directions for solution. Parasuraman and Mouloua are to be commended for creating a book with an understanding of the present and a vision for the future. The products of "the era of clumsy automation" will ultimately suffer the same fate as the World War II systems, largely at the hands of the authors of these chapters, their students and colleagues. History will describe the next generation of systems as "the era of effective, symbiotic, supportive automation." Operator complacency will dwell no more, and there will be a ringing reply to the question, "Who's in charge here?" The human operator, who else?

Finally, I reflect that when I was a graduate student, more years ago than I like to admit, something called "the systems approach" had emerged from post-war systems engineering, operations research, and human factors. Every author, in every paper it seemed, found it necessary, almost as if driven by some scientific or political doctrine, to state that what was needed was a systems approach (in contrast to a component-wise approach) to problems.

Today is it difficult to understand why this needed to be said at all, in fact if one were to write such a thing, it would be an immediate target for a reviewing editor's blue pencil. Why expound on the obvious, the editor would ask?

Let us hope that the time will not be distant before the same can be said of "human-centered automation."

Hughes D., and Dornheim, M. A. (Eds.) A series of articles by various authors on pilot-computer interfaces. *Aviation Week and Space Technology,* January 30 and February 6, 1995.

Wiener, E. I., and Curry, R. E. (1980). Flight-deck automation: Promises and problems. *Ergonomics, 23,* 995–1011.

Preface

There is perhaps no facet of modern society in which the influence of automation technology has not been felt. Whether at work or in the home, while traveling or while engaged in leisurely pursuits, human beings are becoming increasingly accustomed to using and interacting with sophisticated computer systems designed to assist them in their activities. Among these are diagnostic aids for physicians, automated teller machines for bank customers, flight management systems for pilots, navigational displays for drivers, and so on. As we approach the 21st century, it is clear that even more powerful computer tools will be developed, and that many of the more complex human–machine systems will not function without the added capabilities that these tools provide.

The benefits that have been reaped from this technological revolution have been many. At the same time, however, automation has not always worked as advertised; thus there is a real concern that the problems raised by automation could outweigh the benefits. Whatever the merits of any particular automation technology, however, it is clear that automation does not merely supplant human activity but also transforms the nature of human work. Understanding the characteristics of this transformation is vital for successful design of new automated systems. In general, the implementation of complex, "intelligent" automated devices in such domains as aviation, manufacturing, medicine, surface transportation, shipping, and nuclear power has brought in its wake significant challenges for human factors, cognitive science, and systems engineering.

The influence of automation technology on human performance has often been investigated in a fragmentary, isolated manner, with investigators conducting disconnected studies in different domains. Independent workshops and conferences have been held, for example, on automation in air traffic control or automation in anesthesiology. There has been little contact between these endeavors, although the principles gleaned from one domain may well have implications for the other. Also, with a few exceptions, the research has tended to be empirical and only weakly theory driven. In recent years, however, various groups of investigators have begun to examine human performance in automated systems in general, and have begun to develop theories of human interaction with automation technology. Our goal in this book is to present these theories and to assess the impact of automation on different aspects of human performance. The contributors to this volume were asked to examine automation and human performance from the dual perspective of theory and multiple domains of application. By presenting both basic and applied research, we hope to highlight the general principles of human–computer interaction in several domains in which automation technologies are widely implemented. The major premise of this approach is that a broad-based, theory-driven approach will have significant implications for the effective design of automation technology in specific work environments.

We have divided the book into four parts. Part I covers broad theoretical perspectives and concepts in automation research. The opening chapter by Woods provides a general theoretical framework for "decomposing" the complexity of human–automation interaction and poses the research challenges that must be met in the future. In chapter 2, Riley outlines an empirically supported theory of automation usage patterns by human operators. The theoretical and empirical bases of an alternative approach to implementation of automation, *adaptive* automation, are described by Scerbo in chapter 3. Flach and Bennett in chapter 4 present an original theoretical framework, based on the concept of *representation,* for the design of user interfaces to automated systems.

Part II consists of eight chapters devoted to assessing the impact of automation on different aspects of human performance. The domains of human performance covered include: monitoring (Parasuraman, Mouloua, Molloy, & Hilburn, chap. 5); mental workload (Kantowitz & Campbell, chap. 6; Kramer, Trejo, & Humphrey, chap. 7); situational awareness (Endsley, chap. 8); vigilance (Warm, Dember, & Hancock, chap. 9); decision making (Mosier & Skitka, chap. 10); and supervisory control (Coury & Semmel, chap. 11). The final chapter in this section, by Bowers, Oser, Salas, and Cannon-Bowers, discusses aspects of team performance in automated systems.

The eight chapters in Part III discuss issues related to human performance in different domains in which automation technologies have been introduced. Historically, air travel is one area in which many automation innovations were first introduced. Aviation is also the area where much human factors research on automation has been carried out. Accordingly, the first three chapters in Part III are devoted to aviation automation—both airborne, in the cockpit (Sarter, chap. 13; Rogers, Schutte, & Latorella, chap. 14), and on the ground, in air traffic management (Hopkin & Wise, chap. 15). Because the aim of this book was to examine human–automation interaction across domains of application, several systems other than those found in aviation are also discussed. The next two chapters describe automation in different modes of transportation: motor vehicles on the road (Hancock, Parasuraman, & Byrne, chap. 16), and maritime operations (Lee & Sanquist, chap. 17). The remaining three contributions discuss automation in medical systems (Guerlain & Smith, chap. 18), quality control and maintenance (Drury, chap. 19), and oil and gas pipeline operations (Meshkati, chap. 20).

Part IV consists of two chapters that look to the future. In chapter 21, Sheridan speculates on the future relationship between humans and automation. Finally, in chapter 22, Hancock also discusses this relationship in the context of his theme of understanding the "teleology," or grand purpose in design, of automation technology.

Many of the chapters in this book derive from papers presented at the First Automation Technology and Human Performance Conference, held in Washington, D.C. on April 7–9, 1994. This conference was conceived and organized by us, in collaboration with the following members of the Cognitive Science Laboratory, Catholic University of America: Charles A. Adams, Evan A. Byrne, Pamela Greenwood, David Hardy, Brian Hilburn, Robert Molloy, and Sangeeta Panicker. The assistance of the following members of the Scientific Committee is also gratefully acknowledged: Peter Hancock (University of Minnesota), Harold Hawkins (Office of Naval Research), James Howard, Jr. (Catholic University of America), Alan Pope (NASA Langley Research Center), Indramani L. Singh (Banaras Hindu University), and Joel S. Warm (University of Cincinnati). The conference received financial support from the following institutions and agencies: NASA Langley Research Center, Hampton, Virginia; Office of Naval Research, Arlington, Virginia; and the School of Arts and Sciences, The Catholic University of America. Additional support for preparation of this volume was received from NASA Langley Research Center, the National Institute on Aging, and the Office of Naval Research. Finally, we express our gratitude to the following persons for their assistance in the preparation of this book: William J. Bramble Jr., Guillermo Navarro, Carolyn Inzana, Eric Gruber, and Jacqueline A. Duley.

Raja Parasuraman
Washington, DC

Mustapha Mouloua
Orlando, FL

Contributors

Kevin B. Bennett   309 Oleman Hall, Psychology Department, Wright State University, Dayton, OH 45419.

Clint A. Bowers   Team Performance Lab, Department of Psychology, University of Central Florida, Orlando, FL 32816-1390.

Evan A. Byrne   Cognitive Science Laboratory, The Catholic University of America, Washington, DC 20064.

John L. Campbell   Battelle HF Transportation Center, 4000 NE 41st Street, Seattle, WA 98105.

Janis Cannon-Bowers   US Naval Air Warfare Center-TSD, 12350 Research Parkway, Code 4961, Orlando, FL 32826-3224.

Bruce G. Coury   Johns Hopkins University, Applied Physics Laboratory, RSI/Rm. 2-246, Laurel, MD 20723-6099.

William N. Dember   Department of Psychology, ML 376, University of Cincinnati, Cincinnati, OH 45221-0376.

Colin G. Drury   State University of New York at Buffalo, Department of Industrial Engineering, 342 Belt Hall, Buffalo, NY 14260.

Mica R. Endsley   Texas Technical University, Department of Industrial Engineering, Lubbock, TX 79409.

John M. Flach   Wright State University, 309 Oelman Hall, Psychology Dept., Dayton, OH 45435.

Stephanie A. E. Guerlain   210 Baker Systems, Ohio State University, 1971 Neil Avenue, Columbus, OH 43210.

P. A. Hancock   HF Research Lab, University of Minnesota, 141 Mariucci Arena-Ops., 1901 Fourth Street SE, Minneapolis, MN 55455.

Brian G. Hilburn   National Aerospace Lab, Anthony Fokkerweg, Amsterdam, Netherlands.

V. David Hopkin   Human Factors Consultant, Peatmoor, 78 Crookham Road, Fleet, Hants, GU13 0SA, Great Britain.

Darryl G. Humphrey   Department of Psychology, Wichita State University, 1845 Fairmont Street, Wichita, KS 67208.

Barry H. Kantowitz   Battelle HF Transportation Center, P.O. Box C-5395, Seattle, WA 98105.

Arthur F. Kramer   University of Illinois, 603 E. Daniel Street, Psychology Department, Champaign, IL 61820.

Kara A. Latorella   NASA Langley Research Center, Crew/Vehicle Integration Branch, M/S 152, Hampton, VA 23681-0001.

John D. Lee   Battelle HF Transportation Center, P.O. Box 5395, 4000 NE 41st Street, Seattle, WA 98115.

Najmedin Meshkati   Institute of Safety and Systems Management, University of Southern California, Los Angeles, CA 90089-0021.

Robert Molloy   Cognitive Science Laboratory, Catholic University of America, Washington, DC 20064.

Kathleen Mosier   NASA Ames Research Center, MS 262-4, Moffett Field, CA 94035.

Mustapha Mouloua   Center for Applied Human Factors in Aviation, University of Central Florida, Orlando, FL 32816-1780.

Jodi Heintz Obradovich   Cognitive Systems Engineering Laboratory, Department of Industrial, Welding and Systems Engineering, The Ohio State University, 210 Baker Systems, 1971 Neil Avenue, Columbus, OH 43210.

Randall J. Oser   US Naval Air Warfare Center-TSD, 12350 Research Parkway, Code 4961, Orlando, FL 32826-3224.

Raja Parasuraman   Cognitive Science Laboratory, Catholic University of America, Washington, DC 20064.

Victor Riley   Honeywell, Incorporated, HTC, 3660 Technology Drive, MN65-2600, Minneapolis, MN 55418.

William H. Rogers   NASA Langley Research Center, Crew/Vehicle Integration Branch, M/S 152, Hampton, VA 23681-0001.

Sally Rudmann   School of Allied Medical Professionals, The Ohio State University, 535 School of Allied Medicine, 1583 Perry Street, Columbus, OH 43210.

Eduardo Salas   US Naval Air Warfare Center-TSD, 12350 Research Parkway, Code 4961, Orlando, FL 32826-3224.

Thomas F. Sanquist   Battelle Human Affairs Center, 4000 NE 41st Street, Seattle, WA 98105.

Nadine B. Sarter   Ohio State University, 210 Baker Systems Building, 1971 Neil Avenue, Columbus, OH 43210-1210.

Ralph D. Semmel   The Johns Hopkins University, Applied Physics Laboratory, RSI/Room 2-238, Laurel, MD 20723-6099.

Mark W. Scerbo   Old Dominion University, Psychology Department, Norfolk, VA 23529.

Paul C. Schutte   NASA Langley Research Center, Crew/Vehicle Integration Branch, M/S 152, Hampton, VA 23681-0001.

Thomas B. Sheridan   Massachusetts Institute of Technology, 77 Massachusetts Avenue #3-346, Cambridge, MA 02139.

Linda J. Skitka   Department of Psychology, M/C 285, University of Illinois at Chicago, Chicago, IL 60680.

Jack W. Smith   Laboratory for Knowledge Based Medical Systems, Department of Computer Science, The Ohio State University, 395 Dreese Lab, 2015 Neil Avenue, Columbus, OH 43210.

Philip J. Smith   Cognitive Systems Engineering Laboratory, Department of Industrial, Welding and Systems Engineering, The Ohio State University, 210 Baker Systems, 1971 Neil Avenue, Columbus, OH 43210.

Patricia Strohm   Western Reserve Care System, Northside Medical Center Blood Bank, 500 Gypsy Lane, Youngstown, OH 44501.

Leonard J. Trejo   Department of Psychology, University of Illinois, 603 East Daniel St., Champaign, IL 61820.

Joel S. Warm   University of Cincinnati, Department of Psychology, Cincinnati, OH 45221.

John A. Wise   Center for Aviation/Aerospace Research, Embry-Riddle Aeronautical University, 600 S. Clyde Morris Boulevard, Daytona Beach, FL 32114-5416.

David D. Woods   Ohio State University, 1971 Neil Avenue, I & SE Department, Columbus, OH 43210.

I Theories and Major Concepts

1 Decomposing Automation: Apparent Simplicity, Real Complexity

David D. Woods
The Ohio State University

INTRODUCTION

We usually focus on the perceived benefits of new automated or computerized devices. Our fascination with the possibilities afforded by technology in general often obscures the fact that new computerized and automated devices also create new burdens and complexities for the individuals and teams of practitioners responsible for operating, troubleshooting, and managing high-consequence systems. First, the demands may involve new or changed tasks such as device setup and initialization, configuration control, or operating sequences. Second, cognitive demands change as well, creating new interface management tasks, new attentional demands, the need to track automated device state and performance, new communication or coordination tasks, and new knowledge requirements. Third, the role of people in the system changes as new technology is introduced. Practitioners may function more as supervisory controllers, monitoring and instructing lower-order automated systems. New forms of cooperation and coordination emerge when automated systems are capable of independent action. Fourth, new technology links together different parts that were formerly less connected. As more data flows into some parts of a system, the result is often data overload. Coupling a more extensive system more tightly together can produce new patterns of system failure. As technology change occurs we must not forget that the price of new benefits is often a significant increase in operational complexity. Fifth, the reverberations of technology change, especially the new burdens and complexities, are often underappreciated by the advocates of technology change. But their consequences determine when, where, and how technology change will succeed.

My colleagues and I have been studying the impact of technology change on practitioners—those people who do cognitive work to monitor, diagnoses and manage complex systems—pilots, anesthesiologists, process plant operators, space flight controllers (e.g., Woods, Johanessen, Cook, & Sarter, 1994). In these investigations we have seen that technology change produces a complex set of effects. In other words, automation is a wrapped package—a package that consists of changes on many different dimensions bundled together as a hardware/software system. When new automated systems are introduced into a field of practice, change is precipitated along multiple dimensions. In this chapter I examine the reverberations that technology change produces along several different dimensions:

- Automation seen as more autonomous machine *agents*. Introducing automated and intelligent agents into a larger system in effect changes the team composition. It changes how human supervisors *coordinate* their activities with those of the machine agents. Miscommunications and poor feedback about the activities of automated subsystems have been part of accident scenarios in highly automated domains.
- Automation seen as an increase in *flexibility*. As system developers, we can provide users with high degrees of flexibility through multiple options and modes. We also have the ability to place multiple virtual devices on one physical platform so that a single device will be used in many contexts that can differ substantially. But do these flexibilities create new burdens on practitioners, burdens that can lead to predictable forms of error?
- Automation seen as more *computerization*. Technology change often means that people shift to multifunction computer-based interfaces as the means for acquiring information and utilizing new resources. Poor design of the computer interface can force users to devote cognitive resources to the interface itself, asking questions such as: Where is the data I want? What does the interface allow me to do? How do I navigate to that display? What set of instructions will get the computer to understand my intention? Successful computer interfaces (e.g., visualization, direct manipulation) help users focus on their task without cognitive resources (attention, knowledge, workload) being devoted to the interface per se.
- Automation seen as an increase in *coupling* across diverse parts and agents of a system. Tighter coupling between parts propagates effects throughout the system more rapidly. This can produce efficiency benefits by reducing transfer costs, but it also means that problems have greater and more complex effects, effects that can propagate

quickly. But when automated partners are strong, silent, clumsy, and difficult to direct, then handling these demands becomes more difficult. The result is coordination failures and new forms of system failure.

• Much technology change is justified, at least in part, based on *claims* about the impact of technology on human performance — the new system will "reduce workload," "help practitioners focus on the important part of the job," "decrease errors," and so on. But these claims often go unexamined. A number of studies have examined the impact of automation on the cognition and behavior of human practitioners. These studies, many of which are discussed in other chapters of this book, have shown repeatedly that systems introduced to aid practitioners in fact created new complexities and new types of error traps.

• The success of new technology depends on how it affects the people in the field of practice. The dimensions addressed earlier represent some of the ways that technology change can have surprising impacts on human and system performance. By closely examining the reverberations of technology change we can better steer the possibilities of new technology into fruitful directions.

HOW TO MAKE AUTOMATED SYSTEMS TEAM PLAYERS

Heuristic and algorithmic technologies expand the range of subtasks and cognitive activities that can be automated. Automated resources can, in principle, offload practitioner tasks. Computerized systems can be developed that assess or diagnose the situation at hand, alerting practitioners to various concerns and advising practitioners on possible responses.

Our image of these new machine capabilities is that of a machine alone, rapt in thought and action. But the reality is that automated subtasks exist in a larger context of interconnected tasks and multiple actors. Introducing automated and intelligent agents into a larger system changes the composition of the distributed system of monitors and managers and shifts the human's role within that cooperative ensemble (Hutchins, 1994). In effect, these "intelligent" machines create joint cognitive systems that distribute cognitive work across multiple agents (Hutchins, 1990; Roth, Bennett, & Woods, 1987). It seems paradoxical, but studies of the impact of automation reveal that design of automated systems is really the design of a new human–machine cooperative system. The design of automated systems is really the design of a team and requires provisions for the coordination between machine agents and human practitioners (e.g., Layton, Smith, & McCoy, 1994).

However, research on human interaction with automation in many domains, including aviation and anesthesiology, has shown that automated systems often fail to function as team players (Billings, 1991; Malin et al., 1991; Sarter & Woods, 1994b). To summarize the data, automated systems that are strong, silent, clumsy, and difficult to direct are not team players. Automated systems are:

Strong when they can act autonomously.

Silent when they provide poor feedback about their activities and intentions.

Clumsy when they interrupt their human partners during high workload or high criticality periods, or when they add new mental burdens during these high-tempo periods.

Difficult to direct when it is costly for the human supervisor to instruct the automation about how to change as circumstances change.

Systems with these characteristics create new problems for their human partners and new forms of system failure.

"Strong" automation refers to two properties of machine agents. In simpler devices, each system activity was dependent on immediate operator input. As the power of automated systems increases, machine agents, once they are instructed and activated, are capable of carrying out long sequences of tasks without further user interventions. In other words, automated systems can differ in *degree of autonomy* (Woods, 1993). Automated systems also can differ in *degree of authority*. This means that the automated system is capable of taking over control of the monitored process from another agent if it decides that intervention is warranted based on its perception of the situation and its internal criteria (Sarter & Woods, 1994a).

Increasing autonomy and authority create new monitoring and coordination demands for humans in the system (Norman, 1990; Sarter & Woods, 1995; Wiener, 1989). Human supervisors have to keep track of the status and activities of their automated partners. For example, consider the diagnostic situation in a multi-agent environment, when one notices an anomaly in a process being monitored (Woods, 1994). Is the anomaly an indication of an underlying fault, or does the anomaly indicate some activity by another agent in the system, unexpected by this monitor? In fact, in a number of different settings, we observe that human practitioners respond to anomalies by first checking for what other agents have been or are doing to the process jointly managed (Johannesen, Cook, & Woods, 1994).

When machine agents have high autonomy, they will act in the absence of immediate user input. Human practitioners have to *anticipate* how the automated system will behave as circumstances change. Depending on the complexity of the system and the feedback about system activities, this may be difficult. As one commentator has put it, the most common questions people ask about their automated partners are: What is it doing? Why is it doing that? What will it do next? How in the world did we get into that mode? (Wiener, 1989). These questions are indications of coordination breakdowns—what has been termed "automation surprises." Automation surprises are situations where automated systems act in some way outside of the expectations of their human supervisors. Data from studies of these surprises in aviation and medicine (Moll van Charante, Cook, Woods, Yue, & Howie, 1993; Norman, 1990; Sarter & Woods, 1994b) indicate that poor feedback about the activities of automated systems to their human partners is an important contributor to these problems.

Autonomy and authority are properties that convey an agentlike status on the system from the point of view of human observers. This raises an important point. Automated systems have two kinds of interpretations. Based on knowledge of underlying mechanisms, an automated system is deterministic and predictable. However, those who monitor or interact with the system in context may perceive the system very differently. For example, with the benefit of knowledge of outcome and no time pressure, one can retrospectively show how a system's behavior was deterministic. But as system complexity increases, and depending on the feedback mechanisms available, predicting the system's behavior in context may be much more difficult.

A user's perception of the device depends on an interaction between its capabilities and the feedback mechanisms that influence what is observable about system behavior in relation to events in the environment. What feedback is available depends on the "image" the device presents to users (Norman, 1988). When a device is complex, has high autonomy and authority, and provides weak feedback about its activities (what has been termed "low observability"), it can create the image of an animate agent capable of independent perception and willful action. We refer to this as the *perceived animacy* of the automated system. In effect, the system, although determinate from one perspective, seems to behave as if it were an animate agent capable of activities independent of the operator (Sarter & Woods, 1994a).

Flightdeck automation on commercial transport jets illustrates how autonomy combined with low observability can create the perception of animacy (Sarter & Woods, 1994b). Pilots sometimes experience difficulties with tracking system behavior in situations that involve indirect mode transitions. In these situations, the system changes its behavior independent of any immediate pilot instructions. The system acts in response to reaching a preset target (e.g., leveling off at a target altitude) or because an envelope protection threshold is crossed. In other words, based on the programmed mechanisms, the system "realizes" the need for a mode change, carries it out without requesting pilot consent, and provides only weak feedback about the change or the implications of the change for future aircraft behavior. It is in this type of situation that pilots are known to ask questions such as: What is it doing? Why did it do this? What will it do next? (Wiener, 1989). These are questions one asks about another agent with an agenda of its own and an agent that does not communicate very well.

Much work in this area has noted that poor feedback on system status and behavior is at the heart of automation surprises. But what does it mean to say "poor feedback?" When we take a close look at the data provided to the operators of many advanced systems, it becomes quite clear that the amount of data available to the human is increasing. All of the necessary data to build a picture of their automated partner's activities is present in general. But the effectiveness of this data depends on the cognitive work needed to turn it into a coherent interpretation in context.

Effective feedback depends on more than display formats; it is a relation among the system's function, the image the system presents to outsiders, and the observer embedded in an evolving context (Woods, 1995). As a result, it is better to refer to interface and feedback issues in terms of *observability*. This term captures the fundamental relationship among thing observed, observer, and context of observation that is fundamental to effective feedback. Observability depends on the cognitive work needed to extract meaning from the data available. We, as researchers, need to make progress on better ways to measure this property of cognitive systems.

Because automated systems are deterministic, if one has complete knowledge of how a system works, complete recall of the past instructions given to the system, and total awareness of environmental conditions, then one can project accurately the behavior of the automated partner. However, as the system becomes more complex, projecting its behavior also becomes more cognitively challenging. One has to have an accurate model of how the system works, one has to call to mind the portions of this knowledge that are relevant for the current situation, one has to recall past instructions that may have occurred some time ago and may have been provided by someone else, one has to be aware of the current and projected state of various parameters that are inputs to the automation, one has to monitor the activities of the automated system, and one has to integrate all of this information and knowledge together in order to project how the automation will behave in the future. As a result, an automated system can look very different from the perspective of a user in context as compared to an analyst taking a bird's-eye view with knowledge of outcome. The latter will see how the system's behavior was a direct and natural result of previous instructions and current state; the former will see a system that appears to do surprising things on its own. This is the paradox of perceived animacy of automated systems that have high autonomy and authority but low observability (Fig. 1.1). This situation has strong implications for error analysis and incident reconstruction (Woods et al., 1994).

The trend in automation seems to be for greater increases in system autonomy and authority, whereas feedback mechanisms are stagnant at best. The result appears to be that "strong and silent" automation is on the increase (Norman, 1990). Yet the research to date has revealed that there are latent dangers of powerful yet silent automation (e.g., Cook, Woods, & Howie, 1992).

Designing automated systems is more than getting that machine to function autonomously, it is also making provisions for that automated agent to coordinate its activity with other agents. Or, perhaps more realistically, it is making provisions so that other human agents can see the assessments and activity of the automated agent so that these human practitioners can perform the coordination function by managing a set of partially autonomous subordinate agents (see Billings, 1991; Woods et al., 1994).

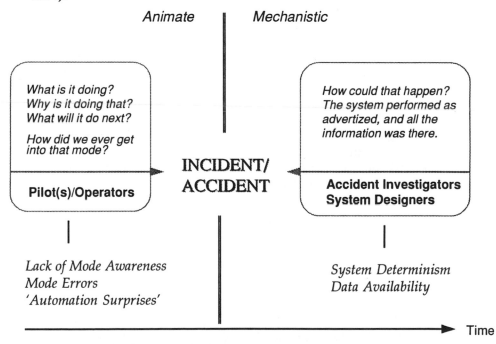

FIG. 1.1. A paradox associated with perceived animacy. Automated systems that have high autonomy and authority but low observability appear to behave as if they are animate agents capable of activities independent of the operator. However, such systems are deterministic, and their behavior is predictable if one has complete and available knowledge of how the system works, complete recall of the past instructions given to the system, and total awareness of the situation and environmental conditions.

FLEXIBILITY: BURDENSOME OR INSTRUMENTAL?

Flexibility and customizability are central to the perceived advantages of the growth in technological powers (Woods, 1993). New automated systems are often flexible in the sense that they provide a large number of functions and options for carrying out a given task under different circumstances. For example, the computers on commercial jet flightdecks provide at least five different mechanisms at different levels of automation for changing altitude. This customizability is construed normally as a benefit that allows practitioners to select the mode or option best suited to a particular situation. However, it also creates a variety of new demands.

To utilize highly flexible systems, the practitioner must learn about all of the available options, learn and remember how to deploy them across the variety of real operational contexts that can occur, and learn and remember the interface manipulations required to invoke different modes or features. Monitoring and attentional demands are also created as practitioners must keep track of which mode is active. All of this represents new burdens on the practitioner to set up and manage these capabilities and features.

If the new tasks and workload created by such flexible systems tend to congregate at high-workload and high-criticality periods, the result is a syndrome called *clumsy automation* by Wiener (1989). Clumsy automation is a form of poor coordination between the human and machine in the control of dynamic processes where the benefits of the new technology accrue during workload troughs and the costs or burdens imposed by the technology (i.e., additional tasks, new knowledge, forcing the user to adopt new cognitive strategies, new communication burdens, new attentional demands) occur during periods of peak workload, high-criticality, or high-tempo operations (Cook & Woods, 1995; Sarter & Woods, 1994b). Significantly, deficits like this can create opportunities for new kinds of human error and new paths to system breakdown that did not exist in simpler systems (Woods et al., 1994).

The result is that we need to understand the difference between two types of flexibility in cognitive artifacts: (a) flexibilities that serve to increase practitioners' range of adaptive response to the variability resident in the field of activity, and (b) flexibilities that simply create new burdens on practitioners, especially at high-tempo or high-criticality periods (Woods, 1993).

PROPERTIES OF THE COMPUTER SHAPE PRACTITIONER COGNITION AND BEHAVIOR

Today, technological change is transforming the workplace through the introduction and spread of new computer-based systems. Thus, automation can be seen in part as computerization. But there are a variety of properties of the computer as a medium that shape practitioner cognition and behavior in predictable but problematic ways.

Computer-based information technology allows designers to combine multiple features, options, and functions onto a single physical platform. The same physical device can be designed to operate in many different contexts, niches, and markets simply by taking the union of all the features, options, and functions that are needed in any of these settings. In a sense, the computer medium allows one to create multiple virtual devices concatenated onto a single physical device. After all, the computer medium is multifunction — software can make the same keys do different things in different combinations or modes, or provide soft keys, or add new options to a menu structure; the CRT or other visual display unit (VDU) allows one to add new displays that can be selected if needed to appear on the same physical viewport.

But to do this pushes the designer to proliferate modes and thereby create the potential for mode errors, to proliferate displays hidden behind the narrow viewport and create navigation problems, to assign multiple functions to controls so that users must remember complex and arbitrary sequences of operation. In other words, the modularity of the computer medium helps designers follow Norman's (1988) tongue-in-cheek advice on how to do things *wrong* in designing computer-based devices. Such systems appear on the surface to be simple because they lack large numbers of physical display devices and controls; however, underneath the placid surface of the CRT workstation a variety of clumsy features may exist that produce operational complexities.

Computerization also has tremendously advanced our ability to collect, transmit, and transform data. In all areas of human endeavor, we are bombarded with computer-processed data. But our ability to digest and interpret data has failed to keep pace with our abilities to generate and manipulate greater and greater amounts of data. Thus, we are plagued by data overload. User interface technology has allowed us to concentrate this expanding field of data into one physical platform, typically a single visual display unit (VDU). Users are provided with increased degrees of flexibility for data handling and presentation in the computer interface through window management and different ways to display data. The technology provides the capability to generate tremendous networks of computer displays as a kind of virtual perceptual field viewable through the narrow aperture of the VDU. These changes effect the cognitive demands and processes associated with extracting meaning from large fields of data (Woods, 1995).

We have demonstrated in several studies how characteristics of computer-based devices influence cognition and behavior in ways that increase the potential for erroneous actions and assessments. In one case (Cook & Woods, in press), a new operating-room patient-monitoring system was studied in the context of cardiac anesthesia. This and other similar systems integrate what was previously a set of individual devices, each of which displayed and controlled a single sensor system, into a single CRT display with multiple windows and a large space of menu-based options for maneuvering in the space of possible displays, options, and special features. The study consisted of observing how the physicians learned to use the new technology as it entered the workplace.

By integrating a diverse set of data- and patient-monitoring functions into one computer-based information system, designers could offer users a great deal of customizability and options for the display of data. Several different windows could be called, depending on how the users preferred to see the data. However, these flexibilities all created the need for the physician to interact with the information system—the physicians had to direct attention to the display and menu system and recall knowledge about that system. Furthermore, the computer keyhole created new interface management tasks by forcing serial access to highly interrelated data and by creating the need to periodically declutter displays to avoid obscuring data channels that should be monitored for possible new events.

The problem occurs because of a fundamental relationship, the escalation principle: the greater the trouble in the underlying system or the higher the tempo of operations, the greater the information processing activities required to cope with the trouble or pace of activities (Woods et al., 1994). For example, demands for monitoring, attentional control, information, and communication among team members (including human–machine communication) all tend to go up with the tempo and criticality of operations. This means that the burden of interacting with the display system tends to be concentrated at the very times when the practitioner can least afford new tasks, new memory demands, or diversions of his or her attention away from patient state to the interface per se.

The physicians tailored both the system and their own cognitive strategies to cope with this bottleneck. In particular, they were observed to constrain the display of data into a fixed, spatially dedicated default organization rather than exploit device flexibility. They forced scheduling of device interaction to low-criticality self-paced periods to try to minimize any need for interaction at high-workload periods. They developed stereotypical routines to avoid getting lost in the network of display possibilities and complex menu structures.

These are all standard tactics people use to cope with complexities created by the clumsy use of computer technology (Woods et al., 1994). We have observed that pilots, space flight controllers, as well as physicians cope with new burdens associated with clumsy technology by learning only a subset of stereotypical methods, underutilizing system functionality. We have observed these practitioners convert interface flexibility into fixed, spatially dedicated displays to avoid interacting with the interface system during busy periods. We have observed these practitioners escape from flexible but complex modes of automation and switch to less automated, more direct means of accomplishing their tasks when the pace of operations increases. This adaptation or tailoring process occurs because practitioners are responsible not just for the device operation, but also for the larger performance goals of the overall system. As a result, practitioners tailor their activities to insulate the larger system from device deficiencies (Cook & Woods, in press; Woods et al., 1994).

HIGHLY COUPLED SYSTEMS BREAK DOWN IN NEW WAYS

The scale of and degree of coupling within complex systems creates a different pattern for disaster where incidents develop or evolve through a conjunction of several small failures, both machine and human (Perrow, 1984; Reason, 1990). There are multiple contributors; all are necessary, but they are individually insufficient to produce a disaster. Some of the multiple contributors are latent; that is, conditions in a system that produce a negative effect but whose consequences are not revealed or activated until some other enabling condition is met. This pattern of breakdown is unique to highly coupled systems and has been labeled the *latent failure model of complex system breakdown* (Reason, 1990).

Computerization and automation couple more closely together different parts of the system. Increasing the coupling within a system has many effects on the kinds of cognitive demands to be met by the operational system. And increasing the coupling within a system changes the kinds of system failures one expects to see (Perrow, 1984; Reason, 1990). The latent failure model for disaster is derived from data on failures in highly coupled systems.

Automation and computerization increase the degree of coupling among parts of a system. Some of this coupling is direct, some is based on potential failures of the automation, and some is based on the effects of automation on the cognitive activities of the practitioners responsible for managing the system. For example, higher coupling produces more side effects to failures. A failure is more likely to produce a cascade of disturbances that spreads throughout the monitored process. Symptoms of faults may appear in what seem to be unrelated parts of the process (effects at a distance). These and other effects can make fault management and diagnosis much more complicated.

Highly coupled processes create or exacerbate a variety of demands on cognitive functions (Woods, 1988). For example, increased coupling creates:

- New knowledge demands (e.g., knowing how different parts of the system interact physically or functionally).
- New attentional demands (e.g., deciding whether or not to interrupt ongoing activities and lines of reasoning as new signals occur).
- More opportunities for situations to arise with conflicts between different goals. New strategic trade-offs can arise as well. Creating or exacerbating conflicts and dilemmas produces new forms of system breakdown (see Woods et al., 1994).

Automation may occur in the service of stretching capacity limits within a system. But these efficiency pressures may very well create or exacerbate double binds that practitioners must face and resolve. These pressures may also reduce margins, especially by reducing the error tolerance of the system and practitioners' ability to recover from error and failures. Characteristics such as error tolerance and the degree of observability through the computer interface can change the ability of practitioners to buffer the system in the face of contingencies and complications (Woods et al., 1994).

Technology change often facilitates greater participation by formerly remote individuals. People, who represent different but interacting goals and constraints, now can interact more directly in the decision-making process. Coordination across these diverse people and representatives and cooperation among their interacting interests must be supported.

Overall, changes in automation, through increased coupling, make systems more vulnerable to the latent failure type of system breakdown where multiple contributors come together in surprising ways (see also Hollnagel, 1993; Woods et al., 1994). Thus, increases in level of automation can change the kinds of incidents, their frequency, and their consequences in ways that can be very difficult to foresee. Interestingly, the signature of failure in tightly coupled systems is often misperceived and labeled as simply another case of human error.

TECHNOLOGY CHANGE TRANSFORMS OPERATIONAL AND COGNITIVE SYSTEMS

These effects of technology change run counter to conventional wisdom about automation. There are two broad themes that run throughout the previous discussion.

First, changes in level of automation *transform* systems. Technology change is an intervention into an ongoing field of activity (Flores, Graves, Hartfield, & Winograd, 1988; Winograd & Flores, 1987). When developing and introducing new technology, one should realize that technology change represents new ways of doing things; it does not preserve the old ways with the simple substitution of one medium for another (e.g., paper for computer-based; hardwired for digital; automatic for manual).

Marketing forces tout the universal benefits of all types of new technology without reservations. However, the difference between skillful and clumsy use of technological powers lies in understanding how automation can transform systems. For example, where and when does it create new burdens? How does the keyhole property of the computer shape practitioner cognition in ways that reduce error and failure recovery? What are the new patterns of system breakdown? What is the new cooperative or joint human–machine system created by more automation? How does this cooperative system function when complicating factors arise at and beyond the margins of normal routines?

Understanding the potential transformations allows one to identify and treat the many postconditions necessary for skillful use of technological possibilities. To do this one must unwrap the automation package. In doing so we must come to recognize that new technology is more than object in itself. When we design new automated and computerized systems we are concerned with more than just a hardware and software object. We are also designing:

- A dynamic *visualization* of what is happening and what may happen next that provides practitioners with feedback about success and failure and about activities and their effects.
- A *tool* that helps practitioners respond adaptively to the many different circumstances and problems that can arise in their field of activity.
- A *team* of people and machine agents that can coordinate their assessments and activities as a situation escalates in tempo, difficulty, and criticality.

APPARENT SIMPLICITY, REAL COMPLEXITY

Conventional wisdom about automation makes technology change seem simple. Automation is just changing out one agent (a human) for another. Automation provides more options and methods. Automation frees up people for other more important jobs. Automation provides new computer graphics and interfaces. However, the reality of technology change, as revealed through close examination of device use in context, is that technological possibilities often are used clumsily, resulting in strong, silent, difficult-to-direct systems that are not team players.

The discussion of how technology change transforms systems points out the irony present in conventional claims about the effects of automation. The very characteristics of computer-based devices that have been shown empirically to complicate practitioners' cognitive activities and contribute to errors and failures are generally justified and marketed on the grounds that they reduce human workload and improve human performance. Technologists assume that automation will automatically reduce skill requirements, reduce training needs, produce greater attention to the job, and reduce errors.

New technology can be used skillfully to increase skilled practice and to produce more reliable human–machine systems, but not through wishful thinking or superficial claims about the impact of new technology on human–machine systems. Understanding or predicting the effects of technology change requires close examination of the cognitive factors at work in the operational system. Studying and modeling joint human–machine cognitive systems in context is the basis for skillful as opposed to clumsy use of the powers of technology (Woods et al., 1994).

ACKNOWLEDGMENTS

The preparation of this manuscript was supported, in part, under a Cooperative Research Agreement with NASA–Ames Research Center (NCC 2–592; Technical Monitor: Dr. Everett Palmer) and under a grant with the FAA Technical Center (Grant No. 93-G-036; Technical Monitor: John Zalenchak).

REFERENCES

Billings, C. E. (1991). *Human-centered aircraft automation philosophy* (NASA Technical Memorandum 103885). Moffett Field, CA: NASA–Ames Research Center.

Cook, R. I., & Woods, D. D. (1995). Adapting to new technology in the operating room. *Human Factors,* in press.

Cook, R. I., Woods, D. D., & Howie, M. B. (1992). Unintentional delivery of vasoactive drugs with an electromechanical infusion device. *Journal of Cardiothoracic and Vascular Anesthesia, 6,* 1-7.

Flores, F., Graves, M., Hartfield, B., & Winograd, T. (1988). Computer systems and the design of organizational interaction. *ACM Transactions on Office Information Systems, 6,* 153-172.

Hollnagel, E. (1993). *Human reliability analysis: Context and control.* London: Academic Press.

Hutchins, E. (1990). The technology of team navigation. In J. Galegher, R. Kraut, & C. Egido (Eds.), *Intellectual teamwork: Social and technical bases of cooperative work* (pp. 191-220). Hillsdale, NJ: Lawrence Erlbaum Associates.

Hutchins, E. (1994). *Cognition in the wild*. Cambridge, MA: MIT Press.

Johannesen, L., Cook, R. I., & Woods, D. D. (1994). Cooperative communications in dynamic fault management. In *Proceedings of the 38th Annual Meeting of the Human Factors and Ergonomics Society* (pp. 225–229). Santa Monica, CA: Human factors and Ergonomics Society.

Layton, C., Smith, P. J., & McCoy, E. (1994). Design of a cooperative problem-solving system for enroute flight planning: An empirical evaluation. *Human Factors, 36*(1), 94–119.

Malin, J., Schreckenghost, D., Woods, D., Potter, S., Johannesen, L., Holloway, M., & Forbus, K. (1991). *Making intelligent systems team players: Case studies and design issues* (NASA Technical Report, TM-104738). Houston, TX: NASA Johnson Space Center.

Moll van Charante, E., Cook, R. I., Woods, D. D., Yue, Y., & Howie, M. B. (1993). Human–computer interaction in context: Physician interaction with automated intravenous controllers in the heart room. In H. G. Stassen (Ed.), *Analysis, design and evaluation of man-machine systems 1992* (pp. 000–000). New York: Pergamon Press.

Norman, D.A. (1988). *The psychology of everyday things*. New York: Basic Books.

Norman, D.A. (1990). The "problem" of automation: Inappropriate feedback and interaction, not "over-automation." *Philosophical Transactions of the Royal Society of London, B 327*, 585–593.

Perrow, C. (1984). *Normal accidents. Living with high-risk technologies*. New York: Basic Books.

Reason, J. (1990). *Human error*. Cambridge, England: Cambridge University Press.

Roth, E. M., Bennett, K. B., & Woods, D. D. (1987). Human interaction with an "intelligent" machine. *International Journal of Man-Machine Studies, 27*, 479–525.

Sarter N. B., & Woods D. D. (1994a). Decomposing automation: Autonomy, authority, observability and perceived animacy. In M. Mouloua & R. Parasuraman (Eds.), *Human performance in automated systems: Current research and trends* (pp. 22–27). Hillsdale, NJ: Lawrence Erlbaum Associates.

Sarter N. B., & Woods D. D. (1994b). Pilot interaction with cockpit automation II: An experimental study of pilots' mental model and awareness of the Flight Management System (FMS). *International Journal of Aviation Psychology, 4*, 1–28.

Sarter, N. B., & Woods, D. D. (1995). "How in the world did we get into that mode?" Mode error and awareness in supervisory control. *Human Factors 37*(1), 5–19.

Wiener, E. L. (1989). *Human factors of advanced technology ("glass cockpit") transport aircraft* (Technical Report 117528). Moffett Field, CA: NASA–Ames Research Center.

Winograd, T., & Flores, F. (1987). *Understanding computers and cognition*. Reading, MA: Addison-Wesley.

Woods, D. D. (1988). Coping with complexity: The psychology of human behavior in complex systems. In L. P. Goodstein, H. B. Andersen, and S. E. Olsen (Eds.), *Tasks, errors, and mental models* (pp. 128–148). New York: Taylor and Francis.

Woods, D. D. (1993). Price of flexibility in intelligent interfaces. *Knowledge-Based Systems, 6*(4), 189–196.

Woods, D. D. (1994). Cognitive demands and activities in dynamic fault management: Abductive reasoning and disturbance management. In Stanton, N. (Ed.), *The human factors of alarm design* (pp. 63–92). London: Taylor and Francis.

Woods, D. D. (1995). Towards a theoretical base for representation design in the computer medium: Ecological perception and aiding human cognition. In J. Flach, P. Hancock, J. Caird, & K. Vicente (Eds.), *An ecological approach to human–machine systems I: A global perspective* (pp. 157–188). Hillsdale NJ: Lawrence Erlbaum Associates.

Woods, D. D., Johannesen, L. J., Cook, R. I., & Sarter, N. B. (1994). *Behind human error: Cognitive systems, computers, and hindsight*. Dayton, OH: Crew Systems Ergonomic Information and Analysis Center (CSERIAC).

2 Operator Reliance on Automation: Theory and Data

Victor Riley
Honeywell Technology Center

INTRODUCTION

On June 30, 1994, an Airbus A330 crashed during a test flight, killing all seven on board. The test was being performed to evaluate how well the aircraft's autopilot system performed with an engine out, simulated hydraulic failure, and rear center of gravity just after takeoff. According to the investigating committee, the crew appeared overconfident and did not intervene in time to prevent the accident. They calculated that if the test pilot had retaken manual control four seconds earlier than he actually did, the accident would have been avoided (Sparaco, 1994).

As illustrated by this accident, the decision to rely or not rely on automation can be one of the most important decisions an operator of a complex system can make. Indeed, the decision has been a critical link in the chains of events that have led to many incidents and accidents in aircraft, railroad, ship, process control, medical, and power plant operations. In these cases, the operators may rely too much on the automation and fail to check up on it or monitor its performance, or they may defeat the automation because they have high, possibly misplaced, confidence in their own abilities to perform the tasks manually. Several aircraft accidents illustrate the former problem; the Chernobyl nuclear power plant accident is one example of the latter.

Until recently, little has been known about what factors influence the decision to use or not use automation and what types of bias to which this decision may be subject. A better understanding of these factors and biases may help system developers anticipate the conditions under which operators may underrely or overrely on automation and guide the development of training methods and user interfaces to encourage rational automation use.

At a time when automation was being given increasing authority in complex systems, Sheridan and Farrell (1974) expressed concern about the changing roles of human operators and automation, and included operator trust in the automation as one of the fundamental elements of supervisory control tasks. Several investigators have since investigated human trust in automated systems. Muir (1987) performed a literature review on trust and developed a theory of trust in automation that combined two previous theories of trust between humans. Muir (1989) went on to perform two experiments using a process control simulation to demonstrate that operator trust in automation could be measured using a subjective scale, that operators could distinguish between faulty and reliable components of a system, and that a subjective measure of trust in a component correlated with the operator's use of that component.

Lerch and Prietula (1989) examined how attributions of qualities to agents affected operator trust in the agent. They also found evidence that it was harder to recover trust in an agent after a failure than to build trust in it initially, in support of one of the hypotheses proposed by Muir (1987). Will (1991) performed a study with petroleum engineers to determine the extent to which they relied on an expert system to perform well analysis. The expert system was intended to examine the well data and recommend a solution, but the experimenters had introduced a fault that would cause the system to make the wrong recommendation. Initially, all subjects used the system and expressed confidence in the result. One expert subject later redid the analysis by hand and discovered the error. The only significant difference between the expert and novice subjects was their expressed opinion of the utility of the system, with experts saying they would have done just as well without the system (without knowing that the answer they had reached was wrong).

Riley (1989) suggested that an operator's decision to rely on automation may not depend only on the operator's level of trust in the system, but rather on a more complex relationship among trust, self-confidence, and a number of other factors. The major thrust of this argument was that if the operator had more confidence in his or her own ability to do the task than trust in the automation, the operator was likely to do the task manually, whereas if the operator's trust in the automation was higher than the operator's self-confidence, the operator was likely to rely on the automation. However, this relationship was mediated by other factors, including the operator's level of workload and the level of risk associated with the situation. For example, a higher level of risk may exaggerate the effects of trust and self-confidence, or it might bias an operator toward manual control because of the operator's fiduciary responsibility for the process.

The theory is shown in Fig. 2.1, where arrows represent hypothesized influences between factors. The "reliance" factor represents the probability that an operator will use automation and is influenced by the operator's self-confidence and level of trust in the automation. Trust, in turn, is influenced by the actual reliability of the automation and a "duration" factor, which is meant to account for increasing stability of the operator's opinion of the automation as the operator gains experience with it.

Muir's (1989) results provide support for the proposed relationship among automation accuracy, trust in automation, and reliance. In addition, Lee (1992) (also Lee & Moray, 1992) performed an extensive series of studies to investigate these relationships and the relationship among trust, self-confidence, and automation use. Using an extension of the process control simulation used by Muir (1989), Lee provided evidence that the combination of trust in automation and self-confidence can influence automation use. He also found a very high level of variance due to individual differences in automation use and task performance strategy. However, the relationships proposed in Fig. 2.1 among workload, self-confidence, and automation use have not received as much support because little workload-related research has looked at the question of automation use. Parasuraman, Molloy, and Singh (1993) demonstrated that an element of workload is necessary for the development of automation related "complacency." Harris, and Arthur (1993) attempted to determine whether giving subjects advance notice of workload increases would prompt them to turn automation on as a workload management strategy, but were unable to

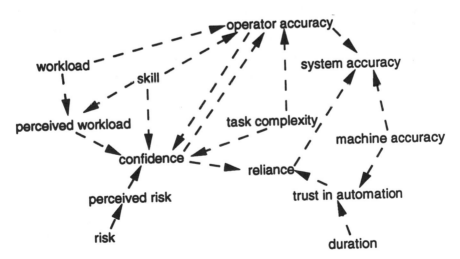

FIG. 2.1.   An initial theory of operator reliance on automation. Arrows indicate the hypothesized directions of influence.

demonstrate a significant effect, partly due to a high degree of between-subject variance. Taken with the previously discussed findings, this evidence provides support for some of the relationships proposed in Fig. 2.1 while leaving others in question. The rest of this chapter summarizes the results of a series of studies to investigate these and further relationships.

AN INVESTIGATION OF FACTORS
THAT INFLUENCE AUTOMATION RELIANCE
USING A SIMPLE COMPUTER GAME

The experiments summarized here made use of a simple computer-based testbed (Riley, 1994). The primary purpose of this testbed was to enable independent manipulations of subject workload, the difficulty or uncertainty associated with the task that could be automated, automation reliability, and the riskiness of decisions. The ability to investigate the independent effects of these factors was a critical requirement for the testbed. In real operations, such as flight and process control, it is often difficult or impossible to separate factors. For example, when the level of risk in a situation rises, workload and task uncertainty often rise at the same time. This makes it difficult to attribute automation use decisions observed in real operations to specific factors. Isolating and testing these factors in the laboratory might shed light on how they may operate in the real world.

Another objective was to minimize the possibility that subjects' automation use decisions would depend on, or be influenced by, their different strategies in performing the task. In complex systems, many different system management strategies can be taken, and automation use decisions may be a fundamental part of these strategies. This was illustrated by Lee (1992). By using a very simple task with few possible strategies, automation use differences due to subject differences can be separated from task strategy differences.

The testbed used two tasks, one of which could be turned over to automation at the subject's discretion. The first task required the subject to categorize a character as either a letter or a number. The level of uncertainty of this task could be controlled by introducing characters that were neither letter nor number at some known rate, and scoring these characters randomly as if they were one or the other. This task could be turned over to an automated "aid" that performed the classifications for the subject. A distraction task required the subject to correct random disturbances of a marker from a target location. Workload could be controlled by varying the probability that the marker would move in a given trial. This task could not be automated, and scoring on the categorization task was contingent on the placement of the marker over the target at the end of each trial. Each trial lasted 1.75 seconds and trials were forced paced: If the subject did not respond to a trial, it was counted as wrong and the next trial began on schedule. All experiments summarized here ran for 2,050 trials (about an hour). A depiction of the screen used in the experiments is shown in Fig. 2.2.

Automation reliability was controlled by setting the probability that the automation would make a correct classification. Pretesting showed that subjects were approximately 85% accurate in performing the task manually, so automation was set at 90% accuracy in three of the experiments for the case when it was working and 50% (chance performance) when it failed; this would make the automation's performance resemble that of a good human operator, yet make it difficult to discriminate between own reliability at manual control and automation reliability. The approximation of expected subject accuracy and the remaining uncertainty due to random wrong answers was thought to reduce the influence of rational comparisons of own and automation accuracy in automation use decisions, making these decisions more open to influence from the factors of interest (workload, automation failures, uncertainty, and risk). Three of the experiments contained two automation failures in isolation and one failure in the

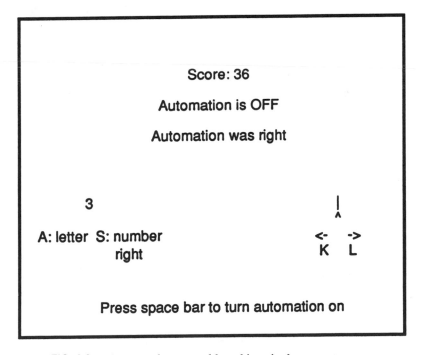

FIG. 2.2. The screen layout used by subjects in the computer game.

presence of higher workload and higher uncertainty, and workload and uncertainty also varied, both two times over the hour-long timeline. The primary dependent variable in three of the experiments was the proportion of subjects who used the automation during each trial. Because of the wide confidence limits for a proportion score, at least 30 subjects took part in each experiment.

Thirty University of Minnesota students enrolled in undergraduate psychology courses took part in Experiment One. The primary purpose of this experiment was to estimate the magnitudes of influence from automation reliability, task uncertainty, and workload on automation use decisions. A $25 award was offered to the subject who posted the highest score in the game, to provide an incentive and something of value that might be lost due to error, a necessary element of the risk manipulation to be explored later.

Fig. 2.3 shows the profiles of automation accuracy, workload, and task uncertainty over the course of Experiments One, Three, and Four. Experiment Two used a similar automation profile which will be described later. As shown, the experiment started with automation accuracy at 90%, workload at 40%, and task uncertainty at 10%. Because the experiment returned to this combination of conditions periodically, this will be referred to as the "normal" condition. After about 6 minutes, the workload

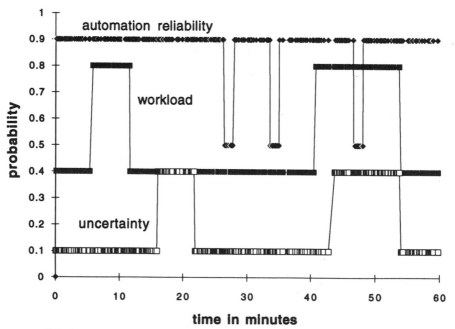

FIG. 2.3.   The profiles of independent variables used in the computer game.

manipulation occurred, with workload rising to 80% for about 6 minutes. The normal condition then returned for about 4½ minutes, followed by the uncertainty manipulation where a variety of new uncategorizable characters appeared. In this condition, there was a 40% chance that the subject would be given a nonnumber/nonletter character, such as an asterisk or pound sign, reducing the expected level of manual task performance by 15%. The uncertainty manipulation lasted for about 6 minutes, followed by a return to the normal condition for 4½ more minutes before the first automation failure. This failure lasted for about 1¾ minutes, then the normal condition returned for about 6 more minutes before the automation failed again for 1¾ minutes. Following this failure, the normal condition returned for about 5½ minutes, then a combined manipulation occurred, started by a workload increase that was augmented, after about 3 minutes, by an increase in uncertainty. After about 3 minutes of the workload and uncertainty combination, the automation failed again for 1¾ minutes, then recovered. After about 6 more minutes, both workload and uncertainty returned to the normal level, which was maintained for the last 6 minutes of the game phase.

Fig. 2.4 shows the proportion of students who used the automation over the course of Experiment One in relation to the manipulation profiles. One

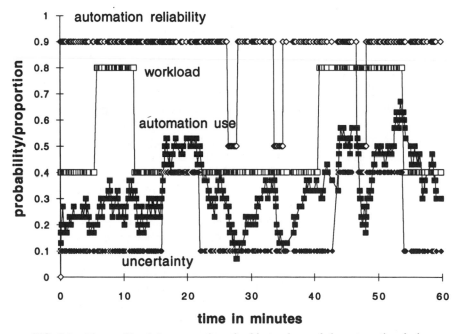

FIG. 2.4.   The profile of the proportion of subjects who used the automation during each trial of Experiment One.

unexpected characteristic of the student automation use profile was its low overall level. In the normal condition, automation use tended to return to about a 35% level, although there was a gradually rising trend over the course of the experiment. Even during the uncertainty manipulation, when even the most skilled subject would expect to make a large number of errors, only about half the students relied on the automation. Although counter to an optimal strategy, this apparent bias toward manual control is consistent with observations by Lee (1992). Another surprising finding was that there was no apparent reluctance to use the automation after failures. Contrary to predictions based on Muir's (1987) investigation of human trust, subjects did not delay turning the automation back on when it appeared to recover after failure events, and automation use after each failure was not less than before the failure. However, the amount of dithering in the profile prevented precise comparisons across the profile, so the dynamics of trust in automation use were explored in more detail in the second experiment.

The independent parameters in Experiment One were very highly auto-correlated. To illustrate, consider that with three automation failures, the expected reliability of the automation changed only six times over 2,050 trials, and there were only 14 state changes (trials during which one of the independent variables changed value) overall. For this reason, a full regression analysis could not be performed to estimate parameter significance levels. Instead, the timeline was divided into segments corresponding to system states (combinations of automation reliability, workload, and task uncertainty), and the proportion of subjects using the automation in each segment was estimated by averaging across the last 10 data points in each segment. This allowed the profile to stabilize after each change of conditions and provided a good estimate of the stable value of automation use in the segment. A regression analysis of this estimator against workload, uncertainty, automation reliability, and the number of automation failures showed that uncertainty and reliability were both significant ($p = .0008$ and $.00004$, respectively), but that workload was not ($p = .96$). The regression produced a good fit, with $R = 0.94$ ($F = 17.45$, $p = .0003$). The Durbin-Watson statistic of this regression was 2.6, indicating that there was no serial correlation present.

One of the interesting results of Experiment One was the list of reasons given by subjects for choosing whether or not to rely on the automation. A comparison of total automation use and average lateness of automation use by reason given for using automation revealed differences in the strategies used by subjects. For example, subjects who cited fatigue tended to use the automation less and later than did other subjects, suggesting that they followed a manual control strategy for much of the experiment but turned the automation on late. Subjects who cited a high level of self-confidence in doing the task themselves used the automation very little, and those who cited the uncertainty manipulation used it differently from those who cited workload or errors. These results suggest that large individual differences exist in automation use strategies, and that the theory of automation use shown in Fig. 2.1 may not apply to individuals, but rather across a group; individuals may use much simpler strategies influenced by small numbers of factors.

The amount of dithering in the automation use profile prevented reliable comparisons that would have shed light on some interesting questions, such as whether it was more difficult to recover trust in the automation after failures than to gain trust in it initially. Experiment Two was intended to better understand the influences and dynamics of trust in automation and to separate the element of trust from possible uncertainty about automation states. Unlike the other three experiments, the workload and uncertainty levels remained constant and automation was 100% accurate when it worked and 0% when it failed. This was intended to reduce the subject's uncertainty about whether the automation or subject was better at the task. The dependent measure was the response time to state changes: how long it took for subjects to turn the automation on initially, off in response to the first failure, on in response to the first recovery, and so on. Three conditions were run with 17 subjects in each condition. As in Experiment One, subjects were students in undergraduate psychology courses at the University of Minnesota.

The three conditions differed only in the amount of information that subjects were given about the automation prior to playing the game. In the Trust/State condition, subjects had no prior information about the automation, so when an automation failure was encountered, subjects would be uncertain both about whether the automation had entered a partially or fully unreliable state and how long the state change would last; thus, both trust and state uncertainty would influence automation use decisions in this condition. In the Trust condition, subjects were told that the automation could only get all the answers right or wrong at a time. This was intended to eliminate the subjects' suspicion that the automation may have entered a partially reliable state, leaving trust (the subjects' projection of the automation's accuracy into the future) as the sole remaining influence. In the None condition, subjects were also told how long the automation would stay failed and recovered in each state transition, so neither trust nor state uncertainty would influence decisions. Differences among the three conditions would reveal the contributions of each element (state uncertainty and trust) to subject automation use decisions.

The response time data exhibited a large amount of skew, so a log transformation was applied to normalize the data for illustration. However, other characteristics of the data prevented the use of analysis of variance: because automation was fully under subject control, there were many cases in which a subject did not turn automation on or off prior to a state change, so no response was required for that state change. This meant that some of the cells in the data matrix were empty. For these reasons, nonparametric techniques were applied to estimate the significance of response time differences. Fig. 2.5 shows the normalized response times for reference; the normalized data correspond better to the significance levels of the differences diagnosed by the nonparametric measures than do the raw data.

The results of Experiment Two demonstrated that both state uncertainty and trust affect automation use decisions, but only early in the subjects' experience with the automation. Prior to the first failure, subjects in the None condition turned the automation on significantly faster than did those in the other two conditions, suggesting that low trust in the automation delayed subject use of the automation. However, trust was developed by flawless automation behavior over the first 20 minutes of the timeline, so high trust delayed subject responses to the first failure. Contrary to popular theories of trust, however, subjects did not show any reluctance to use the automation again after the first failure; in the Trust condition, subjects turned the automation on following the failure significantly faster than they turned it on initially, which opposes the hypothesis that trust would be harder to recover following failures than to gain initially, and there was no

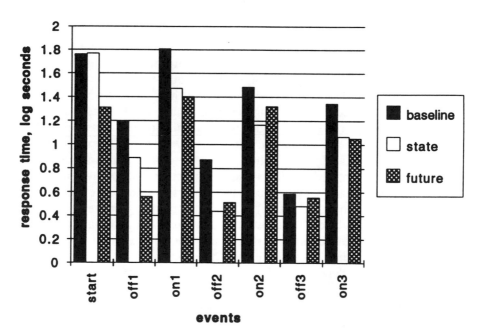

FIG. 2.5. Normalized response times to automation failure and recovery events in Experiment Two.

difference between on responses from subjects in the Trust and None conditions, suggesting that trust did not play a role in this response. After the first failure and recovery, differences between the conditions disappeared, suggesting that subjects learned to recognize automation states and anticipate its behaviors. This demonstrates that system-specific information can replace other factors as influences on automation use decisions.

One interesting feature of the response times was the large amount of skew observed. Histograms of all response times showed a cluster of responses early, then a scattering of responses late. This suggested that some of the subjects were responding directly to the manipulations whereas others turned the automation on or off in response to other factors, such as fatigue, boredom, or distraction. This was confirmed by the reasons given by subjects for making automation use decisions following the completion of the experiment. The large range of individual differences demonstrated here agrees with Lee (1992) and Harris et al. (1993) and suggests that large subject pools are required for automation studies.

Experiments One and Two provided important data regarding the fundamental influences on automation use decisions, but the use of university students and the artificial nature of the experiment environment prevent generalizing these conclusions to real systems. The first step toward drawing such generalizations is to determine whether the operators of real systems, with high levels of training and experience with advanced automation, exhibit the same automation use characteristics as did the students. After understanding the behaviors of real system operators in this highly controlled environment, we should be better able to interpret their behaviors in real systems.

Experiment Three was intended to carry out this first step. Experiment Three replicated Experiment One, but commercial airline pilots, who are trained and highly experienced with advanced automation, served as subjects. Although there was no evidence to suggest that pilots would use the automation any differently from the students, the fact that pilots use automation so extensively in real systems and the importance of their automation use decisions made this a topic of great interest. Thirty-four pilots from a major airline took part, and the conditions of Experiment One were replicated as closely as possible, except for two differences: First, the pilots were offered a $100 award for best performance due to their higher expected income than the students'; and second, the pilots were run using a laptop computer whereas the students were run on a desktop machine. The types of stimuli used and a comparison of the visual characteristics of the two displays suggested that any display differences would not produce task performance differences.

The automation use profile produced by the pilots is shown along with that produced by the students in Fig. 2.6. The pilot profile virtually

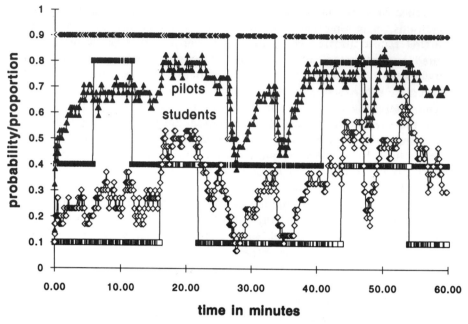

FIG. 2.6.   The profile of the proportion of pilots who used the automation during each trial of Experiment Three, compared with student use of automation in Experiment One. The pilots used the automation much more than the students did, but the dynamic characteristics of automation use were very similar.

duplicated the student profile in its dynamic behaviors, but the pilots' overall level of automation use was much higher and showed a substantial bias in favor of automated control. The proportions were separated by an average of 34% ($p < .01$), and the greater number of trials over which automation was used by the pilots was highly significant ($p < .00005$, two-tailed test). Secondary tests supported no explanations for this difference due to age, willingness to accept risk, attitudes toward automation, or other differences between the groups. This result suggests that some aspect of pilot experience or training biases them in favor of automation use. However, the dynamic characteristics of pilot automation use almost replicated those of student use; workload was insignificant ($p = .90$), uncertainty significant ($p = .01$), and automation reliability highly significant ($p = 8.14\text{E-}7$), using the same method of taking the last 10 data points in each segment as point estimators for regression analysis.

One of the surprising aspects of the pilot automation use profile was its high level during automation failures. Fully one third of the pilots continued to use the automation throughout the failure periods. However, those pilots who did turn the automation off in response to failures did so somewhat faster than the students did, arguing against the possibility that those pilots who did not turn the automation off simply did not notice the failure. These results, however, do not imply that pilots will overrely on automation on the flightdeck; this particular experiment used a highly artificial environment, and the potential loss of a $100 award does not compare with the consequences of errors committed on the flightdeck. To investigate whether differences in risk, defined as the likelihood and consequences of error, might influence automation use, the final experiment was developed with an additional penalty for categorization errors. The expectation going into this experiment was that when the consequences of error increased, subjects would want to assume manual control, preferring errors of commission over errors of omission (letting the automation have control and make the costly errors).

Experiment Four replicated Experiment Three, but each categorization error produced a 5% probability of the loss of 10% of the subject's current point total. Over the course of 2,050 trials, each subject was expected to lose 10% of his or her current points between 10 and 15 times. As in Experiment Three, subjects in Experiment Four were offered a $100 award for the highest score posted on the game. Experiment Four subjects were pilots from the same airline as those in Experiment Three. Thirty-one pilots took part.

The automation use proportion produced by subjects in Experiment Four is shown in Fig. 2.7 along with the profile from Experiment Three. The Experiment Four profile matched that produced in Experiment Three until after the second automation failure. Subjects in the higher risk condition then took much longer to turn the automation back on following the failure and ended up using the automation at a much lower rate after the third failure. The ending difference between the proportions was about 20%, which was significant ($p = .05$). Otherwise, the close match between the two profiles prior to the second failure reinforced the results of Experiment Three, with the same high reliance on automation throughout failures.

Other questions of interest were also explored in the four experiments. It was thought that subjects may use automation in relation to their own level of manual proficiency; if a subject were able to accurately assess his or her own level of accuracy doing the task manually and compare that with the automation's apparent accuracy, that would influence the decision to rely on automation. As stated earlier, the levels of automation accuracy were selected to make this comparison difficult, to make the reliance decision more open to influence from the other factors of interest. To determine whether subject proficiency still played a part in the decision, the correlation between manual accuracy and automation use was examined, and no relationship was found. In addition, a measure of subject self-confidence was constructed by giving subjects the opportunity to perform additional trials under manual control and receive double the additional number of

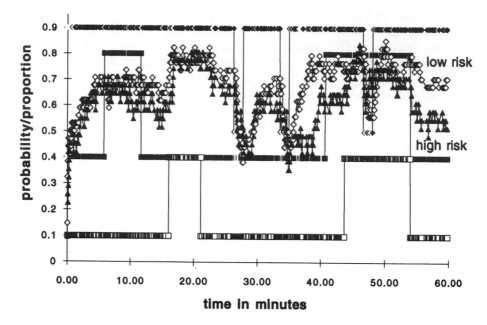

FIG. 2.7.   The profile of the proportion of pilots who used the automation during each trial of Experiment Four compared with the automation use profile from Experiment Three. Pilots in the higher-risk condition tended to use the automation less than did those in the lower-risk condition after the second automation failure.

points if they performed those trials perfectly. There was no correlation between actual manual performance accuracy during the game phase and the number of additional trials selected, indicating that subjects were not good at estimating their own proficiency.

However, behavior in this additional trials task could also have been influenced by risk-taking attributes, and these attributes might also account for other differences of interest, such as between the student and pilot populations and between the low- and high-risk conditions in the pilot experiments. To examine this possibility, a subjective and an objective measure of risk taking were incorporated into all experiments. No differences in risk-taking attitudes were found between any of the groups using either measure.

Finally, it was thought that pilots and students might have different attitudes toward automation, and that this difference would account for behavior differences. To examine this possibility, the Complacency Potential Rating Scale developed by Singh, Molloy, and Parasuraman (1993) was administered to all subjects. No differences were found between groups using this scale.

CONCLUSIONS

Taken together, these results provide firm support for more of the relationships proposed in Fig. 2.1 and for some new ones. Fig. 2.8 shows a revised theory of automation use, incorporating these results. Dashed lines indicate those relationships that were originally hypothesized but not supported by the evidence (they were examined in this series of studies but the results are not reported here; see Riley, 1994, for a complete treatment), whereas solid lines indicate relationships for which evidence was gained in these and other existing studies. Fatigue and learning about system states now replace the duration factor in the first theory, and learning is shown as influencing trust and reliance directly. The influence of learning on trust is necessary to account for the effect that experience with the automation has on trust (as demonstrated in Experiment Two), but the direct influence of learning on reliance replaces the influence of trust on reliance as subjects are better able to recognize system states and anticipate system behaviors rationally. The results also reaffirm the wide range of individual differences in automation use. Therefore, the theory shown in Fig. 2.8 should not be considered as applying to an individual's automation use decisions, but rather to represent automation use behaviors produced across a group. It is likely that an individual uses a much simpler strategy than this theory suggests, and that individual strategies are influenced by a few factors or small subsets of this overall theory.

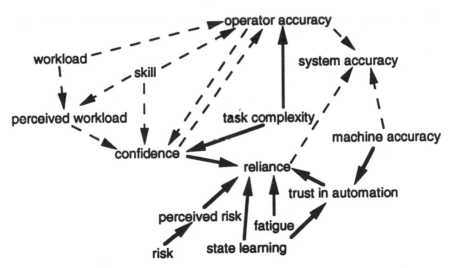

FIG. 2.8.   The revised theory of automation use. Dotted arrows show hypothesized relationships that have not been confirmed by experimental evidence, whereas solid lines represent those relationships supported by evidence from these studies.

All of this suggests several conclusions for real-world systems. First, different people may be susceptible to different types of inappropriate automation use behaviors based on their self-confidence in doing the task, their level of trust in the automation, their responses to fatigue, and their incorporation of these and other factors into their decision-making processes. Uncovering these biases may be a valuable aid to personnel selection and training, but perhaps a better approach would be to encourage more rational automation use strategies through training. For example, if operators know they are more likely to rely on automation when fatigued, they should compensate for this bias by increasing their monitoring of the system and perhaps intervening earlier if the automation appears to fail. Second, the strong effect of learning suggests that training should concentrate on helping operators better recognize system states and anticipate system behaviors. This may replace less rational influences that may lead to less appropriate automation use decisions.

Until now, the use of automation has been either dictated by procedures or left to operator discretion without a good understanding of the various factors that can influence these decisions, even though they may be among the most important decisions an operator can make. With automation increasing in sophistication, complexity, and authority, it may be time to attempt to systematize the decision to use or not use automation and reduce the likelihood that an individual will employ possibly faulty biases and influences.

ACKNOWLEDGMENTS

The author wishes to express his gratitude to Dr. Gordon Legge of the University of Minnesota for his ideas and recommendations, Dr. Beth Lyall and Andreas Petridis for administering Experiments Three and Four, and John Zalenchak of the Federal Aviation Administration for suggesting and supporting Experiments Three and Four.

REFERENCES

Harris, W., Hancock, P., & Arthur, E. (1993). The effect of taskload projection on automation use, performance, and workload. In R. Jensen & D. Neumeister (Eds.). *Proceedings of the Seventh International Symposium on Aviation Psychology* (pp. 8909–f). Columbus: The Ohio State University.

Lee, J. (1992). *Trust, self confidence, and operators' adaptation to automation.* Unpublished doctoral thesis, University of Illinois, Champaign.

Lee, J., & Moray, N. (1992). Trust, control strategies and allocation of function in human–machine systems. *Ergonomics, 35,* 1243–1270.

Lerch, F., & Prietula, M. (1989). How do we trust machine advice? In G. Salvendy & M. Smith (Eds.), *Designing and using human–computer interfaces and knowledge-based systems* (pp. 410–419). Amsterdam: Elsevier Science.

Muir, B. (1987). Trust between humans and machines, and the design of decision aids. *International Journal of Man-Machine Studies, 27,* 527–539.

Muir, B. (1989). *Operators' trust in and use of automatic controllers in a supervisory process control task.* Unpublished doctoral thesis, University of Toronto.

Parasuraman, R., Molloy, R., & Singh, I. (1993). Performance consequences of automation-induced "complacency." *The International Journal of Aviation Psychology, 3*(1), 1–23.

Riley, V. (1989). A general model of mixed-initiative human–machine systems. Paper presented at the Human Factors Society 33rd Annual Meeting, Denver, CO.

Riley, V. (1994). *Human use of Automation.* Unpublished doctoral dissertation, University of Minnesota, Minneapolis.

Riley, V., Lyall, E., & Wiener, E. (1994). *Analytic methods for flight-deck automation design and evaluation, phase two report: Pilot use of automation* (Contract report DTFA01-91-C-00039). Washington, DC: Federal Aviation Administration.

Sheridan, T., & Farrell, W. (1974). *Man–machine systems: Information, control, and decision models of human performance.* Cambridge, MA: MIT Press.

Singh, I., Molloy, R., & Parasuraman, R. (1993). Automation-induced "complacency": Development of the Complacency-Potential Rating Scale. *The International Journal of Aviation Psychology, 3*(2), 111–121.

Sparaco, P. (1994). A330 crash to spur changes at airbus. *Aviation Week and Space Technology, 141*(6), 20–22.

Will, R. (1991). True and false dependence on technology: Evaluation with an expert system. *Computers in Human Behavior, 7,* 171–183.

3 Theoretical Perspectives on Adaptive Automation

Mark W. Scerbo
Old Dominion University

INTRODUCTION

Adaptive automation is a form of automation that is flexible or dynamic in nature. In adaptive systems, decisions regarding the initiation, cessation, and type of automation are shared between the human operator and machine intelligence. Unlike more traditional forms of automation, adaptive automation can adjust its method of operation based on changing situational demands. Although still in its infancy, researchers and developers alike have begun to espouse the virtues and promise of adaptive automation (Morrison, Gluckman, & Deaton, 1991; Rouse, 1988).

Numerous theories, models, and platforms for delivering adaptive automation have already been proposed. The literature also contains the concerned voices of those who see the potential danger in adaptive automation evolving from a technological impetus instead of a human-centered design philosophy (Billings & Woods, 1994; Wiener, 1989). This is fortunate, for as we shall see, the ultimate success of adaptive automation may rest more with what we know about ourselves than with what we know about technology. This chapter describes the theoretical basis of adaptive automation and surveys research and development efforts aimed at validation of this approach to automation design.

AUTOMATION

Automation can be thought of as the process of allocating activities to a machine or system to perform (Parsons, 1985). These can be entire activities or portions thereof. As Davis and Wacker (1987) suggested, an activity is automated when it can be performed without human assistance under normal operating conditions. A truly automatic device is typically one that can both detect changes in the environment and affect the environment. A thermostat and the automatic pilot in a cockpit are both examples of automatic devices. Each can detect deviations from some reference point and can take action to return the system to that reference point. What distinguishes the automatic pilot from the thermostat is the number and variety of inputs, the complexity of the processing, and the number and variety of responses the device can make.

Wickens (1992) described three general classes of automation that serve different purposes. First, automation can perform functions that are beyond the ability of humans. For example, the faceted appearance of Lockheed's F-117 stealth fighter makes the aircraft almost impossible to fly. Automated systems, however, "interpret" the control stick movements of the pilot in order to control the unusual aerodynamics of the aircraft. Second, automation can perform tasks that humans do poorly. An example would be the autoexposure system in a camera. This feature eliminates the need for photographers to make estimates about the amount of light in a scene as well as having to calculate the proper shutter speed and lens opening. Third, automation can assist humans by performing undesirable activities. For instance, an automatic transmission depresses the clutch and shifts gears for those of us who find that aspect of driving burdensome.

Benefits of Automation

There are numerous benefits to automation. As machines, devices, or systems become capable of performing more and more activities, there are fewer for the human to do. This is ideal in situations in which the operator is overloaded with activities. In fact, several researchers have reported that some of the more successful applications of automation are in cases where it relieves the human from having to deal with nuisance or housekeeping activities (Gaba, 1994; Rouse, 1991).

Automation can also increase the flexibility of operations or permit control of more complex systems (Woods, 1994). On the other hand, automation can attenuate the variability associated with human performance and thereby significantly reduce errors.

Within the aviation industry, cockpit automation has made it possible to reduce flight times, increase fuel efficiency, navigate more effectively, and extend or improve the pilot's perceptual and cognitive capabilities (Wiener, 1988). In medicine, the availability of automatic ventricular fibrillation in the field has made it possible to save lives (Thompson, 1994).

Costs of Automation

The benefits derived from automation come at a price. Several researchers have indicated that automation brings with it a different set of problems (Billings, 1991; Wiener & Curry, 1980; Woods, 1994). As noted previously, when tasks become highly automated there may be fewer activities for the operator to perform and his or her role shifts from that of an active participant to one of a passive monitor. The unfortunate consequence of this changing role is that humans are not well suited to monitor sources of information for extended periods of time (Parasuraman, 1986; Warm, 1984). Parasuraman and his colleagues demonstrated that the ability to detect automation failures deteriorates under automatic as opposed to manual operating conditions (Parasuraman, Molloy, & Singh, 1993; Parasuraman, Mouloua, & Molloy, 1994).

Another problem concerns workload. At first glance, it seems that automation should help reduce mental workload. However, evidence shows that this does not necessarily happen. Instead, Woods (1994) argued that automation merely changes how work is accomplished. In fact, Wiener (1989) claimed that in some instances the introduction of automation may even increase workload. He cautioned that too often automated systems may operate well under periods of low workload and become a burden during high-workload periods.

Still, another problem often associated with automation concerns the maintenance of skills. Wickens (1992) argued that manual skills may deteriorate in the presence of long periods of automation. Along similar lines, others argued that automation removes the operator from the "loop" leading to decreases in situation awareness (Sarter & Woods, 1992). Consequently, overreliance on automation may make the operator less aware of what the system is doing at any given moment, leaving the operator ill-equipped to deal with a failure of automation.

In complex systems, Woods (1994) claimed that automation can lead to incongruent goals among subsystems. Also, when subsystems are highly coupled, it may be more difficult to isolate the locus of a problem if the system as a whole should fail.

Finally, there appears to be a good deal of skepticism about automation among its users. Confidence, both in oneself and the automation, impact on its usage (Lee & Moray, 1992, Muir, 1987). Extensive experience with the automation is needed for operators to assess its reliability. In his work with pilots, Rouse (1991) learned that indices of reliability had to be much higher than 95% for automatic systems to be considered useful.

Thus, it appears that some degree of confidence in automation is necessary for users to embrace it. However, too much confidence brings with it another set of problems. Once a sense of trust has developed and an automatic system has become an accepted method of operation, there lies the potential danger that individuals will become overly reliant on the automation. This may lead operators to be less willing to evaluate or even monitor the automated activities, a situation that has been described as automation induced "complacency" (Parasuraman, Molloy, & Singh, 1993).

Consequently, there are both benefits and costs associated with automation. Woods (1994) suggested that it may not be appropriate to view automation in terms of costs and benefits. Rather, he argued that automation transforms the nature of work. Offloading a task to an automated system to perform does not leave a gap in the operator's responsibilities. Instead, it changes his or her responsibilities, often requiring a redistribution of resources.

LEVELS OF AUTOMATION

Dual Mode

The simplest form of automation operates in two modes, manual and automatic. For example, a thermostat automatically maintains a desired temperature. If it is broken in the dead of winter, you must manually turn on the heating system each time you begin to feel cold and turn it off when you are warm.

Surprisingly, little is known about the effects of automation on human performance. In fact, in 1985 an entire issue of *Human Factors* was devoted to the topic of automation, but offered little comparative data regarding performance on manual and automatic tasks.

Recently, however, researchers have begun to examine how aspects of automation impact performance by themselves and in concert with other activities. Several studies have addressed monitoring efficiency under manual and automatic modes. For example, Parasuraman, Molloy, and Singh (1993) had subjects perform a systems monitoring task in which they were to watch a set of gauges for periodic deviations from prescribed levels and reset them. Subjects were quite capable of performing the task by itself or in conjunction with two other tasks (a compensatory tracking task and a resource management task). In another condition, however, the system monitoring task was automated, that is, the system corrected the deviations when they occurred. The subjects were responsible for monitoring the automation in addition to performing the tracking and resource management tasks. In this condition, the automated task failed to take corrective action periodically. Parasuraman and his colleagues found that under these conditions, monitoring performance declined dramatically after only 20 minutes. These findings indicate that operators may be unable or unwilling to continue monitoring automation while performing other tasks. Further, monitoring behavior appears to become increasingly inefficient with longer periods of automation.

In another study, Parasuraman, Mouloua, Molloy, and Hilburn (1992) again asked subjects to monitor automation. However, in this experiment the monitoring task was interrupted with a brief period requiring manual control. The investigators found that overall monitoring performance improved under these conditions and remained fairly stable for the rest of the study. Parasuraman et al. (1992) argued that a temporary return to manual operations may act as a countermeasure against poor monitoring behavior induced by automation.

In addition, it appears that there may be benefits to shorter periods of automation. Parasuraman, Hilburn, Molloy, and Singh (1991) found no costs associated with 10-minute cycles of manual and automated control. On the contrary, when subjects had to work at three tasks simultaneously, they found that performance on any of the nonautomated tasks benefited from automation on the other two tasks. Similar results were reported by Gluckman, Carmody, Morrison, Hitchcock, and Warm (1993), who found that tracking performance was superior in the context of an automated fuel management task.

To date, little is known about how the frequency and duration of cycles of automation affect performance. In a recent study, Scallen, Duley, and Hancock (1994) found superior performance on a tracking task that cycled between manual and automated modes every 15 seconds as compared to every 60 seconds. By contrast, Hilburn, Molloy, Wong, and Parasuraman (1993) observed a progressive deterioration in tracking performance with increases in the frequency of shifts between automated and manual modes. It is important to note that the pattern of shifts between modes in the Hilburn et al. study was derived from operator performance. It is, therefore, unlikely that the timing parameters surrounding mode shifts in this study were as consistent as those in the Scallen et al. (1994) study, which followed a precise temporal schedule. Thus, the benefits accruing shorter cycles reported by Scallen et al. may be attributable, in part, to a more predictable schedule of mode shifts.

Other investigators have been concerned about potential carryover effects emerging after operating under periods of automation (Morris & Rouse, 1986). This refers to the notion of an "automation deficit," that is, a degradation in manual performance after a period of automated control of a task. Glenn et al. (1994) explored this issue by asking subjects to perform a compensatory tracking task, a tactical assessment task, and a communication task simultaneously over three 10-minute periods. One of the tasks was automated during the second period. The researchers compared performance in Period 3 with that of Period 1, but found no evidence of improvements or deficits attributable to automation in the second period. Parasuraman, Bahri, Molloy, and Singh (1991) also failed to find any positive or negative residual effects after 10-minute periods of automation.

These findings contradict those of Ballas, Heitmeyer, and Perez (1991a), who reported an automation-related deficit in response time on a tactical assessment task. This effect, however, was observed on the first response after return to manual operation. Thus, it is possible that any automation-related carryover effects are short-lived. Furthermore, Ballas, Heitmeyer, and Perez (1991b) later indicated that residual effects may be tied to interface design.

Multiple Modes

In more complex systems, there may be several different levels and/or modes of operation. The level of automation can vary along a continuum from none at all (i.e., manual operations) to fully automatic. Billings (1991) distinguished among seven levels of automation. Differences along the continuum reflect different levels of autonomy. At the low end of the spectrum, activities and functions are under control of the operator. At the other end, systems are capable of executing complete functions and monitoring their own actions. Somewhere between these two extremes are systems that might perform some portions of an activity or make recommendations to the operator about a course of action.

Rouse and Rouse (1983) distinguished three unique modes of automation based on how the technology might assist the operator. For example, entire tasks can be allocated to the system to perform. Or, a task could be partitioned so that the system and operator would each be responsible for controlling some portion of the task. In the third mode, a task might be transformed to make it easier for the operator (e.g., changing the format used to present information to the operator). The allocation and partition modes would be high in autonomy because the system has control over all or some part of the task. Transformations, however, are low in autonomy because the operator is still responsible for performing the task.

In complex systems, it is not uncommon to find multiple modes of automation. The Flight Management System (FMS) found in the most advanced commercial aircraft is responsible for eight unique kinds of functions including navigation, guidance, system monitoring, and management of its own systems (Billings, 1991). The Honeywell MD-11 FMS can be configured for at least 12 separate modes of operation (Billings, 1991).

Complex systems with multiple modes of automation like the FMS present operators with a real challenge. First, they are difficult to learn. Pilots must learn the boundaries of each mode of operation and how to fly under each. Second, they may increase workload because the pilot must not only be cognizant of the operating procedures associated with each mode, but also be aware of which mode is active at all times. Indeed, Sarter and Woods (1994a) claimed that in the presence of multiple modes of automation, flying becomes a task of orchestrating a "suite of capabilities" for different sets of circumstances.

Recently, Endsley and Kiris (1994) examined the impact of different levels of automation on performance. The investigators had subjects use an expert system to help them make decisions about automobile navigation. Groups of subjects were asked to use one of four systems that differed in level of autonomy. Systems at the low end either recommended a course of action or sought concurrence on a suggested action. On the high end, systems made decisions and acted on them. Subjects worked through a set of four problems after which the system failed and they were required to solve two more problems on their own. Performance on the last two problems was compared to that of a "manual" group that made all of their decisions without use of an expert system. The investigators found that the decision times of subjects working with more autonomous systems were longer than those in the manual condition. Their findings suggest that higher levels of autonomy remove the operator from the task at hand and can lead to poorer performance during automation failures.

ADAPTIVE AUTOMATION

Recently, there has been a growing interest in the merits of automation that is dynamic or adaptive in nature (Hancock & Chignell, 1987; Morrison, Gluckman & Deaton, 1991; Rouse, 1976). In adaptive automation, the level or mode of automation or the number of systems that are automated can be modified in real time. Furthermore, both the human and the machine share control over changes in the state of automation. Parasuraman, Bahri, Deaton, Morrison, and Barnes (1992) argued that adaptive automation represents an optimal coupling of the level of automation to the level of operator workload. This, of course, presupposes that levels of operator workload can be specified to permit suitable adjustments in automation.

Adaptive Mechanisms

Perhaps the most critical challenge facing developers of adaptive automation concerns how changes among modes or levels of automation will be accomplished. Morrison and Gluckman (1994) discussed several potential candidates for triggering changes among levels of automation. One candidate is the operator's own performance. The operator's interactions with the system interface would be monitored and evaluated against some standard to determine when to change levels of automation. At present, there is much debate about how to derive the standards against which performance is to be assessed. Presumably, one could create a database of human performance information that would then be accessed to evaluate operator performance on line. However, such an approach would be of limited utility because the system would be entirely reactive. Furthermore, it is unlikely that a comprehensive database could ever be assembled.

A more promising alternative lies with operator performance modeling. This approach enables the system to generate standards derived from models of human performance. More important, performance models provide the capability for a more proactive system. For example, Geddes (1985) and his colleagues (Rouse, Geddes, & Curry, 1987–1988) offered a model to invoke automation based on information about the current state of the system, external events, and expected operator actions. The operator's intentions are predicted from patterns of his or her activity. In a well-structured environment (i.e., one with a rigid script of goals and plans), the system is capable of changing operation modes to meet future demands.

Hancock and Chignell (1987, 1988) also proposed an adaptive aiding mechanism. In their model, tasks or subtasks are allocated to the human or the system based on both current and future levels of operator workload. Unique to this approach is the idea that current levels of workload are determined in part by deviations from an ideal state. The prioritization and scheduling of activities needed to achieve desired states as well as the activities themselves all contribute to current estimates of workload. The discrepancy between current and expected states coupled with current levels of performance and the predicted levels of performance needed to reach the desired state drive the adaptive allocation of tasks. (Other models have also been proposed. For a more thorough review of operator performance models, see Parasuraman, Bahri, Deaton, Morrison, & Barnes, 1992; Rouse, 1988).

Morrison and Gluckman (1994) also proposed using biopsychometrics to invoke adaptive automation. Under this method, physiological signals that reflect autonomic or brain activity and presumably changes in workload might serve as the trigger for shifting among the automation modes. The advantage to biopsychometrics is that such measures can be obtained continuously with little or no impact on operator performance. Several physiological measures have been shown to reflect operator workload and are potential candidates for triggering adaptive automation, including heart rate variability, eye movements, and event-related brain potentials (Byrne & Parasuraman, in press; see also Kramer, Trejo, & Humphrey, chap. 7, this volume).

The third possibility for invoking adaptive automation outlined by Morrison and Gluckman (1994) is to monitor the activities of the mission itself. For example, Barnes and Grossman (1985) advocated the development of systems that would monitor ongoing activities for the occurrence of critical events. Detection of such events would, in turn, activate the automation. These critical events might be emergency situations or predetermined periods of high workload. Alternatively, the priority and sequencing of functions or the allocation of activities between the pilot and the cockpit might be scheduled to take place at prespecified points in the mission. This method of invocation may be the most immediately accessible of the three described by Morrison and Gluckman. Unfortunately, as Parasuraman, Bahri, Deaton, Morrison, & Barnes (1992) indicated, activity monitoring systems are not very sophisticated and are only loosely coupled to operator workload or performance.

System Responsiveness

As noted earlier, adaptive automation is still in its infancy, and as a result there are many issues that will need to be addressed before such systems become truly viable. One of these issues is system responsiveness. Early human factors work with computers revealed that system response time was an important determinant of both usability and user acceptance (Bailey, 1982). The issue is likely to be more critical with adaptive automation. Successful adaptive automation will require the proper amount of automation at the proper times. Under high-workload conditions, this may necessitate instantaneous system response. Proponents of operator performance models argue that system response times such as these are unlikely to be achieved without the ability to predict future workload demands (Greenstein & Revesman, 1986; Rouse, Geddes, & Curry, 1986).

Timing

Although predictive modeling is necessary for ideal system response times, it is not sufficient. As noted previously, developers of adaptive automation will also need to be concerned about cycles of automation and the frequency of changes among states. Furthermore, the timing parameters surrounding fluctuations in automation have yet to be worked out. There are two strategies that might be employed. One would be to match as closely as possible the changes in task demands with changes in automation. This procedure would maximize the chances of having the appropriate mode of automation in operation at any moment. However, this strategy might be inappropriate under conditions of rapidly fluctuating workload demands. Clearly, as the time spent under any mode of automation grows shorter, a point will be reached where the operator can no longer be effective. Furthermore, rapidly shifting modes of automation would impose an additional burden on the individual to maintain awareness of a constantly changing set of unique operational parameters. The alternative strategy would be to keep the number of mode changes to a minimum. This strategy would produce greater discrepancies between task demands and the appropriate automation mode at any moment, but would reduce the need to keep track of a rapidly fluctuating set of operating procedures. Ultimately, a compromise between theses strategies must be reached. The optimal choice, however, is likely to depend on the nature of the task as well as the individual.

AUTHORITY AND INVOCATION

Operator Authority

Adaptive automation makes it possible for a system to have autonomous control over changes among modes of automation. This has led some researchers to ask whether a system should have that authority.

There are many who argue that the operator should always have authority over the system (Billings & Woods, 1994; Malin & Schrenckeng-host, 1992). Pilots, too, have argued that because they have responsibility for the aircraft, themselves, and any passengers they should have the authority to initiate changes in automation (Billings, 1991).

Although research in this area is scarce, there is performance-based evidence to support operator-initiated invocation of automation. For instance, Harris, Hancock, Arthur, and Caird (1991) had subjects perform a resource management task, a systems monitoring task, and a compensatory tracking task simultaneously. Under one condition, the subjects performed all tasks manually, whereas in a second condition, the tracking task was automated. In a third condition, the subjects were encouraged to invoke automated tracking to help maintain optimal performance on the other two tasks. The results showed that subjects were more efficient at the resource management task when they had control over invoking automation. Hilburn et al. (1993) also reported an advantage for operator-invoked automation on a similar suite of tasks.

System Authority

There are also arguments to support system-initiated control over automation invocation. For instance, the operator may need to change automation modes at the precise time he or she is too busy to make the change (Sarter

& Woods, 1994b; Wiener, 1989). Recently, Harris, Goernert, Hancock, and Arthur (1994) gathered evidence that validates this concern. These investigators had subjects perform a systems monitoring task, a resource management task, and a tracking task simultaneously. The tracking task was under operator control in one condition and under system control in another. In the operator-controlled condition, subjects were asked to use the automation whenever they encountered difficulties with the other two tasks. Throughout the course of the experiment there were predetermined changes in task difficulty. Half of these task changes were preceded with a warning. Harris and his colleagues found that there was little difference in use of automation between conditions. However, there was an important interaction between automation and warning conditions in the resource management task. Specifically, subjects were less able to manage resources in the operator-initiated condition when workload increases came without warning. These data support the idea that operators may be unprepared to adequately manage automation in the context of abrupt, unexpected changes in workload.

Still another concern is whether the operator is the best judge of when and if automation is needed (Morris & Rouse, 1986). Harris, Hancock, and Arthur (1993) observed that as subjects became more fatigued under multitask conditions, they showed a tendency to use less automation. Moreover, the decline in automation use was accompanied by a decline in performance on a nonautomated task. These researchers argued that operator-initiated automation is unlikely to be of any value if it is not used precisely when it is needed.

Finally, there are situations in which it would be very beneficial for the system to have authority over automation invocation. If an operator's life or other lives were in danger or the continued operation of a system were to result in serious damage, clearly one would want the system to intervene and circumvent the threat or minimize the potential damage. For example, it is not uncommon for many of today's fighter aircraft to sustain higher levels of G forces than the pilot can withstand physically (Buick, 1989). Despite the presence of anti-G protective equipment, high enough G forces can still render the pilot of an armed and fast-moving aircraft unconscious for periods of up to 12 seconds (Whinnery, 1989). Thus, it is not surprising that situations such as these have been held up as ideal candidates for applications of adaptive automation (Whinnery, 1988).

RESEARCH AND DEVELOPMENT
IN ADAPTIVE AUTOMATION

The concept of adaptive automation originated with work in artificial intelligence in the 1970s. Specifically, efforts were aimed at developing adaptive aids to facilitate decisions about allocating tasks between a human and computer (Rouse, 1976, 1977). Later, efforts were directed at applying the adaptive aiding ideas to aviation systems (Hammer & Rouse, 1982).

In the mid-1980s, the Defense Advanced Research Projects Agency (DARPA), Lockheed Aeronautical Systems Company, McDonnell Aircraft Company, and what was then Wright Research and Development Center combined efforts to utilize state-of-the-art intelligent systems to assist pilots of advanced fighter aircraft. This effort was called the Pilot's Associate program, and its primary objective was to provide the pilot with "the right information, in the right format, in the right location, at the right time" (Judge, 1991, p. 86).

The Pilot's Associate (PA) is a consortium of cooperative knowledge-based systems capable of monitoring and assessing events and generating plans to respond to problems (Hammer & Small, 1995). Situation assessment systems monitor events in the external environment as well as the status of internal systems. This information can be evaluated and presented to the pilot. Planning systems inform the pilot of potential actions the system might take. These actions range from short-term tactics (e.g., adjusting the throttle) to changes in the overall mission (e.g., modifying the route). In addition, the PA includes an intelligent interface capable of selecting the appropriate information to present to the pilot at the appropriate time. The intelligent pilot–vehicle interface is a complex communications system capable of conversing with the pilot in different modes and at different levels of detail (Rouse, Geddes, & Hammer, 1990).

The PA program was a successful showcase of current intelligent system capabilities. Much knowledge of human information processing was needed to demonstrate the adaptive automation concepts. The initial stages of design drew heavily on the current models of human information processing and performance. Interestingly, many had to be abandoned by the second phase because they were too restrictive in a more contextually rich information environment (see Hammer & Small, 1995).

The original objective of the PA program was to demonstrate how artificial intelligence might be used to assist fighter pilots. It was never, however, conceived of as a plan for how adaptive automation ought to be implemented in a fighter cockpit. Recently, the Navy initiated its Adaptive Function Allocation for Intelligent Cockpits (AFAIC) program (Morrison, Gluckman, & Deaton, 1991). The primary objective of this effort is to understand the human performance issues surrounding adaptive automation and to develop guidelines for its implementation. Morrison, Cohen, and Gluckman (1993) outlined a threefold classification scheme for understanding how various strategies for implementing adaptive automation affect performance (see Fig. 3.1). On one dimension of the taxonomy are the factors which either invoke or change modes of automation; critical external events, changes in human performance and changes in physiology.

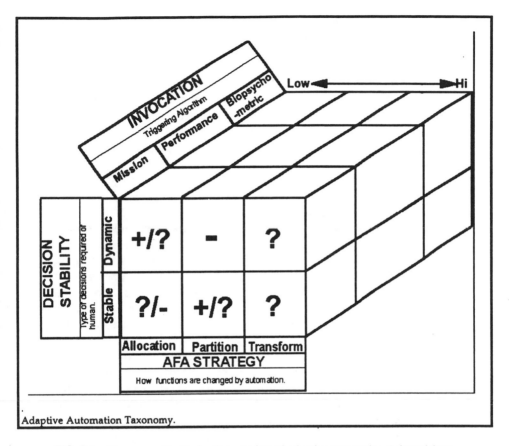

FIG. 3.1. Taxonomy for the implementation of adaptive automation (adapted from Morrison, Cohen, & Gluckman, 1993).

The second dimension is based on Rouse and Rouse's (1983) ideas about how functions might be changed. Entire tasks or portions thereof can either be allocated to the system, to the pilot, or transformed. The third dimension of the taxonomy reflects the complexity of the decisions and actions required of automated tasks. Stable decisions are those with consistent stimuli and actions, whereas dynamic decisions have more varied inputs and a wider range of actions. Efforts to validate the taxonomy are still underway, although the AFAIC program has generated some preliminary guidelines for implementing adaptive automation (see Morrison, 1993; Morrison & Gluckman, 1994).

IMPLEMENTING ADAPTIVE AUTOMATION

Although adaptive automation is still very much a concept that exists in laboratories, that has not kept researchers and designers from speculating about what form it ought to take and the human performance issues associated with it. Consider some of the unique issues that are likely to arise in systems equipped with adaptive automation. The operator will have to learn the capabilities and limitations of the system, and the system, in turn, will have to learn those of the operator. The operator may have to log extensive hours of training with the system to achieve optimal levels of performance. The operator and the system are likely to have unique responsibilities, but will collaborate on many others. Moreover, the system should be capable of assuming some of the operator's responsibilities during periods of high workload or anticipating the operator's actions and adjusting its own behavior accordingly. Perhaps, most importantly, the operator and system will have to be able to exchange information freely and effortlessly.

Activities like collaboration, backing up one another, and communication suggest the notion of teamwork. Indeed, several researchers have recently appealed to team concepts when referring to advanced automation. Hammer and Small (1995), for instance, viewed the Pilot's Associate as "an electronic crew member." Woods (1994) described advanced automation as a human-machine cooperative system with cognitive work distributed across agents and discusses the need for "automated agents" to coordinate activities among all agents.

Others have been more forthright in their conceptualizations. Malin et al. (1991) and Malin and Schreckenghost (1992) considered the design issues necessary to make an intelligent system a team player in space operations. Space teams have common goals. The individuals on those teams must coordinate, collaborate, and communicate effectively. Team members perform tasks both individually and cooperatively. In addition, members have some ability to back up one another's specialties.

Malin and her colleagues argued that an intelligent system may be considered a team member if certain criteria are met (Malin & Schreckenghost, 1992). An intelligent system can be included as a team member if it: is reliable, can communicate effectively, can coordinate activities, and can be guided by a coach. The system must be able to do what is necessary without causing harm or interference. Furthermore, the system must also be capable of being repaired; that is, the information within the system, its activities, and its ability to reason must be modifiable. Second, the system must communicate effectively by providing only as much information as is necessary when it is needed. Third, the system must coordinate activities to optimize performance among all team members and to minimize the potential for interference. Team members must be able to monitor the system and the system must be able to monitor the activities of the team members. This requires the need to exchange information about goals, plans, intentions, and beliefs. Finally, teams are managed or coached. The humans not only have their own responsibilities but must also manage the collective activities of all members, including the system. Thus, the human must be fully aware of the capabilities and limitations of the system, what it is doing at any given time, and how the system impacts the activities of the other team members.

The criteria outlined by Malin and her colleagues are the foundation for a set of guidelines aimed at developing intelligent systems that might behave more like a human in a team context. Scerbo (1994), on the other hand, argued that an understanding of team dynamics should be the force guiding the development of adaptive automation.

A team can be considered as two or more individuals who interact dynamically and interdependently to achieve a common goal (Salas, Dickinson, Converse, & Tannenbaum, 1992). Team members are dependent on one another. Furthermore, there is a need for individual members to coordinate and synchronize their activities and to communicate among one another.

It is probably no accident that those characteristics that define human teams (e.g., coordination, communication, etc.) found their way into the "team player" criteria proposed by Malin et al. (1992). Accordingly, Scerbo (1994) believed that an examination of the team literature might provide important insights about interactions with adaptive automation.

To fully appreciate team behavior, one has to understand those factors that affect team performance. Nieva, Fleishman, and Rieck (1978) discussed four such factors. First, there are the member resources. These are the skills, abilities, and knowledge that individual members contribute to the team. Second, there are task characteristics that dictate how the individual and interdependent activities will impact on the performance of the team as a whole. Third, there are also team characteristics. These refer to those aspects of the team itself that impact performance. Fourth, the member resources and the task and team characteristics are each affected by external conditions. These might be environmental or organizational in nature.

These factors have been embodied in a model of team behavior that emphasizes both individual and collective contributions to performance (Fleishman & Zaccaro, 1992). According to this model, there are seven categories of team functions: orientation, resource distribution, timing, response coordination, systems monitoring, motivation, and procedure maintenance. Orientation functions describe those activities associated with the acquisition and distribution of information. This may include details about team goals or the overall mission, potential problems, constraining influences, and the prioritization of activities. Information can either be collected from external sources and distributed to team members or gathered from team members and disseminated externally. Resource distribution functions refer to those activities necessary to match member resources with appropriate tasks and to distribute work equitably. Timing functions determine the pace of work at the individual and team levels. Response coordination functions refer to the sequencing and timing of member responses. The systems monitoring functions address how progress is measured at the individual and team levels. Work may be adjusted or tasks reallocated after performance is evaluated. Motivational functions refer to the establishment of team objectives and performance norms as well as the processes necessary to motivate members to meet those objectives. These functions also include measures needed to balance individual competition against team objectives and resolve performance-relevant disputes. The last category, procedure maintenance, refers to monitoring individual and team activities for compliance with standard procedures or organizational policies.

Fleishman and Zaccaro's (1992) taxonomy was derived from many years of studying humans in team situations. Scerbo (1994), however, argued that there are analogs for these functions in the adaptive automation arena. Consider orientation functions. With respect to aviation, the pilot and the aircraft would need to know the capabilities and limitations of one another. The tasks needed to complete a mission would have to be allocated between the pilot and the aircraft. Resource distribution functions, for example, in an aircraft with adaptive automation refer to task allocation and workload distribution. Systems monitoring functions for teams are similar to performance assessment functions in adaptive automation. Procedure maintenance in teams and adaptive automation would also be similar. The one area where the taxonomy does not have a suitable analog in adaptive automation concerns motivational functions. These functions do not readily translate into performance issues for adaptive automation. However, the need for pilots to get to know and trust the automation as they would another human team member should not be overlooked.

Scerbo (1994) used this taxonomy as a framework to identify those team functions addressed by the two most ambitious research efforts in adaptive automation to date, the PA and the AFAIC programs. He found that in both programs, considerable effort had been directed at developing methods to allocate functions between the pilot and the cockpit and to distribute pilot workload more uniformly (i.e., resource distribution functions). Systems monitoring functions have also been studied, however, efforts to assess pilot performance have received more emphasis than that of the cockpit or the combined performance of the two. The PA program has also addressed many orientation functions in order to build an intelligent planning system. The PA program also considered numerous response coordination functions in the creation of their intelligent interface system.

Scerbo (1994) also noted that several functions have yet to be addressed. Perhaps the most important oversight is the need to monitor the collective performance of the pilot and the aircraft. Overall mission effectiveness will ultimately be determined by how well the pilot and aircraft operate as a team. Neither program has directly addressed the timing functions, although it is possible that the PA considered pacing and task sequencing in their planning system. The timing functions, however, become more critical when activities must be performed in parallel. In fact, Amalberti and Deblon (1992) claimed that under emergency situations, flying becomes an exercise in scheduling and prioritizing activities. It is unclear whether either program addressed procedure maintenance functions; however, monitoring the pilot's and cockpits' compliance with standard protocols and policies could probably be handled with little difficulty. Also, neither program has considered motivational functions. However, research in this area might make a compelling argument for adaptive automation. Helmreich and Foushee (1993) indicated that the social dynamics of the flight crew have contributed to its share of aviation mishaps. In situations like these adaptive automation may have an advantage over human teams. Issues such as individual competition, individual acceptance of team norms, compliance with authority, or disregard for protocol would be meaningless for adaptive automation.

That there are many issues surrounding adaptive automation still to be explored should not be taken as short-sightedness on the part of either the PA or AFAIC programs. To their credit, they have addressed many important team functions without formally acknowledging a pilot–cockpit team. Instead, the gaps identified by Scerbo (1994) should be viewed as support for approaching adaptive automation from a broader, team perspective.

FUTURE DIRECTIONS

It is clear that automation is becoming increasingly sophisticated and the issues surrounding it more complex. Some researchers and developers have begun to describe future incarnations of automation as "electronic team members" (Emerson, Reising, Taylor, & Reinecke, 1989; Malin et al., 1992; Scerbo, 1994). It is likely that as both the systems and the interactions among people and systems become more complex, drawing on the human factors and cognitive engineering literatures will no longer be sufficient to address the myriad of issue that will arise. In fact, Hammer and Small (1995) came to a similar conclusion in their review of PA. They discovered that traditional human factors knowledge did not adequately address many of the human performance issues that arose during the course of the project. Hammer and Small concluded that "an examination of how humans decide to share tasks and information may be a more fruitful area in which to develop a theory of human–associate interaction" (p. 29).

In the following section, I raise several issues that are apt to increase in importance if we are to realize the potential of adaptive automation. I continue to use the team analogy because it provides a broader perspective of how work is accomplished. The following topics should not be considered exhaustive, but merely representative of the kind of issues that need to be considered for successful applications of adaptive automation.

Mental Models

A mental model is an individual's cognitive representation of how a system operates. Mental models enable an individual to describe, explain, and make predictions about system operations (Rouse & Morris, 1986; Wickens, 1992).

Several researchers have argued that it is advantageous for operators to form accurate mental models of complex systems (Carroll & Olson, 1988; Wickens & Flach, 1988). Doing so enables them to anticipate future system states, formulate plans, and troubleshoot effectively (Wickens, 1992). In fact, Wickens (1992) and Woods (1994) claimed that adding automation increases the complexity of systems. Thus, the need for an accurate mental model may be all the more important for automated systems.

There has also been a growing interest among some researchers studying team performance in the potential value of mental models in their work (Cannon-Bowers, Salas, & Baker, 1991; Rouse, Cannon-Bowers, & Salas, 1992). An accurate mental model may be particularly important to teams that must act on information that is integrated from separate sources and different members. Cannon-Bowers et al. (1991) argued that team effectiveness is apt to be determined by the degree to which all members work from a common mental model.

It should be noted that the utility of mental models has not gone unquestioned. Norman (1983), for instance, argued that mental models require time to evolve, and they may be incomplete and even imprecise. Also, mental models about particular systems may be structured from knowledge of other systems, from prototypes, or even from faulty information. In his observations of people using calculators, Norman found numerous instances of "superstitious" behavior that he argued was the result of an improper mental model.

These problems are likely to become more serious in the context of adaptive systems. This is because adaptive automation adds another layer of complexity to mental model formation. In addition to knowledge about how a system operates under various conditions, the mental model must now include self-evaluative information. In other words, a complete understanding of an adaptive system will require knowledge of one's own behavior, how it affects the system, and how the system reacts to it.

There is little in the human factors literature that addresses this type of mental model. Again, it may be necessary to consider other avenues of research outside the traditional vein of our discipline.

At this level of complexity, the relationship between the user and the system becomes truly interdependent. Interactions with adaptive systems begin to resemble interpersonal interactions. Therefore, an understanding of how individuals come to know one another's, strengths, weaknesses, and behavior patterns may provide some clues to mental model construction of adaptive systems. In this regard, theories about the development of relationships, communication patterns and self-disclosure, and the development of team skills may prove valuable.

Although research along these lines may hold some promise for understanding how mental models of adaptive systems might be formulated, one should not underestimate the challenge that lies ahead. Current knowledge of interpersonal behavior is incomplete, at best. Moreover, our ability to understand the intentions and actions of others is generally poor. Therefore, one should not be surprised to learn that operators will find it quite difficult to form a representation of an adaptive system, if it can be done at all.

Training and Practice

As noted previously, learning a complex system with multiple modes of automation is a lengthy process. As it is, pilots often indicate that they would welcome additional training (Orlady, 1993). Introducing adaptive automation will only increase training requirements. Organizations utilizing adaptive systems should understand the need for and make available additional time to train their users. It has been said that incomplete knowledge of operations leads to questions about what the system is doing and why the system is behaving the way it is (Wiener, 1989). Systems appear to take on a life of their own and act independently (Sarter & Woods, 1994a).

Again, if one adopts a team perspective, then learning a system with adaptive automation is not unlike learning to work with a new team member. Research shows that training can improve team performance, particularly if it is directed at communication or coordinated activities (Salas et al., 1992). Practice sessions allow members to come to know the boundaries of their teammate's abilities. This can benefit overall team performance in two ways. First, it allows team members to capitalize on one another's strengths and address each other's weaknesses. The second advantage of team practice is that as players come to know the skills and abilities of one another, the quality of their communication changes. For example, Foushee, Lauber, Baetge, and Acomb (1986) observed that flight crews that had spent longer periods of time together communicated more, verbalized more of their intentions, and were more likely to acknowledge one another's communications. Moreover, this pattern of communication was also associated with better performance when compared to other crews who were unfamliar with one another.

Team members must also learn to coordinate their activities. This is one of the primary differences between individual and team performance. Within an individual, the brain "knows" what the right and left hand are doing and, in some instances, may require few if any attentional resources to coordinate their activities (see Schneider & Shiffrin, 1977). In a team context, the quality of any executive or managerial process to coordinate individual activities is grossly inferior to what the human brain can accomplish. Consequently, an enormous amount of effort and resources are needed to monitor, plan, and schedule the activities of individual members to achieve overall objectives. This requires that goals and objectives be clearly stated, that plans and intentions be communicated, and that team members have the ability to observe one another's performance. Extensive practice enables team members to build models of one another's behavior and provide information and assistance at the times they are needed (Kanki & Palmer, 1993).

Communication

The success of adaptive systems will in large part be determined by the interface. I use the term *interface* to include all methods of information exchange. Most researchers agree that effective communication among team members is critical for overall success (Kanki & Palmer, 1993; Streufert & Nogami, 1992). Individual team members each possess knowledge and information that the other members do not. Thus, individuals must share information with one another in order to make decisions and carry out actions. Team members must also coordinate their activities, thereby requiring each to communicate their intentions.

To ensure the successful exchange of information, humans use any and all available means to communicate with one another. Consider the wide range of options that individuals have available to them. They can use spoken and written language. They can draw diagrams and pictures. They may also use hand gestures, facial expressions, and eye contact. In fact, they may even resort to physical contact. All of these are important forms of communication. Individuals constantly make decisions about when to use which form or forms of communication to best express themselves in different situations. It follows that an adaptive system designed with only one method of information exchange (e.g., an alphanumeric interface) is apt to severely limit interaction with that system. It would be analogous to asking team members to restrict their repertoire of communication to passing notes back and forth.

Designers of adaptive systems must, at the outset, attempt to include as many information formats as possible (e.g., text, graphics, voice, and video). There are at least two advantages to this strategy. First, multiple modes allow information to flow more freely because users can communicate more naturally. Second, workload can be reduced by eliminating the requirement to translate all information into only one or two formats.

FINAL THOUGHTS

The purpose of this chapter has been to survey recent developments and research on a new form of technology—adaptive automation. Particular attention was paid to the hopes and concerns surrounding adaptive automation and its potential impact on human performance. Before closing, however, I revisit an idea raised earlier. Woods (1994) argued that automation must be viewed from the broader context in which it is applied. Given that a great deal of automation is introduced on the job, it might be instructive to consider a more global view of work in order to gain a better understanding of the impact of automation.

There is a natural flow to the progression of work in organizations. The introduction of automation or new tools or procedures disrupts this flow. Some activities may be performed faster. Some activities may no longer be needed. New activities may be added. Productivity may even fall off temporarily until employees can fully incorporate the new methods or technology into their routines. But the flow of work continues.

What this perspective shows is that automation is neither inherently good nor bad. It does, however, change the nature of work, and, in doing so, solves some problems while creating others (Wiener & Curry, 1980). Both the developers and consumers of automation need to be aware of this. Designers must not incorporate automation into systems without considering how it will impact on other activities in the flow of work. This may require extensive testing to ensure that the desired benefits of automation exceed the unintentional consequences of automation. Likewise, organizations must prepare their employees for the restructuring of work brought about by automation and provide proper training for them to adapt to the new processes.

Adaptive systems represent the next step in the evolution of automation, and it is a big step. This type of technology will have a profound impact on how work is performed. Thus, it is not surprising to find the current literature filled with cries of caution. Many have argued that automation, in general, necessitates a human-centered approach to design (Billings, 1991; Rouse, 1991; Wickens, 1992; Wiener, 1988). Adaptive automation, on the other hand, may require a social-centered approach. As noted earlier, a team perspective is likely to provide important insights into human inter-action with adaptive automation. However, social and organizational issues may also have some bearing on the successful operation and acceptance of adaptive automation. This may be particularly true when multiple users must interact with the same adaptive system.

At present, adaptive automation is still in its conceptual stages. Although prototypes do exist, it will take many years for the technology to mature. Fortunately, this gives designers, cognitive engineers, and psychologists a chance to begin studying the many issues that surround adaptive automation before implementation of the technology is widespread. We have a real opportunity at this point in time to guide the development of adaptive automation from an understanding of human requirements instead of from restrictions imposed by current technological platforms.

REFERENCES

Amalberti, R., & Deblon, F. (1992). Cognitive modelling of fighter aircraft process control: A step towards an intelligent on-board assistance system. *International Journal of Man–Machine Studies, 36,* 639–671.

Bailey, R. W. (1982). *Human performance engineering: A guide for system designers.* Englewood Cliffs, NJ: Prentice-Hall.

Ballas, J. A., Heitmeyer, C. L., & Perez, M. A. (1991a). Interface styles for adaptive automation. *Proceedings of Sixth International Symposium on Aviation Psychology* (pp. 96–101). Columbus, OH: The Department of Aviation, The Aviation Psychology Laboratory, The Ohio State University.

Ballas, J. A., Heitmeyer, C. L., & Perez, M. A. (1991b). Interface styles for the intelligent cockpit: Factors influencing automation deficit. *Proceedings of American Institute of Aeronautics & Engineers Computing in Aerospace, 8* (pp. 657–667).

Barnes, M., & Grossman, J. (1985). *The intelligent assistant concept for electronic warfare systems* (NWC TP 5585). China Lake, CA: NWC.

Billings, C. E. (1991). *Human-centered aircraft automation philosophy: A concept and guidelines* (Technical Memorandum no. 103885). Moffett Field, CA: NASA.

Billings, C. E., & Woods, D. D. (1994). Concerns about adaptive automation in aviation systems. In M. Mouloua & R. Parasuraman (Eds.), *Human performance in automated systems: Current research and trends* (pp. 264–269). Hillsdale, NJ: Lawrence Erlbaum Associates.

Buick, F. (1989). *+Gz protection in the future—review of scientific literature* (DCIEM 89-RR-47). Downsview, Ontario, Canada: Defense and Civil Institute of Environmental Medicine.

Byrne, E. A., & Parasuraman, R. (in press). Psychophysiology and adaptive automation. *Biological Psychology.*

Cannon-Bowers, J. A., Salas, E., & Baker, C. V. (1991). Do you see what I see? Instructional strategies for tactical decision making teams. *Proceedings of the 13th Annual Interservice/Industry Training Systems Conference* (pp. 214–220). Washington, DC: National Security Industrial Association.

Carroll, J. M., & Olson, J. R. (1988). Mental models in human–computer interaction. In M. Helander (Ed.), *Handbook of human-computer interaction* (pp. 45–65). Amsterdam: North Holland.

Davis, L. E., & Wacker, G. J. (1987). Job design. In G. Salvendy (Ed.), *Handbook of human factors* (pp. 431–452). New York: Wiley.

Emerson, T., Reising, J. M., Taylor, R. M., & Reinecke, M. (1989). *The human-electronic crew: Can they work together?* (WRDC-TR-89-7008). Wright-Patterson Air Force Base, OH: Wright Research and Development Center.

Endsley, M. R., & Kiris, E.O. (1994). The out-of-the-loop performance problem: Impact of level of automation and situation awareness. In M. Mouloua & R. Parasuraman (Eds.), *Human performance in automated systems: Current research and trends* (pp. 50–56). Hillsdale, NJ: Lawrence Erlbaum Associates.

Fleishman, E. A., & Zaccaro, S. J. (1992). Toward a taxonomy of team performance functions. In R. W. Swezey & E. Salas (Eds.), *Teams: Their training and performance* (pp. 31–56). Norwood, NJ: Ablex.

Foushee, H. C., Lauber, J. K., Baetge, M. M., & Acomb, D. B. (1986). *Crew factors in flight operations III: The operational significance of exposure to short-haul air transport operations* (NASA Technical Memorandum 88322). Moffet Field, CA: NASA-Ames Research Center.

Gaba, D. M. (1994). Automation in anesthesiology. In M. Mouloua & R. Parasuraman (Eds.), *Human performance in automated systems: Current research and trends* (pp. 57–63). Hillsdale, NJ: Lawrence Erlbaum Associates.

Geddes, N. D. (1985). Intent inferencing using scripts and plans. In *Proceedings of the First Annual Aerospace Applications of Artificial Intelligence Conference* (pp. 160–172). Wright-Patterson Air Force Base, OH: U.S. Air Force.

Glenn, F., Barba, C., Wherry, R. J., Morrison, J., Hitchcock, E., & Gluckman, J.P. (1994). Adaptive automation effects on flight management task performance. In M. Mouloua & R. Parasuraman (Eds.), *Human performance in automated systems: Current research and trends* (pp. 33–39). Hillsdale, NJ: Lawrence Erlbaum Associates.

Gluckman, J. P., Carmody, M. A., Morrison, J. G., Hitchcock, E. M., & Warm, J. S. (1993). Effects of allocation and partitioning strategies of adaptive automation on task performance and perceived workload in aviation relevant tasks. In *Proceedings of the Seventh International Symposium on Aviation Psychology* (pp. 150–155). Columbus, OH: The Department of Aviation, The Aviation Psychology Laboratory, The Ohio State University.

Greenstein, J., & Revesman, M. (1986). Development and validation of mathematical models of human decision making for human–computer communication. *IEEE Transactions on Systems, Man, and Cybernetics, 16*, 148–154.

Hammer, J. M., & Rouse, W. B. (1982). Design of an intelligent computer-aided cockpit. *Proceedings of the 1982 IEEE Conference on Systems, Man, and Cybernetics* (pp. 449–452). New York: IEEE.

Hammer, J. M., & Small, R. L. (1995). An intelligent interface in an associate system. In W. B. Rouse (Ed.), *Human/technology interaction in complex systems* (Vol. 7, pp. 1–44). Greenwich, CT: JAI Press.

Hancock, P. A., & Chignell, M. H. (1987). Adaptive control in human–machine systems. In P. A. Hancock (Ed.), *Human factors psychology* (pp. 305–345). North Holland: Elsevier.

Hancock, P. A., & Chignell, M. H. (1988). Mental workload dynamics in adaptive interface design. *IEEE Transactions on Systems, Man, and Cybernetics, 18*, 647–658.

Harris, W. C., Goernert, P. N., Hancock, P. A., & Arthur, E. (1994). The comparative effectiveness of adaptive automation and operator initiated automation during anticipated and unanticipated taskload increases. In M. Mouloua & R. Parasuraman (Eds.), *Human performance in automated systems: Current research and trends* (pp. 40–44). Hillsdale, NJ: Lawrence Erlbaum Associates.

Harris, W. C., Hancock, P. A., & Arthur, E. J. (1993). The effect of taskload projection on automation use, performance, and workload. In *Proceedings of the Seventh International Symposium on Aviation Psychology* (pp. 178–184). Columbus, OH: The Department of Aviation, The Aviation Psychology Laboratory, The Ohio State University.

Harris, W. C., Hancock, P. A., Arthur, E., & Caird, J. K. (1991, September). *Automation influences on performance, workload, and fatigue.* Paper presented at the 35th Annual Meeting of the Human Factors Society, San Francisco, CA.

Helmreich, R. L., & Foushee, H. C. (1993). Why crew resource management? Empirical and theoretical bases of human factors training in aviation. In E. L. Wiener, B. G. Kanki, & R. L. Helmreich (Eds.), *Cockpit resource management* (pp. 3–45). San Diego, CA: Academic Press.

Hilburn, B., Molloy, R., Wong, D., & Parasuraman, R. (1993). Operator versus computer control of adaptive automation. In *Proceedings of the Seventh International Symposium on Aviation Psychology* (pp. 161–166). Columbus, OH: The Department of Aviation, The Aviation Psychology Laboratory, The Ohio State University.

Judge, C. L. (1991). Lessons learned about information management within the Pilot's Associate program. In *Proceedings of the Sixth International Symposium on Aviation Psychology* (pp. 85–89). The Department of Aviation, The Aviation Psychology Laboratory, The Ohio State University, Columbus, OH.

Kanki, B. G., & Palmer, M. T. (1993). Communication and crew resource management. In E. L. Wiener, B. G. Kanki, & R. L. Helmreich (Eds.), *Cockpit resource management* (pp. 99–136). San Diego, CA: Academic Press.

Lee, J. D., & Moray, N. (1992). Trust, control strategies and allocation of function in human–machine systems. *Ergonomics, 35*, 1243–1270.

Malin, J. T., & Schreckenghost, D. L. (1992). Making intelligent systems team players: Overview for designers (NASA Technical Memorandum 104751). Houston, TX: Johnson Space Center.

Malin, J., Schreckenghost, D., Woods, D., Potter, S., Johannesen, L., Holloway, M., & Forbus, K. (1991). *Making intelligent systems team players: Case studies and design issues. Vol. 1: Human–computer interaction design; Vol. 2: Fault management system cases* (NASA Technical Report 104738). Houston, TX: Johnson Space Center.

Morris, N. M., & Rouse, W. B. (1986). *Adaptive aiding for human–computer control: Experimental studies of dynamic task allocation* (Tech. Report AAMRL-TR-86-005). Wright-Patterson Air Force Base, OH: Armstrong Aerospace Medical Research Laboratory.

Morrison, J. G. (1993). *The Adaptive Function Allocation for Intelligent Cockpits (AFAIC) Program: Interim research and guidelines for the application of adaptive automation* (Technical Report NAWCADWAR-93031-60). Warminster PA: Naval Air Warfare Center, Aircraft Division.

Morrison, J. G., Cohen, D., & Gluckman, J. P. (1993). Prospective principles and guidelines for the design of adaptively automated crewstations. In J. G. Morrison (Ed.), *The adaptive function allocation for intelligent cockpits (AFAIC) program: Interim research and guidelines for the application of adaptive automation* (Technical Report No. NAWCADWAR-93031-60). Warminster, PA: Naval Air Warfare Center, Aircraft Division.

Morrison, J. G., & Gluckman, J. P. (1994). Definitions and prospective guidelines for the application of adaptive automation. In M. Mouloua & R. Parasuraman (Eds.), *Human performance in automated systems: Current research and trends* (pp. 256–263). Hillsdale, NJ: Lawrence Erlbaum Associates.

Morrison, J. G., Gluckman, J. P., & Deaton, J. E. (1991). *Program plan for the adaptive function allocation for intelligent cockpits (AFAIC) program* (Final Report No. NADC-91028-60). Warminster, PA: Naval Air Development Center.

Muir, B. M. (1987). Trust between humans and machines, and the design of decision aids. *International Journal of Man–Machine Studies, 27*, 527–539.

Nieva, V. F., Fleishman, E. A., & Rieck, A. M. (1978). *Team dimensions: Their identity, their measurement, and their relationships.* Washington, DC: ARRO.

Norman, D. A. (1983). Some observations on mental models. In D. Gentner & A. Stevens (Eds.), *Mental models* (pp. 7–14). Hillsdale, NJ: Lawrence Erlbaum Associates.

Orlady, H. W. (1993). Airline pilot training today and tomorrow. In E. L. Wiener, B. G. Kanki, & R. L. Helmreich (Eds.), *Cockpit resource management* (pp. 447–477). San Diego, CA: Academic Press.

Parasuraman, R. (1986). Vigilance, monitoring, and search. In K. Boff, L. Kaufman, & J. Thomas (Eds.), *Handbook of perception and performance* (Vol. 2, pp. 43.1–43.9). New York: Wiley.

Parasuraman, R., Bahri, T., Deaton, J. E., Morrison, J. G., & Barnes, M. (1992). *Theory and design of adaptive automation in aviation systems* (Progress Report No. NAWCADWAR-92033-60). Warminster, PA: Naval Air Warfare Center, Aircraft Division.

Parasuraman, R., Bahri, T., Molloy, R., and Singh, I. (1991). Effects of shifts in the level of automation on operator performance. *Proceedings of the Sixth International Symposium on Aviation Psychology* (pp. 102–107). Columbus, OH: The Department of Aviation, The Aviation Psychology Laboratory, The Ohio State University.

Parasuraman, R., Hilburn, B., Molloy, R., & Singh, I. (1991). *Adaptive automation and human performance: III. Effects of practice on the benefits and costs of automation shifts.* (Tech. Rep. CSL-N91-2). Washington, DC: The Catholic University of America, Cognitive Science Laboratory.

Parasuraman, R., Molloy, R., & Singh, I. L. (1993). Performance consequences of automation-induced "complacency." *International Journal of Aviation Psychology, 3*, 1–23.

Parasuraman, R., Mouloua, M., & Molloy, R. (1994). Monitoring automation failures in human–machine systems. In M. Mouloua & R. Parasuraman (Eds.), *Human performance in automated systems: Current research and trends* (pp. 45–49). Hillsdale, NJ: Lawrence Erlbaum Associates.

Parasuraman, R., Mouloua, M., Molloy, R., & Hilburn, B. (1992). *Training and adaptive automation II: Adaptive manual training* (Technical Report CSL-N92-2). Washington, DC: Cognitive Science Laboratory, Catholic University of America.

Parsons, H. M. (1985). Automation and the individual: Comprehensive and comparative views. *Human Factors, 27*, 99–111.

Rouse, W. B. (1976). Adaptive allocation of decision making responsibility between supervisor and computer. In T. B. Sheridan & G. Johannsen (Eds.), *Monitoring behavior and supervisory control* (pp. 295–306). New York: Plenum.

Rouse, W. B. (1977). Human–computer interaction in multi-task situations. *IEEE Transactions on Systems, Man, and Cybernetics, SMC-7*, 384–392.

Rouse, W. B. (1988). Adaptive aiding for human/computer control. *Human Factors, 30*, 431–443.

Rouse, W. B. (1991). *Design for success: A human centered approach to designing successful products and systems.* New York: Wiley.

Rouse, W. B., Cannon-Bowers, J. A., & Salas, E. (1992). The role of mental models in team performance in complex systems. *IEEE Transactions on Systems, Man, and Cybernetics, 22*, 1296–1308.

Rouse, W. B., Geddes, N. D., & Curry, R. E. (1986). An architecture for intelligent interfaces: Outline of an approach to supporting operators of complex systems. *Human Computer Interaction, 3*, 87–122.

Rouse, W. B., Geddes, N. D., & Curry, R. E. (1987–1988). An architecture for intelligent interfaces: Outline of an approach to supporting operators of complex systems. *Human-Computer Interaction, 3*, 87–122.

Rouse, W. B., Geddes, N. D., & Hammer, J. M. (1990). Computer-aided fighter pilots. *IEEE Spectrum, 27*(3), 38–41.

Rouse, W. B., & Morris, N. M. (1986). On looking into the black box: Prospects and limits in the search for mental models. *Psychological Bulletin, 100*, 349–363.

Rouse, W. B., & Rouse, S. H. (1983). *A framework for research on adaptive decision aids* (Technical Report AFAMRL-TR-83-082). Wright-Patterson Air Force Base, OH: Air Force Aerospace Medical Research Laboratory.

Salas E., Dickinson, T. L., Converse, S. A., & Tannenbaum, S. I. (1992). Toward an understanding of team performance and training. In R. W. Swezey & E. Salas (Eds.), *Teams: Their training and performance* (pp. 3–29). Norwood, NJ: Ablex.

Sarter, N. B., & Woods, D. D. (1992). Pilot interaction with cockpit automation: Operational experiences with the Flight Management System. *International Journal of Aviation Psychology, 2*, 303–322.

Sarter, N. B., & Woods, D. D. (1994a). Decomposing automation: Autonomy, authority, observability and perceived animacy. In M. Mouloua & R. Parasuraman (Eds.), *Human performance in automated systems: Current research and trends* (pp. 22–27). Hillsdale, NJ: Lawrence Erlbaum Associates.

Sarter, N. B., & Woods, D. D. (1994b). Pilot interaction with cockpit automation II: An experimental study of pilots' mental model and awareness of the Flight Management System (FMS). *International Journal of Aviation Psychology, 4*, 1–28.

Scallen, S. F., Duley, J. A., & Hancock, P. A. (1994). Pilot performance and preference for cycles of automation in adaptive function allocation. In M. Mouloua & R. Parasuraman (Eds.), *Human performance in automated systems: Current research and trends* (pp. 154–160). Hillsdale, NJ: Lawrence Erlbaum Associates.

Scerbo, M. W. (1994). Implementing adaptive automation in aviation: The pilot–cockpit team. In M. Mouloua & R. Parasuraman (Eds.), *Human performance in automated systems: Current research and trends* (pp. 249–255). Hillsdale, NJ: Lawrence Erlbaum Associates.

Schneider, W., & Shiffrin, R. M. (1977). Controlled and automatic human information processing I: Detection, search, and attention. *Psychological Review, 84*, 1–66.

Streufert, S., & Nogami, G. (1992). Cognitive complexity and team decision making. In R.W. Swezey & E. Salas (Eds.), *Teams: Their training and performance* (pp. 127–151). Norwood, NJ: Ablex.

Thompson, J. M. (1994). Medical decision making and automation. In M. Mouloua & R. Parasuraman (Eds.), *Human performance in automated systems: Current research and trends* (pp. 68–72). Hillsdale, NJ: Lawrence Erlbaum Associates.

Warm, J.S. (1984). *Sustained attention in human performance.* Chichester, England: Wiley.

Whinnery, J. E. (1988). *Considerations on aircraft autorecovery based on +Gz-induced loss of consciousness* (Technical report NADC-88091-60). Warminster, PA: Naval Air Warfare Center.

Whinnery, J. E. (1989). Observations on the neurophysiologic theory of acceleration (+Gz) induced loss of consciousness. *Aviation, Space, and Environmental Medicine, 6*, 589–593.

Wickens, C. D. (1992). *Engineering psychology and human performance* (2nd ed.). New York: HarperCollins.

Wickens, C. D., & Flach, J. M. (1988). Information processing. In E.L. Wiener & D.C. Nagel (Eds.), *Human factors in aviation* (pp. 111–155). San Diego, CA: Academic Press.

Wiener, E. L. (1988). Cockpit automation. In E. L. Wiener & D. C. Nagel (Eds.), *Human factors in aviation* (pp. 433–461). San Diego, CA: Academic Press.

Wiener, E. L. (1989). *Human factors of advanced technology ("glass cockpit") transport aircraft* (Technical Report 117528). Moffett Field, CA: NASA Ames Research Center.

Wiener, E. L., & Curry, R. E. (1980). Flight-deck automation: Promises and problems. *Ergonomics, 23*, 995–1011.

Woods, D. D. (1994). Automation: Apparent simplicity, real complexity. In M. Mouloua & R. Parasuraman (Eds.), *Human performance in automated systems: Current research and trends* (pp. 1–7). Hillsdale, NJ: Lawrence Erlbaum Associates.

4 A Theoretical Framework for Representational Design

John M. Flach
Kevin B. Bennett
Wright State University

INTRODUCTION

In complex systems, much of the designers' understanding of the work space is now imbedded in automatic control systems designed to implement the functional goals for the system. However, this understanding is generally incomplete. For this reason, human operators are often included to complete the design. An important role of the human operator is to respond to problems and contingencies that could not have been anticipated in the design stage. For example, there are many accidents where errors concatenated in ways that could not have been anticipated by the designers (e.g., Perrow, 1984). Also, for workspaces such as advanced tactical flight, the workspace itself is of such complexity that novelty is to be expected. Thus, events that arise in these workspaces go beyond the rule based capacity of automated systems. Adapting to these events requires creative, "productive" thinking—an achievement that has yet to be captured fully in automated systems. A human, armed with an appropriate representation, is often uniquely capable of this achievement. The catch, and the challenge for interface design, is to provide the appropriate representation. How do we evaluate the appropriateness of a representation?

The importance of representation has been a recurrent theme in the problem solving literature. The following quote from Wertheimer (1959) summarizes the Gestalt position on the requirement that a representation (i.e., envisioning) reflects the "structural truths" of a problem:

> Thinking consists in envisaging, realizing structural features and structural requirements; proceeding in accordance with, and determined by, these

requirements; . . . that operations be viewed and treated in their structural place, role, dynamic meaning, including realization of the changes which this involves; realizing structural transposability, structural hierarchy, and separating structurally peripheral from fundamental features . . . looking for structural rather than piecemeal truth. (pp. 235-236)

In this chapter, we address theoretical and methodological issues for evaluating interfaces as representations to support productive problem solving. The chapter begins with a discussion of our theoretical framework. This theoretical framework provides a context for understanding the need for representative designs to evaluate interfaces and productive problem solving.

A THEORETICAL FRAMEWORK

In this section, we present our theoretical framework for designing and evaluating interfaces. We attempt to clearly outline the basic tenets of our theory so that the reader can evaluate these ideas critically. It is important to note that in explicating constructs we are dealing with first principles. These are intuitive commitments that are not open to empirical falsification. Lachman, Lachman, and Butterfield (1979) referred to the shared intuitive commitments of a group of scientists within a common paradigm as the conventional rules of science. No matter how rigorous a scientific approach, it is impossible to escape from the requirement to bootstrap the empirical process on a set of intuitive commitments or conventional rules (e.g., commitments about the fundamental nature of causality, time, information, meaning, the dimensionality of space, etc.).

Perhaps the logical place to start our discussion of the theoretical basis for evaluating interfaces is with a definition of interface. The interface is the medium between the human and a work or task environment. In human--machine systems, the medium typically includes displays and controls. However, it is almost never the case that the medium is limited to displays and controls.[1] In addition to the artifactual displays and controls, the interface includes the natural sources of information (e.g., optical flow fields) and natural constraints on action (e.g., biodynamic constraints). It also includes the operators' memory and knowledge base (i.e., mental model or internal representation). The role of operator expertise in determining the functionality of the medium was emphasized by Hutchins,

---

[1]It is not uncommon that the terms *display* and *interface are used interchangeably. We use the term display* to refer to an artifactual representation that is often one local component within the interface.

Hollan, and Norman (1986). They noted that the facility and flexibility of a tool such as a text editor depends critically on the skill of the user (e.g., even a nondirect interface can appear to be direct when the user has extensive experience). Human operators have adapted and learned to perform fluently using some very poorly designed interfaces.

The medium reflects the *interactions* among the physical constraints on action, the information constraints on perception, and the value constraints (e.g., goals and cost functionals) that explicitly or implicitly define a task or work domain. These interacting constraints are illustrated in Fig. 4.1 in a way that emphasizes the belief that these interacting constraints are central to the problem of coordinated or situated action.

Although most of the current diagrams of the human information processing system include a feedback loop, these diagrams typically show stimulus and response as distinct entities—the stimulus entering from the left and the response leaving from the right. This creates the impression that the relation between stimulus and response is remote or arbitrary. Thus, in the laboratory the stimulus is manipulated as an independent variable and the response is measured as a dependent variable. The feedback loop is generally cut so that the experimenter can maintain strict control over the

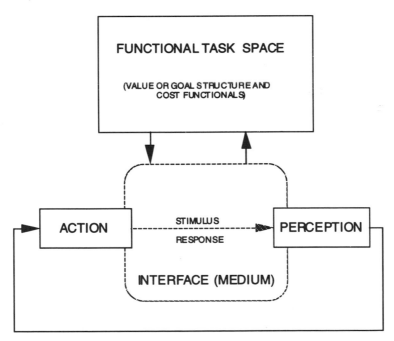

FIG. 4.1. The closed-loop human/machine/environment system. This diagram is intended to emphasize the medium as the center of a coordinated perception-action system.

independent variable, unconfounded by actions of the experimental subject.

Dewey (1896/1972) challenged this approach. He felt that this approach resulted from a failure to appreciate the implication of the closed-loop structure of the system. He argued that "the reflex arc idea, as commonly employed, is defective in that it assumes sensory stimulus and motor response as distinct psychical existences, while in reality they are always inside a co-ordination and have their significance purely from the part played in maintaining and reconstituting the co-ordination" (p. 99). In Fig. 4.1 we have attempted to illustrate the closed-loop system in a way that emphasizes the intimate coupling between stimulus and response within the co-ordination. It should be noted that there is no explicit distinction between human and environment in Fig. 4.1. Value and goal constraints may be internalized by the operator or may be explicit design constraints for the system. Action constraints may arise from the biodynamics of an operator's body, from the controls, or from the plant or vehicle being controlled. Perception constraints reflect the perceptual systems of the operator, the physical sensors of the system and the associated displays, and natural displays such as flow fields. The medium, too, represents both the physical representation as well as the internal representation. Within this diagram there is no distinction between human and environment. The human and environment are integrated throughout the diagram.

The word *interactions*, as used in our definition of medium, is highlighted to emphasize an assumption that will be fundamental to our approach to the design and evaluation of interfaces in the context of functional systems. This is the assumption that the constraints on action, information, and value are not independent. In fact, we make a much stronger claim — meaning arises out of the interactions among these various sources of constraint. Still stronger, we claim that the measure of an interface is the ability of the human agent to make contact with meaning, where contact with meaning is reflected in "structural understanding" and "coordinated action" appropriate to the value system for a specific workspace.

This focus on meaning as a relational property within the medium is in contrast to a more traditional view in which meaning is considered to be a product of information processing. To paraphrase Mace (1977), whereas the traditional information processing view places meaning inside the head, our position considers meaning to result from what the head is inside of. The traditional information processing view is linked to Shannon and Weaver's (1963) statistical notion of information. Information statistics (e.g., bits/second) are excellent for addressing issues of channel capacity but fail to address the issue of correspondence with an external reality. That is, information statistics cannot distinguish between being precisely right (always saying "yes" when yes is the correct answer) and being precisely wrong (always saying "yes" when no is the correct answer). Thus, information statistics do not provide a very effective framework for addressing issues of semantics. There is no basis for addressing the meaning of a message. Although the basic research community has recognized some of the limitations of the information statistic as a performance parameter (e.g., Lachman, Lachman, & Butterfield, 1979), vestiges of this approach remain in terms of the communications channel metaphor. Research has focused on properties of the channel (e.g., capacity, parallel vs. serial processing, locating the bottleneck) and has neglected the semantics of the message that is being communicated over this channel. The result is a tendency to focus on channel capacity or bandwidth as the critical issue for display design and a failure to address the problem of meaning. Our view, in contrast, emphasizes meaning or correspondence as a fundamental issue and channel capacity as only a secondary consideration. Therefore, our theoretical position might be characterized as a meaning processing approach, rather than an information processing approach. In this approach meaning is not the product of processing, but rather is the raw material from which coordinated, productive, or adaptive behaviors are molded. Haken (1988) presented a somewhat similar position:

> The concept of information is a rather subtle one . . . information is linked not only with channel capacity or with orders given from a central controller to individual parts of a system—it can acquire also the role of "medium" to whose existence the individual parts of a system contribute and from which they obtain specific information on how to behave in a coherent, cooperative fashion. At this level, semantics may come in. (p. 23)

It is important to note that there is ample evidence that issues of meaning and bandwidth are not independent. Miller's (1956) famous article summarizes the position that the capacity of working memory depends on the ability to organize or chunk information into meaningful units. A logical implication of Miller's analysis is that the dimensions of meaning that determine the organization of information into chunks cannot be ignored when predicting processing capacity. In fact, we argue that issues of meaning must take precedence in evaluation of the functional bandwidth of the operator when interacting with a work domain through an interface. In particular, the bandwidth of an interface will be greatly affected by the experience and knowledge of the operator and by the organization (i.e., in Gestalt terms the deep structure) of the interface.

The statistical notion of information together with the concept of independent processing stages fitted well within the zeitgeist of a linear world view that has until recently been a dominant perspective of science.

However, science is gaining a new respect for the power and importance of nonlinear systems. In nonlinear systems, interaction is the rule and independence is rare. We believe that human–machine systems are nonlinear and that this fact must be reflected in our research programs.

The claim that the human–machine system is a nonlinear system, where interactions are the rule, has important implications for the distinction between syntax and semantics. The term *syntax* is used here to refer to the form, appearance, or surface structure of an interface (e.g., alphanumeric versus graphical, object versus bar graph, integral versus separable). In the terms of Hutchins, Hollan, and Norman (1986), syntax relates to the construct of articulatory distance; that is, it has to do with the physical form or structure of the interface vocabulary.

The term *semantics* is used here to refer to the meaning of an interface; that is, how it maps on to distinctions within the problem or work space. This relates to what Hutchins et al. referred to as *semantic distance*. However, there may be subtle differences between our notion of semantics and the ideas of Hutchins et al. In discussing semantic distance, Hutchins et al. tended to ask questions such as: "Does the language support the user's conception of the task domain? does it encode the concepts and distinctions in the domain in the same way that the user thinks about them?" (p. 100). Such questions imply that the benchmark for semantic distance is the user's "mental model." For us, the benchmark should reflect a normative analysis of the work domain constraints—what could or what should the user be thinking about?

These contrasting relations between the mental model (or internal representation) and display (external representation) were discussed by Wilson and Rutherford (1989): "When applied within systems design—process control, HCI, or other—two positions seem to be taken. One is that the displays of a process or system must be compatible with operators' internal representations of the system; the other is that the displays themselves ought to determine that certain mental models be built up . . ." (p. 628).

Thus, one perspective is that representation aids should be designed to match the mental models that expert users have developed. In this design approach, experts' understandings and conceptualizations are studied and then translated into representation aids. The second design approach maintains that representation aids should be designed to determine the user's mental model. As Norman (1983) argued:

> People's mental models are apt to be deficient in a number of ways, perhaps including contradictory, erroneous, and unnecessary concepts. As designers, it is our duty to develop systems and instructional materials that aid users to develop more coherent, useable mental models. As teachers, it is our duty to

develop conceptual models that will aid the learner to develop adequate and appropriate mental models. (p. 14)

In order to know what is "adequate and appropriate," the domain itself must be analyzed. The results can then be used to guide the development of representation aids. We advocate this second approach. However, in very complex domains, knowledge elicitation will often be a critical aspect of the domain analysis, and the mental models of domain experts may provide the best available window to the real domain constraints. However, the ultimate design should be framed in terms of an understanding of the ecological constraints. The goal is to provide a representation that allows the cognitive agent to fully utilize the opportunities that the ecology affords. Performance should not be constrained by an under specified or inappropriate mental model. This is a lofty, and perhaps, unattainable goal. It may be very difficult or impossible to fully understand all the implications for the many interacting variables within a complex work domain. However, focusing on mental models does not necessarily reduce the complexity (the expert's mental model will reflect the requisite variety, i.e., full complexity, of the work domain). Further, as scientists, we generally will have greater control and will be capable of more direct analysis of the work domain than of the mental model. The challenge of representational design is to open up and broaden the perspective of the cognitive agent so that all the possibilities within the workspace are accessible.

In designing interfaces to determine mental models rather than to match existing mental models, perceptual skills of the human become critical. Whereas the approach of matching existing mental models uses what the human knows as a leverage against complexity, our approach leverages perceptual skill against the problem of complexity. The goal is to make the domain constraints "directly perceptible." The goal is to support exploration and discovery at the interface, so that the operators can quickly accommodate the actual work constraints into their mental representations of the problem space.

In sum, semantics raises the question of what information to present — what distinctions are important for the operator. Syntax raises the question of how to present that information. Syntax and semantics are conceptually distinct degrees of freedom for interface design. The same information mapping (semantics) can be accomplished in many different forms (syntax). However, in operation, the two dimensions will generally become intimately coupled. The medium literally becomes the message.

This blending of semantics and syntax is implicit in Rasmussen's (1986) observation that operators interact with processes at multiple levels — knowledge-based, rule-based, and skilled-based interactions. The distinction between syntax and semantics holds only for knowledged-based interactions in which the interface functions as a symbol representing features of the external world that the operator must interpret based on a conceptual understanding or mental model. However, in the course of interaction within a work domain, the relation between operator and interface often evolves so that rule- and signal-based interactions emerge. At this stage of skill, surface structures of the display will directly trigger productions and actions. In an important sense, the interface becomes the world. For skill- and rule-based interactions, the operator does not explicitly attend beyond the interface. The mapping to the domain beyond the interface is implicit in the actions and rules that have evolved and survived due to past successes. The capacity for skill- and rule-based interactions emerge with experience and training. These modes will dominate for experienced operators under "normal" operating conditions. However, novel events or system faults will often require a shift to knowledge-based interactions. Because the interface will be required to function at all three levels, semantics and syntax become intimately bound together within any particular interface. The result is that it will be very difficult, if not impossible, to unconfound syntax and semantics when evaluating displays. Prescriptively, the implication of this confounding is that the syntactical structure should be designed to reflect the semantic constraints (i.e., deep structure) of the problem space. This prescription is central to an approach that has been variously referred to as *ecological interface design* (Rasmussen & Vicente, 1989; Vicente, 1991; Vicente & Rasmussen, 1990), *representational aiding* (Woods, 1991), or the *semantic mapping principle* (Bennett & Flach, 1992).

To summarize our position, the claim is that meaning is a central construct when evaluating interfaces. Meaning permeates the problem of display design. It is an emergent property of the interactions between action, information, and values. Meaning reflects constraints arising from the workspace, constraints on action (in AI terms the application of operators), and the value or cost functionals by which actions and solutions are scored. Within displays, semantics (meaning) and syntax (form) are intimately coupled. A representation will be effective to the extent that form (syntax) reflects function (i.e., meaning or semantics). Finally, it is humans' capacity to pick up meaning, to think productively, and to achieve a structural understanding that make them such valuable components in complex systems.

REPRESENTATIVE DESIGN

Internal validity refers to the soundness of inference, the logical consistency of a methodology. The focus on internal validity is a defining attribute that differentiates scientific reasoning from other forms of knowing and argument. The primary threats to internal validity are confounds. Concern for internal validity is a primary motive behind experimental reductionism. That is, an approach in which the phenomenon of interest is parsed so that each dimension can be examined in isolation from other confounding components. This approach can be a very successful approach to the extent that systems are linear. That is, to the extent that performance of the whole is the sum of the components. However, if the system is nonlinear, if there are emergent properties, then parsing must be done with great care.

It is difficult, if not impossible, to escape reductionism in experimental approaches. In order to have control, to avoid confounds, and to reduce problem complexity to a manageable level, experiments are generally reduced versions of the phenomenon of interest. The issue, with respect to evaluating interfaces, and with respect to experimental science in general, is not whether or not to use reductionist methodologies; the issue is what are the most effective ways to parse the problem so that the phenomenon is simplified, but not broken. That is, our methodologies must simplify in a way that preserves critical emergent properties.

Hammond (1993) contrasted two approaches to the parsing problem — Wundt's choice and Brunswik's choice. Wundt's choice was based on the assumption that the deep structure of basic causal relationships was obscured by the surface features of the environment. Thus, Wundt argued that "by experiment . . . we strip the phenomenon of all its accessory conditions, which we can change at will and measure" (Wundt, quoted by Hammond, 1993, pp. 206–207). Wundt's choice evolved into the traditional information processing approach in which the deep structure was characterized in terms of a series of information processing stages that are assumed to function relatively independently. Thus, these stages provide a natural partitioning by which cognitive processing can be reduced for experimental evaluation.

Brunswik, however, believed that understanding the environment was fundamental to the problem of cognition. As Hammond (1993) observed, Brunswik believed that the "irregular, uncertain, confusing environment is the environment of interest, not the sanitized environment of the psychophysics laboratory or the perception laboratory of illusions and other 'impoverished stimulus' conditions" (p. 208). Hammond continued, "Brunswik's choice led to a design that includes a *formal* representation of all those conditions toward which a generalization is intended; representative design thus refers to the logical requirement of representing in the experiment, or study, the conditions toward which the results are intended to generalize" (p. 208). Brunswik called his methodological approach *representative design*.

The important implication of representative design for experimental methodologies is that the researcher must look to structural properties of the task environment in order to make decisions about how to partition the phenomenon of interest so that controlled experimentation is possible. For Brunswik, these structural properties were modeled in terms of the Lens model. The Lens model predicts the "achievement" of the organism as a relation between "ecological validity" and "cue utilization." Ecological validity refers to the probabilistic mapping between the environment and the medium of perception and cue utilization refers to the integration over the medium necessary to make an inference about the corresponding environment. For Brunswik, ecological validity provided both a guide for experimental design and a normative limit for achievement. The normative characteristic of ecological validity is clear as Brunswik (1956) wrote that "one of the most important aspects of functional theory concerns the relationship between ecological validity and utilization. *Ideally, cues should be utilized in accordance with their validity*" (p. 141, italics added).

Gibson (1979) also considered ecological validity to be central to the problem of perception:

> First, the environment must be described, since what there is to be perceived has to be stipulated before one can even talk about perceiving it. Second, the information available for perception in an illuminated medium must be described. This is not just light for stimulating receptors but the information in the light that can activate the system. . . . Third, (and only here do we come to what is called psychology proper), the process of perception must be described. This is not the processing of sensory inputs, however, but the extracting of invariants from the stimulus flux. (Gibson, 1979, p. 2)

For Gibson, the mapping from the medium to the objects of the world was lawful not probabilistic. The existence of lawful or invariant structural relationships in the optic array is a central premise of the concept of direct perception.

The trend toward more representative, naturalistic, and ecological approaches to cognition can also be seen in memory research. Early memory research was dominated by Wundt's choice, as typified in Ebbinghaus' research program that treated meaning and knowledge as nuisance variables to be controlled out. However, the current trend is toward more naturalistic, context-rich approaches as typified by Bartlett's (1932) and more recently Neisser's (1976; Neisser & Winograd, 1988) work, which considers meaning and knowledge to be critical variables for remembering. Bahrick and Karis (1982) noted that "scientists have become more aware of their obligations to be responsive to the problems of society, and the lack of ecological relevance in most memory research has become a matter of explicit concern" (p. 427).

The importance of considering the environment when parsing the problems of cognition was also found to be important for early research in artificial intelligence. Simon (1981) illustrated this in his parable of the ant. In this parable, he pointed out that the structure of the beach (i.e., the problem space) over which the ant locomotes provides critical information for modeling the ant's trajectory. Thus, research in artificial intelligence typically begins with a formal specification of the state space or problem space. The state space shows the critical dimensions of a problem and the possible paths from the initial condition to the goal. Like the concept of ecological validity, the state space provides a framework for bringing normative considerations to research on problem solving. For example, it allows us to compare the solution paths of humans to the "shortest" path through the problem space. Such normative considerations, if possible, could also be important for evaluating interfaces. Although AI researchers have taken great care to study the task constraints, they have had the luxury to choose their tasks. Some might argue that tasks such as cryptoarithmatic, missionaries and cannibals, and the tower of Hanoi are not representative of the kinds of tasks that operators face when trying to manage complex sociotechnical systems such as a chemical processing plant.

Pew (1994) also made a strong case for the need to consider the environment when evaluating human–machine systems and when designing displays to support effective situation awareness (SA):

> The SA requirements are the essential elements of information and knowledge needed to cope with each unique situation. Since virtually all measurements in human factors are relative, we argue that measuring SA implies having a standard, a set of SA requirements, if you will, against which to compare human performance. Such a standard must encompass an *abstract ideal,* a *physically realizable ideal*, and a *practically realizable ideal*. The abstract ideal includes the full set of information and knowledge that would make a contribution to accomplishing a particular goal. This is an abstract ideal because it is unconstrained by the design of the crew station and the information that is actually available to the crew member. Definition of the physically realizable ideal introduces the constraint of a real crew station. It is the information and knowledge that a crew member could obtain, given the current information sources in the workplace, that is, the current suite of displays and controls. It places no constraints on the information processing capacities and limitations of the crew member. Finally, we think in terms of a practically realizable ideal, what any real individual might be able to achieve under the best of circumstances, taking into account typical human performance capacities and limitations. It sets the standard against which to evaluate how well an individual performed given the system he or she had to work with. The definition of the abstract ideal helps us to understand what might be accomplished with better design and implementation. (Pew, 1994, p. 2)

What do these calls for representative experimental design imply for research to evaluate interfaces? We believe the implication is that the interface should not be dissected from the functional work domain. The dissection of interface from its natural work domain destroys the interactions from which meaning emerges within the medium. Thus, it becomes impossible to evaluate the interface in terms of structural truth or in terms of normative models of what the operator ought to know or ought to do given the constraints of a particular problem or work space. This does not mean that we must give up reductionism. The implication is that experimenters must consider the medium in terms of the interactions among task environment, action, and perception. The problem must be parsed in light of those interactions so that functional meaning is preserved in the laboratory. The laboratory environment will always be different from the actual work environment; however, the laboratory environment should be structured so that functional meanings are represented as fully as possible.

A FEW REPRESENTATIVE CASES

In this section, we review several examples from the display literature that illustrate the interplay of semantics and syntax and that highlight the importance for parsing problems within the framework of the task semantics. In reviewing these cases, the details of particular interfaces proposed are far less important than the framework within which the question are asked. The five cases presented here were chosen to illustrate the intimate coupling of semantics and syntax and the dangers that result when experiments or theories treat these as independent dimensions. The first two studies (Sanderson et al. and MacGreger & Slovic) illustrate that the form of a display (i.e., object, integral, separable, configural) cannot be evaluated independently from the structural properties (i.e., deep structure, ecological validity) of the problem being represented. It is the mapping of structure in the representation to the structure of the problem that ultimately determines performance. Bennett and Flach (1992) articulated this in terms of the "semantic mapping principle," and Woods (1991) articulated this in terms of principles for analogical integration.

The second two studies (Vicente and Moray et al.) illustrate that the ruler against which a display is measured (the performance index) must also reflect the semantic constraints of the target domain. It is important not to be seduced by the convenience and apparent generality of generic indices such as performance in a retrospective memory task or other generic task batteries that have only mundane links to the semantics of the task domain. The experimental task should be meaningful with respect to the role of the interface in the target work domain.

The last case (Roth et al.) also demonstrates the importance of semantically based norms (i.e. canonical solutions) for performance measurement. These norms suggest the dimensions that are relevant to task semantics and provide benchmarks against which performance can be scaled.

Another vestige of the information processing approach is a tendency for research to fixate on reaction time (RT) as the critical index of performance. There is the implicit assumption that minimum RT (i.e., maximum information processing rate) is the ideal. Roth et al. illustrated the use of more semantically rich measures of performance.

With regard to time as a performance index, "timing" may be more important than absolute time. Although much has been written on speed-accuracy trade-offs, the implicit assumption has been that speed and accuracy are independent performance dimensions (that, of course, may trade off). This approach fails to acknowledge the possibility for accuracy "in time." This can be most readily seen in music. The quality does not improve if the music is speeded up (even though all the notes are still played accurately). There is an appropriate or optimal rhythm. The construct of "timing" emphasizes the importance of synchrony or coordination with a process (i.e., making the right response at the right time—neither too slow or too fast). Synchrony is a relational construct; that is, synchrony is a measure of the match between temporal constraints in the task domain and temporal constraints on the operators' actions. Synchrony cannot be found in either the operator or the process. It is yet another factor that illustrates the need for constructs that integrate over human and environment.

The other reason to include the Roth et al. study was to emphasize that semantics are not simply a concern for graphical interfaces, but are a concern for any medium of representation intended to support the operator to adapt to the demands of a particular work domain. Graphical displays and expert systems are different in form, but as representations they serve a similar function—to support the operator in managing complex task domains.

Another important factor that interacts with the domain semantics and that has important implications for experimental methodologies is expertise. Both Vicente and Moray et al. manipulated expertise as an independent variable. This is clearly a critical dimension. However, we do not elaborate on this dimension here, as there is ample awareness and discussion of the implications of sampling for the validity and generality of experimental research. Despite this awareness, there has been far too much reliance on college sophomores and manufactured "experts" (i.e., experts created in the lab using 3 to 20 hours of practice in a toy world domain) when evaluating displays. The Moray et al. study is noteworthy in terms of being one of the few studies that has taken the trouble to enlist real experts to evaluate displays for a complex work domain.

Sanderson, Flach, Buttigieg, and Casey (1989)

Earlier research by Wickens and his colleagues (Barnett & Wickens, 1988; Carswell & Wickens, 1987; Casey & Wickens, 1986; Wickens, 1986; Wickens et al., 1985) suggested that object displays (e.g., a triangle) resulted in more effective information integration than separated displays (e.g., bar graphs). However, Sanderson et al. questioned whether the superior performance of object displays was due to the surface feature of "object-ness" or whether it might reflect differences in the deep structure of the representations. That is, differences in the mapping of the representation to the underlying task structure. The task structure was to evaluate the state of a process where two inputs combined (averaged) to produce an output. The subjects were to detect a deviation from the normal scaling of input to output. Sanderson et al. noted that the triangle (object) display had a higher-order feature (the apex angle) that mapped directly to the scaling of input to output. They noted that "in all normal transformations of the triangle for the present system, the angle of the apex varies only between about 85 and 95 deg. Any deviation outside this range immediately signals that the process is in an abnormal state" (p. 185). No such higher-order property was present in the bar graph displays used by Wickens. However, it was possible to create such a higher-order property by simply rearranging the bar graph so that the output was positioned between the two inputs on a common baseline. The result is that the tops of the bar graph would be collinear when the scaling was normal and would deviate from collinearity when a failure occurred. Results showed that the performance advantage for the object display was reversed and superior integration of information was achieved with the bar graph display.

The early research by Wickens et al. confounded semantic and syntactic dimensions of the interface. The conclusion that "objectness" supports more effective information integration ignored the semantic dimension. The Sanderson et al. study is a good illustration of how semantic considerations (the mapping of display constraints to process constraints) supersede syntactic considerations (object versus separated forms). Sanderson et al. concluded that "whether a display is best described as an object or separated display, the crucial determinant of its effectiveness will be how well the significant states (e.g., normal vs. failed) of the system it represents are mapped onto changes in its emergent, or configural, properties" (p. 197).

MacGregor and Slovic (1986)

Whereas the studies discussed in the previous section evaluated an abstract process (the averaging of two inputs to create an output), MacGregor and

Slovic assessed performance in a more naturalistic task of predicting the times for runners to complete a marathon (real data for actual runners were used to predict time in an actual race). Four variables were displayed in four different graphical formats. The formats included a bar graph display, a deviation bargraph display, a spoke display (quadrangle object display), and a face display. The variables displayed included the runner's age, the total number of miles run in the 2 months prior to the marathon, the fastest time in a 10K race, and the runners self-rating of motivation level. The interesting result of the MacGregor and Slovic study was that the face display resulted in either clearly superior performance or performance equivalent to the worst of the other formats depending on the mapping of variables to features in the face. When the most diagnostic variable (fastest 10K time) was mapped to the most salient feature of the face (mouth), then performance with the face display was superior to all other displays. When the most diagnostic variable was mapped to a feature with lower salience (height of eyebrows), performance was no better than the worst of the other formats.

Again, as with the Sanderson et al. study, the MacGregor and Slovic study illustrates that semantics (mapping to the domain) supersede syntax (surface features of the graphic) in determining performance. However, the MacGregor and Slovic study further illustrates how a form such as a face has its own "semantics." A happy face has its own meaning. This semantic affects the salience of elements in the display. Success of the display depends on the mapping of salience to the diagnosticity or importance of domain variables relative to the task or decision being supported. Highly diagnostic variables should be mapped to salient features. MacGregor and Slovic noted that "as integral display designs become more complex and pictorial, the potential for incompatibility between the normative importance of information features and the psychological salience of display features becomes greater, requiring thorough understanding of how display features are perceived and the quality of attention they are given" (p. 198). In the case of face displays and other pictorial formats, the interface functions as a metaphor. Thus, issues of the structural properties of the mapping from the metaphorical to target domains become critical concerns (e.g., Gentner, 1983; Gentner & Gentner, 1983). The range of performance, from best to worst, found with the face display clearly shows how critical the mapping can be. This emphasizes the danger of drawing inferences from effects due to syntax without considering interactions with domain semantics.

Vicente (1992)

Semantic considerations also have important implications for the performance measures that are used to evaluate interfaces. A performance index such as fault diagnosis, in which a display is scored in terms of the operator's ability to identify changes in system state with respect to the design goals for a process, is rich in semantics. However, a performance index in terms of retrospective memory in which a display is scored in terms of the operator's ability to recall the state of all (or a subset of) system variables can be a semantically impoverished measure, particularly if the variables that are to be recalled are chosen arbitrarily.

Vicente used both diagnosis performance and retrospective memory performance to evaluate two displays in the context of a thermo-hydraulic process simulation. One display was a mimic display that showed the physical variables (pump states, flow rates, reservoir levels, heater settings, temperatures, etc.), and a second display was constructed to show additional higher-order process constraints such as the mass and energy balances. The results showed that the addition of the functional information in the second display resulted in clear improvements in diagnosis. However, performance in the retrospective memory probe task was not consistent. When memory performance for all variables was evaluated there was no clear advantage for either display: "Memory was better for whichever display was experienced second" (p. 368). When the performances for physical and functional variables were evaluated independently, the results again showed no relation between diagnosis and recall of physical variables. However, there were significant correlations between diagnosis and the recall of functional variables. The functional variables reflected constraints on the process and were relevant to diagnosing the status of the process.

Moray et al. (1993)

Moray et al. also compared diagnosis performance with retrospective memory in evaluating three display formats for the control of feedwater in a nuclear power plant. The three formats were a single-sensor-single-indicator analog display format, a similar analog format with an added animated pressure-temperature graphic, and an animated graphical display in which all the variables were integrated within a space defined by a Rankine cycle diagram (a temperature/entropy state diagram for a heat engine). Three performance tasks were evaluated. A quantitative retrospective memory recall task required the operators to specify the values for 35 system variables. A qualitative retrospective memory recall task required the operators to answer 21 yes/no questions about the state of the process (e.g., Did the reactor trip during the trial? Did the hotwell level exceed 90% full? Did the generator function normally?). Finally, a diagnosis task required the subjects to specify whether a fault had occurred and to describe the fault. Results showed a clear advantage for the Rankine cycle display in the diagnosis task, but the picture for memory task performance was less clear. There was a disassociation between performance in the diagnosis task and the memory tasks. The authors concluded that "performance on the quantitative recall tests did not correlate with diagnostic performance to a significantly useful extent. Performance on the qualitative recall test correlated more strongly, but not at a level which is sufficient to make it a reliable indirect performance indicator to evaluate displays. The evidence from this project suggests that diagnostic performance itself is the best of the three ways to rank the quality of displays" (p. 56).

The Vicente and Moray et al. studies illustrate, again, the importance of semantic considerations. In evaluating displays, the meter for performance should reflect the semantic constraints of the domain. Semantically neutral tasks, such as global tests of recall for all system variables, will not provide valid indices for scaling the merits of a display.

Roth, Bennett, and Woods (1987)

The previous examples focused on graphical displays. However, there are many other types of representations that can be effective for supporting human problem solving. Roth et al. evaluated an expert system designed to assist technicians in troubleshooting a new generation electromechanical transport system. There are several aspects of this experiment that make it a prototypical example of semantic-based interface evaluation. First, the expert system was an actual system that was in its final stages of development; one that was deemed ready for placement in the field. Second, the experiments were conducted with the actual device and with actual technicians. Third, the six problems that were developed for the experiment varied on semantically relevant dimensions — whether or not knowledge resulting from previous experience with the old technology was relevant to problem solution and the number of competing hypotheses that needed to be ruled out. For each problem the appropriate fault was placed in the device, the technicians were provided with a brief description (much like the trouble report they would normally receive), and were asked to interact with the expert system to diagnose and repair the problem. Each problem session was videotaped, a computer log was recorded, and notes were taken by the experimenters.

The evaluation was semantic-based. The interaction of the technicians and the expert system were observed in the course of solving actual domain problems. A "canonical" solution path was defined that specified the optimal series of expert requests and corresponding technician responses required to reach a correct solution. Deviations from the canonical or most efficient solution path were analyzed for the cause of deviation and for the types of knowledge that were applied by the technicians to bring the problem solving episode back on track. The results indicated that, despite the expert system support, deviations were the rule (observed in 78% of the cases). Because of the traditional "question and answer" format for the expert system, much of the knowledge was hidden within the expert system. Thus, success depended largely on the knowledge and skills of the operator. Roth et al. concluded that "successful performance depended on the ability of the technician to apply knowledge of the structure and function of the device and sensible trouble-shooting approaches. This ability was necessary to follow the under specified instructions, to infer machine intentions, to resolve impasses and to recover from errors (person or machine) that led the machine expert off track" (p. 491).

Summary

The purpose of reviewing these studies was to illustrate the importance of task semantics both to the design of interfaces and to the design of experimental evaluations of interfaces. Subtle changes in the mapping of domain semantics onto display features or in the semantic properties of the tasks operators are asked to perform can have marked consequences for the patterns of performance that are observed. The interactions observed in these studies must be considered very carefully when designing either displays or research programs.

CONCLUSIONS

An active research science cannot be intelligently understood by reference to the rational rules of science alone. It is equally necessary to understanding the paradigm that guides the scientists to do experiments. Without understanding the paradigm, a student may find the experiments unrelated to each other; or the answers the experiments are supposed to provide may seem incomprehensible. The questions the scientists have chosen to ask may seem trivial or exotic, and their controversies may resemble tempests in teapots. However, to the student who grasps the paradigm guiding research, the relationship between theory and experiment will become clearer. The way in which experiments relate to each other will become more evident. The questions scholars in the field have chosen to ask will not seem so arbitrary, and their approach to answering the questions will look more reasonable. (Lachman, Lachman, & Butterfield, 1979, p. 15)

In this chapter, we focused more on elucidating a theoretic framework for evaluating displays than on detailing methodological prescriptions for research. In fact, the detailed methodological implications of this paradigm remain to be worked out. The critical premise of the theoretic framework is that meaning is a central consideration when evaluating interfaces. Every choice that a researcher makes when designing experimental programs to evaluate displays — choice of the dimensions of displays to vary, of the task, of the performance measures, of the subject populations, of the cover story — must be informed by considerations of meaning.

For those who disagree with this central premise, for those who consider meaning to be peripheral (a nuisance variable, an obstacle to generalizability, or an orthogonal dimension to syntactical considerations), our arguments will indeed be perceived as a tempest in a teapot. There is no rational basis to resolve this disagreement.

Others agree in principle that meaning is central, but have pragmatic concerns about the possibility for science in a world where evaluations reflect the specific constraints of particular work domains. The concern focuses on the issue of generalizability. Is it possible to discover general principles? Is it possible to generalize from solutions in one domain (e.g., aviation) to problems in another domain (e.g., process control)? Without such generalizations, is it possible to do science?

In the information processing model, generalization was guided by the boxes in our models. So, for example, the Sternberg (1966) task was used as a "dip stick" to determine whether a particular problem loaded on peripheral (encoding/response generation) or central (working memory) stages of processing. A solution that was effective for relieving memory load in one domain could be generalized to reducing memory load in a second domain.

What is the basis for generalization in a paradigm where meaning is central? Again, much work needs to be done, but we believe that it will be possible to generalize based on structural properties of task domains. Hammond (1993) argued that generalizations are possible with naturalistic approaches to decision making. He noted that such generalization will require the development of formal models of the environment. These formal models will be reductionist, in the sense that many of the environmental details will not be critical to the generalization. So, for Hammond, social judgment theory (SJT) provides a general theory of task environments that is "independent of the substantive materials of any particular judgment task" (p. 212).

Carswell and Wickens (1987, 1990) also identified an important structural property of work domains or tasks that has important general implications for displays. This is the dimension of task integrality. *Integrality* refers to the extent to which the meaningful distinctions within a task space are the measured variables themselves (separable), or whether the meaningful distinctions depend on relations across multiple measured variables (integral). In most complex sociotechnical systems, integrality is the rule. The demands for integration across measured variables in an important motivation behind the increasing interest in graphical displays. Configural graphic displays can be powerful devices for supporting perceptual integration. It is important to note, however, that despite the increased need for integrality, there are still important occasions that demand direct attention to the measured variables. For example, communication between multiple remote operators often depends on the precision of specific values for measured variables (Hansen, 1995). Thus, representations often will have to allow operators to perceive both specific measured variables and higher order relations across those variables. Bennett and Flach (1992) discussed this dual design goal and the implications for configural displays.

The research on manual and process control is another example where such generalizations have been successful. In these research areas, domain constraints such as the order of control and the magnitude of feedback delays provide a basis for bounding problems and generalizing solutions. Models such as McRuer's crossover model (e.g., McRuer & Jex, 1966) and the Optimal control model reflect both cognitive and domain constraints (see Flach, 1990, for an expanded discussion of these models and detailed references). In particular, McRuer's crossover model illustrates that there is no invariant human transfer function. Invariance is only apparent at the level of the human–machine system, and this invariance reflects global constraints on stability in closed-loops systems. These constraints are independent of whether that system is composed of human and machine or of purely electronic or physical components. These constraints reflect properties of the deep structure of control problems. Rather than generalizing based on common elements (common processing stages), we believe that generalizations should be guided by global properties of the human/machine/environment organization. Thus, control theory and dynamical systems theory may help to guide generalizations and to provide the framework for discovering structural, rather than piecemeal truths.

The critical point is that generalizations must be grounded in a theory of meaning. Meaning is a relational property that requires theoretical constructs that integrate over actor and environment. Meaning is not of mind, nor is it of matter. Meaning is about what matters (Flach, 1994)! A good interface must provide a representation of what matters! It must make the distinctions that matter perceivable! A sound methodological approach to display evaluation must do the same thing. It must be representative of the semantic as well as the syntactic constraints on performance. If displays are to support creative problem solving and coordinated adaptation, then evaluations of displays must be meaningful! This is not an antireductionist position. Simplification and control will still be important for insuring internal validity. However, parsing must preserve the essence of the functional human/machine/environment system. The parsing must preserve meaning. If, as research scientists, we sacrifice meaning to achieve control, then we will have nothing to offer in the quest to design representations for visualizing the possibilities of tomorrow.

ACKNOWLEDGMENTS

Our sincerest thanks to Raja Parasuraman and Mustapha Mouloua for their invitation to contribute to this volume. Raja Parasuraman and Kim Vicente gave us critical and useful feedback based on an earlier draft of this chapter. We take full responsibilities for any errors that remain. John Flach received support through a grant from the Air Force Office of Scientific Research during preparation of this chapter. Opinions expressed are those of the authors and do not represent an official position of the Air Force or any other organization.

REFERENCES

Bahrick, H. P., & Karis, D. (1982). Long-term ecological memory. In C. R. Puff (Ed.), *Handbook of research methods in human memory and cognition* (pp. 427–465). San Diego, CA: Academic Press.

Barnett, B. J., & Wickens, C. D. (1988). Display proximity in multicue information integration: The benefits of boxes. *Human Factors, 30*(1), 15–24.

Bartlett, F. C. (1932). *Remembering: A study in experimental and social psychology.* Cambridge, England: Cambridge University Press.

Bennett, K. B., & Flach, J. M. (1992). Graphical displays: Implications for divided attention, focused attention, and problem solving. *Human Factors, 34*(5), 513–533.

Brunswick, E. (1956). *Perception and the representative design of psychological experiments* (2nd ed.). Berkeley, CA: University of California Press.

Carswell, C. M., & Wickens, C. D. (1987). Information integration and the object display. *Ergonomics, 30,* 511–527.

Carswell, C. M., & Wickens, C. D. (1990). The perceptual interaction of graphical attributes: Configurality, stimulus homogeneity, and object integration. *Perception & Psychophysics, 47,* 157–169.

Casey, E. J., & Wickens, C. D. (1986). *Visual display representation of multidimensional systems: The effect of information correlation and display integrality* (Tech. Report CPL-86-2). Urbana–Champaign, IL: Cognitive Psychophysiology Laboratory, University of Illinois.

Dewey, J. (1972). The reflex arc concept in psychology. In J. A. Boydston (General editor) & F.Bowers (Consulting textual editor), *John Dewey: The early works, 1882–1898* (pp. 96–109). Carbondale, IL: Southern Illinois University Press. (Original work published 1896)

Flach, J. M. (1990). Control with an eye for perception: Precursors to an active psychophysics. *Ecological Psychology, 2*(2), 83–111.

Flach, J. M. (1994). Ruminations on mind, matter, and what matters. *Proceedings for the 38th Annual Meeting Human Factors and Ergonomics Society.* Santa Monica, CA: Human Factors and Ergonomics Society.

Gentner, D. (1983). Structural mapping: A theoretical framework for analogy. *Cognitive Science, 7*(2), 155–170.

Gentner, D., & Gentner, D. (1983). Flowing waters or teeming crowds: Mental models of electricity. In D. Gentner & A. L. Stevens (Eds.), *Mental models* (pp. 99-129). Hillsdale, NJ: Lawrence Erlbaum Associates.

Gibson, J. J. (1979). *The ecological approach to visual perception*. Boston, MA: Houghton Mifflin.

Haken (1988). *Information and self organization*. Berlin: Springer-Verlag.

Hammond, K. R. (1993). Naturalistic decision making from a Brunswikian viewpoint: Its past, present, future. In G. A. Klein, J. Orasanu, & C. E. Zsambok (Eds.), *Decision making in action: Models and methods* (pp. 205-227). Norwood, NJ: Ablex.

Hansen, J. P. (1995). Representation of system invariants by optical invariants in configural displays for process control. In P. A. Hancock, J. M. Flach, J. K. Caird, & K. J. Vicente (Eds.), *Local applications in the ecology of human-machine systems* (pp. 208-233).. Hillsdale, NJ: Lawrence Erlbaum Associates.

Hutchins, E. L., Hollan, J. D., & Norman, D. A. (1986). Direct manipulation interfaces. In D. A. Norman & S. W. Draper (Eds.), *User centered system design* (pp. 87-124). Hillsdale, NJ: Lawrence Erlbaum Associates.

Lachman, R., Lachman, J. L., & Butterfield, E. C. (1979). *Cognitive psychology and information processing: An introduction*. Hillsdale, NJ: Lawrence Erlbaum Associates.

Mace, W. M. (1977). James J. Gibson's strategy for perceiving: Ask not what's inside your head but what your head's inside of. In R. E. Shaw & J. Bransford (Eds.), *Perceiving, acting and knowing* (pp. 43-65). Hillsdale, NJ: Lawrence Erlbaum Associates.

MacGregor, D., & Slovic, P. (1986). Graphic representation of judgmental information. *Human-Computer Interaction, 2*, 179-200.

McRuer, D. T., & Jex, H. R. (1967). A review of quasi-linear pilot models. *IEEE Transactions on Human Factors in Electronics, HFE-8*(3), 231-249.

Miller, G. A. (1956). The magical number seven, plus or minus two: Some limits on our capacity for processing information. *Psychological Review, 63*, 81-97.

Moray, N., Jones, B. J., Rasmussen, J., Lee, J. D., Vicente, K. J., Brock, R., & Djemil, T. (1993). *A performance indicator of the effectiveness of human-machine interfaces for nuclear power plants* (NUREG/CR-5977). Washington, DC: USNRC.

Neisser, U. (1976). *Cognition and reality*. San Francisco: Freeman.

Neisser, U., & Winograd, E. (1988). *Remembering reconsidered: Ecological and traditional approaches to the study of memory*. New York: Cambridge University.

Norman, D. A. (1983). Some observations on mental models. In D. Gentner & A. L. Stevens (Eds.), *Mental models* (pp. 7 -14). Hillsdale, NJ: Lawrence Erlbaum Associates.

Perrow, C. (1984). *Normal accidents*. New York: Basic Books.

Pew, R. (1994). Situation awareness: The buzzword of the '90s. *CSERIAC Gateway, 5*(1), 1-4.

Rasmussen, J. (1986). *Information processing and human-machine interaction: An approach to cognitive engineering*. New York: Elsevier

Rasmussen, J., & Vicente, K. (1989). Coping with human errors through system design: Implications for ecological interface design. *International Journal of Man-Machine Studies, 31*, 517-534.

Roth, E. M., Bennett, K. B., & Woods, D. D. (1987). Human interaction with an "intelligent" machine. *International Journal of Man-Machine Studies, 27*, 479-525.

Sanderson, P. M., Flach, J. M., Buttigieg, M. A., & Casey, E. J. (1989). Object displays do not always support better integrated task performance. *Human Factors, 31*(2), 183-198.

Shannon, C. E., & Weaver, W. (1963). *The mathematical theory of communication*. Urbana, IL: University of Illinois Press.

Simon, H. A. (1981). *The sciences of the artificial* (2nd ed.). Cambridge, MA: MIT Press.

Sternberg, S. (1966). High-speed scanning in human memory. *Science, 153*, 652-654.

Vicente, K. J. (1991). *Supporting knowledge-based behavior through ecological interface design* (Tech. Report EPRL-91-1). Urbana–Champaign, IL: Engineering Psychology Research Laboratory and Aviation Research Laboratory, University of Illinois.

Vicente, K. J. (1992). Memory recall in a process control system: A measure of expertise and display effectiveness. *Memory & Cognition, 20*(4), 356–373.

Vicente, K. J., & Rasmussen, J. (1990). The ecology of human–machine systems II: Mediating "direct perception" in complex work domains. *Ecological Psychology, 2*(3), 207–249.

Wertheimer, M. (1959). *Productive thinking*. New York: Harper & Row.

Wickens, C. D. (1986). The object display: Principles and a review of experimental findings (Tech. Report CPL-86-6). Urbana–Champaign, IL: Cognitive Psychophysiology Laboratory, University of Illinois.

Wickens, C. D., Kramer, A., Barnett, B., Carswell, M., Fracker, L., Goettl, B., & Harwood, K. (1985). *Display-cognitive interface: The effect of information integration requirements on display formatting for C3 displays* (Tech. Report EPL-85-3). Urbana–Champaign, IL: Engineering Psychology Research Laboratory and Aviation Research Laboratory, University of Illinois.

Wilson, J. R., & Rutherford, A. (1989). Mental models: Theory and application in human factors. *Human Factors, 31*(6), 617–634.

Woods, D. D. (1991). The cognitive engineering of problem representations. In G. R. S. Weir & J. L. Alty (Eds.), *Human-computer interaction and complex systems* (pp. 169–188). London: Academic Press.

II Assessment of Human Performance in Automated Systems

5 Monitoring of Automated Systems

Raja Parasuraman
Catholic University of America

Robert Molloy
Catholic University of America

Mustapha Mouloua
University of Central Florida

Brian Hilburn
National Aerospace Lab, Amsterdam

INTRODUCTION

The revolution ushered in by the digital computer in the latter half of this century has transformed many of the characteristics of work, leisure, and travel for most people throughout the world. Even more radical changes are anticipated in the next century as computers increase in power, speed, availability, flexibility, and in that elusive concept known as "intelligence." Only a neo-Luddite would want to enter the 21st century without the capabilities that the new computer tools provide; and perhaps even a latter-day Thoreau would not wish to trade in his word processor for pen and paper. And, yet, although we have become accustomed to the rise of computers and as consumers demanded that they perform even greater feats, many have felt a sense of unease at the growth of computerization and automation in the workplace and in the home. Although there are several aspects to this disquiet (see Hancock, chap. 22, this volume), there is one overriding concern: *Who will watch the computers?*

The concern is not just the raw material for science-fiction writers or germane only to the paranoid mind, but something much more mundane. Computers have taken over more of human work—ostensibly leaving humans less to do, to do more in less time, to be more creative in what they do, or to be free to follow other pursuits. For the most part, computers have led to these positive outcomes—they have freed us from the hard labor of repetitive computation and allowed us to engage in more creative pursuits. But in some other cases, the outcomes have not been so sanguine; in these instances, human operators of automated systems may have to work as hard or even harder, for they must now watch over the computers that do their work. This may be particularly true in complex human–machine systems in which several automated subsystems are embedded, such as the commercial aircraft cockpit, the nuclear power station, and the advanced manufacturing plant. Such complex, high-risk systems, in which different system subcomponents are tightly "coupled," are vulnerable to system monitoring failures that can escalate into large-scale catastrophes (Perrow, 1984; Weick, 1988). Editorial writers have rightly called for better understanding and management of these low-probability, high-consequence accidents (Koshland, 1989).

One of the original reasons for the introduction of automation into these systems was to assist humans in dealing with complexity and to relieve them of the burden of repetitive work. The irony (Bainbridge, 1983) is that one source of workload may be replaced by another: Monitoring computers to make sure they are doing their job properly can be as burdensome as doing the same job manually, and can impose considerable mental workload on the human operator. Sheridan (1970) first discussed how advanced automation in modern human–machine systems changes the nature of the task demands imposed on the human operator of such systems. He characterized the role of the human operator in highly automated systems as altered from that of an active, manual controller to a supervisor engaged in monitoring, diagnosis, and planning. Each of these activities can contribute to increased mental workload.

Many of the changes brought about by automation have led to significant system benefits, and it would be difficult to operate many complex modern systems such as nuclear power plants or military aircraft without automation (Sheridan, 1992). Although users of automated systems often express concerns about the trend of "automation for automation's sake" (Peterson, 1984), many automated systems have been readily accepted and found invaluable by users (e.g., the horizontal situation indicator map display used by pilots). At the same time, some other changes associated with automation have reduced safety and user satisfaction, and a deeper understanding of these changes is necessary for successful implementation and operation of automation in many different systems (Mouloua & Parasuraman, 1994; Wickens, 1994; Wiener, 1988). Among the major areas of concern is the impact of automation on human monitoring. Automation of a task for long periods of time increases the demand on the operator to monitor the performance of the automation, given that the operator is expected to intervene appropriately if the automation fails. Because human monitoring can be subject to error in certain conditions, understanding how automation impacts on monitoring is of considerable importance for the design of automated systems. This chapter discusses the interrelationships of automation and monitoring and the corresponding implications for the design of automated systems.

EXAMPLES OF OPERATIONAL MONITORING: NORMAL PERFORMANCE AND INCIDENTS

It has become commonplace to point out that human monitoring can be subject to errors. Although this is sometimes the case, in many instances operational monitoring can be quite efficient. In general, human operators perform well in the diverse working environments in which monitoring is required. These include air traffic control, surveillance operations, power plants, intensive-care units, and quality control in manufacturing. In large part, this probably stems from general improvements over the years in working conditions, and, in some cases (although not generally), from increased attention to ergonomic principles. In one sense, when the number of *opportunities* for failure are considered — virtually every minute for these continuous, 24-hour systems — the relatively low frequency of human monitoring errors is quite striking.

This is not to say that errors do not occur. But often when human monitoring is imperfect, it occurs under conditions of work that are less than ideal. Consider the monitoring performance of personnel who conduct x-ray screening for weapons at airport security checkpoints. These operators are trained to detect several types of weapons and explosives, yet may rarely encounter them in their daily duty periods. To evaluate the efficiency of the security screening, FAA inspectors conduct random checks of particular airline screening points using several test objects corresponding to contraband items, including guns, pipe bombs, grenades, dynamite, and opaque objects. The detection rate of these test objects by airport x-ray screening personnel is typically good, although not perfect, as shown in Fig. 5.1 (Air Transport Association, 1989). However, this is readily understandable given even a cursory evaluation of the working environment of security screening personnel. To say that the work conditions typically do not conform to well-established human factors principles is to point out the obvious — screeners work for long periods in a noisy, distracting environment, sit on uncomfortable chairs viewing monitors under adverse visual conditions (glare, poor contrast, etc.), and have few or no computer aids to help in detection. In addition, the operators are minimally trained, earn low wages, and have few opportunities for career advancement in the security field — conditions that are responsible for the high turnover in these positions. Given these circumstances, the level of detection achieved is what would be anticipated, although it could clearly be improved by more

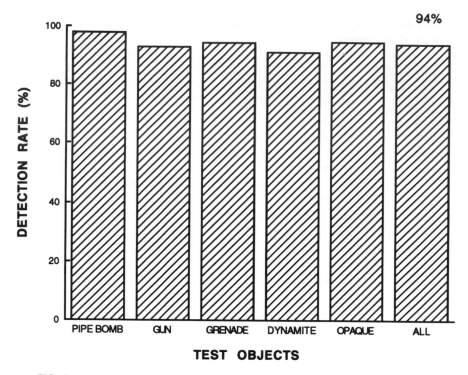

FIG. 5.1. Operational detection rates of test object items by airport x-ray security screening personnel (Air Transport Association, 1989).

attention to ergonomic design. General human engineering principles for optimizing monitoring have been known for years (Craig, 1984; Davies & Parasuraman, 1982).

Unfortunately, established ergonomic principles and guidelines have not been routinely applied to job design so as to improve operator monitoring. Furthermore, the advent of increased levels of automation creates more opportunities for failures of monitoring; as the number of automated subsystems, alarms, decision aids, and so on increases, so does the monitoring load on the human operator. Numerous aviation incidents over the past two decades have involved problems of monitoring of automated systems as one, if not *the* major cause of the incident. An early example is the crash of Eastern Flight 401 in the Florida Everglades, in which the crew, preoccupied with diagnosing a possible problem with the landing gear (itself involving a machine monitoring failure), did not notice the disengagement of the autopilot and did not monitor their altitude (NTSB, 1973).

The following case from the Aviation Safety and Reporting System (ASRS) database provides another example (Singh, Molloy, & Parasuraman, 1993):

The aircraft was at FL370 in Washington center airspace, with the first officer (F/O) flying, using the autopilot. Air-traffic control (ATC) gave a clearance to cross 20 miles west of DQO at FL240. At the top of the descent point, the aircraft began a power-off descent. To comply with the crossing restriction ATC requested an increase in cruise speed to 320 knots when level at FL240. The captain used the Flight Management Computer (FMC) cruise page to re-establish a cruise altitude of FL240 and a new cruise speed of 320 knots. This action eliminated the earlier altitude crossing restriction. Since the F/O was on the progress page and the captain was on the cruise page, the loss of the restriction went unnoticed (the legs page displays altitude restrictions). The aircraft reduced its descent and slowly added power, causing the aircraft to cross the restriction fix 1000 feet high. The problem was noticed just prior to the fix, too late for any action. In filing this report, the captain felt that the crew's confidence in the aircraft to make crossing restrictions that are programmed into the FMC caused the crew to become moderately complacent and not vigilantly monitor altitude. (pp. 112)

In this and many related cases, pilot overreliance on automation has been thought to be a contributing factor. Analyses of ASRS reports have provided evidence of monitoring failures thought to be related to excessive trust in, or overreliance on, automated systems such as the autopilot or flight management system (Lee & Moray, 1992; Mosier, Skitka, & Korte, 1994; Muir, 1988; Riley, 1994; Singh et al., 1993). Misplaced trust in diagnostic expert systems (Will, 1991) and other forms of computer technology (Weick, 1988) has also been widely reported. Mosier et al. (1994) examined a number of similar reports in the ASRS database. They found that 77% of the incidents in which overreliance on automation was suspected involved a probable failure in monitoring.

MACHINE MONITORING

Because case reports and other sources of evidence suggest that human monitoring can be inefficient, particularly after long periods of time at work (Parasuraman, 1987), automating the monitoring function has been proposed as a solution. Some monitoring tasks can be automated; for example automated checklists for preflight procedures (e.g., Palmer & Degani, 1991). Pattern recognition methods are also available for machine detection of abnormal conditions (e.g., of abnormalities in nuclear power plant control) (Gonzalez & Howington, 1977). Automation of some routine monitoring tasks can be a effective strategy given that the detection algorithms and associated software are reliable, and given that the operator can quickly ascertain the appropriate response to the failure from a higher-level display indicator. If the appropriate action is obvious from this indicator, then corrective action can be taken even if the lower-level automated monitor fails.

In general, machine monitoring may be an effective design strategy in particular instances, particularly for lower-level functions. In fact, machine monitoring is used extensively in many complex systems. However, automated monitoring may not provide a *general* solution to the monitoring problem, for at least two reasons, both associated with component proliferation. First, automated monitors can increase the number of alarms, which are already high in many settings (e.g., 866 in the Lockheed L-1011). Human operator response to multiple alarms raises many human factors concerns (Stanton, 1994). Second, to protect against failure of automated monitors, designers may be tempted to put in another system that monitors the automated monitor, a process that could lead to infinite regress. These high-level monitors too can fail, and when failing may disguise the reason for failure. Further automation of monitoring functions would therefore compound the problem of "peripheralization" (Norman et al., 1988) of the human operator from the primary control functions, with negative consequences for human monitoring and diagnosis (Satchell, 1993). Automated warning systems can also lead to reliance on the warning signals as the primary indicator of potential system malfunctions rather than as secondary checks, a problem referred to as "primary-backup inversion" by Wiener and Curry (1980).

WORKLOAD AND AUTOMATION

The view that automation does not necessarily reduce workload was first pointed out some years ago by Edwards (1976). Nevertheless, this early admonition was perhaps not widely heeded. One of the benefits anticipated by designers was that the introduction of automation would reduce operator mental workload. In a number of instances, automation had this desired effect. In addition, on the presumption that reduced workload leads to safer operation, it was thought that automation would reduce human error and improve system safety. The potential fallacy in this line of reasoning was recognized quite early, even by writers in the popular technical press (Bulloch, 1982).

Several lines of evidence indicate that automation does not consistently lead to a reduction in the operator's mental workload. The first source of evidence comes from surveys of commercial pilots and their attitudes toward advanced cockpit automation (Wiener, 1988). Although a significant proportion of pilots agreed with the statement that advanced automation had reduced pilot workload, an equal number disagreed. The general experience in aviation has been that advanced automated devices often do reduce workload, but usually at flight phases where workload is already low, such as cruise; whereas some automation actually increases workload at critical phases, such as takeoff and landing. Thus, automation does not necessarily reduce workload, but shifts the pattern of workload between work phases.

There was a second fallacy in early thinking about the benefits of automation. This was the notion that an operator would have less to do, thereby allowing more time for vigilant monitoring. Vigilance itself was seen as a low-workload task. As noted earlier, the work of Sheridan (1970) and others had already exploded these myths, but they still persist today in some quarters. In fact, in some cases, the human operator may be faced with greater monitoring workload levels with an automated system than existed prior to the automation, despite the fact that the automation was intended to reduce workload. McDaniel (1988) described how this problem can occur in the context of automation in high-performance military aircraft:

> If the automation of a critical function is not perfectly reliable, the pilot will need to monitor it in order to intervene quickly should a malfunction occur. If the pilot continuously monitors the automation, he can intervene in about one second. If he is attending to another task when the malfunction occurs, the reaction time will be several seconds because he must also refresh his awareness of the situation as well as detect that a malfunction has occurred, what has malfunctioned, and what to do about it. In many situations, the malfunctioning aircraft cannot survive even those few seconds. As a result, a pilot dares not perform a second noncritical task rather than monitor the automated critical task. So, while this type of automation permits a useful task to be accomplished, it does nothing to free the pilot's attention resources for other tasks. (p. 837).

Wickens (1992) stated that automating a function increases from one to three the number of decisions the human operator must make in diagnosing a potential system malfunction. For example, consider that an automated system monitors the doors of a commercial aircraft to ensure that they are closed during flight. In the event of a failure indication, the crew must decide whether it reflects a dangerous condition (open door), a failure with the automated monitor, or a malfunction in the display indicator of the automated system. The reasons why automation may increase operator workload are clear.

The paradox is that implementing automation in an attempt to reduce workload may actually result in increased workload, because of the cognitive workload associated with monitoring the automation. Moreover, the workload of monitoring may be considerable, contrary to popular belief. Traditionally, monitoring and vigilance have been considered to be "unstimulating" tasks that do not tax the operator's capacities. This line of reasoning led to the development of the arousal theory of vigilance, which postulated that the level of physiological arousal fell during a vigilance task, leading to the classic vigilance decrement over time (Duffy, 1957). But this view is based on older conceptualizations of arousal in which arousal was linked to underload, monotony, and boredom. Newer conceptualizations of arousal theory have linked arousal to the deployment of attentional resources (e.g., Hancock & Warm, 1989; Matthews, Davies, & Lees, 1990). Recent research on vigilance has also led to a revision of the view that the workload of vigilance is low (Deaton & Parasuraman, 1993; Warm, Dember, & Hancock, chapter 9, this volume). These studies indicate that even superficially "simple" vigilance tasks can impose considerable mental workload, of the level associated with such tasks as problem solving and decision making.

Clearly, then, the notion that automation always reduces workload fails on at least two grounds. First, automation may change the pattern of workload across work segments, but does not always reduce overall workload. Second, even in highly automated systems in which the human operator is engaged primarily in monitoring, workload may not be reduced because the demands of monitoring can be considerable.

FAILURE DETECTION UNDER ACTIVE AND PASSIVE CONTROL

Automation clearly can either decrease or increase mental workload. And monitoring demands can increase with automation. What of monitoring efficiency? Are human operators better at monitoring a system when they also control it manually or when they only monitor an automated control system? There is a large body of research on human vigilance indicating that human monitoring performance is prone to error when monitoring must be performed for long, uninterrupted periods of time (Davies & Parasuraman, 1982; Warm, 1984). However, most of this work has been carried out using simple sensory vigilance tasks that do not approach the complexity of monitoring and search jobs in real systems (Moray, 1984; Parasuraman, 1986). Moreover, despite the general consensus that automation increases monitoring demands, there is surprisingly little empirical work directly comparing human monitoring behavior in automated systems to that of manual performance. Research on vigilance has shown that detection of low-probability events is degraded after prolonged periods on watch (Davies & Parasuraman, 1982). One might predict, therefore, that human operator detection of a failure in the automated control of a task, which is likely to be an improbable event, would be very poor after a prolonged period spent under automation control. However, most vigilance research has been carried out with simple tasks requiring detection of infrequent signals that carry little significance or meaning to the subject (Davies & Parasuraman, 1982), and hence these findings may not apply in the richer, multitask environment of automated systems. Until recently, there were few studies that have specifically examined monitoring performance in automated systems.

The few controlled empirical studies that have been conducted have mostly involved tracking or flight-control tasks in which subjects were required to detect sudden failures in the dynamics of the controlled element (Bortolussi & Vidulich, 1989; Johannsen, Pfendler, & Stein, 1976; Kessel & Wickens, 1982; Wickens & Kessel, 1981; Young, 1969). In one study, subjects either monitored only, or both monitored and controlled, a tracking task. The monitoring task involved a sudden change in the control system dynamics; this represented the "failure" that the subjects were required to detect. In the "manual" condition subjects actively controlled the tracking and monitored at the same time. In the "passive automation" condition subjects only monitored the system dynamics. Speed of failure detection was found to be significantly slower in the automated than in the manual condition—a result attributed to insufficient propriceptive feedback when tracking tasks are under automated control (Wickens & Kessel, 1979). Similar results were obtained with more realistic flight simulation experiments; one study in which subjects were required to detect autopilot failure during a landing approach with simulated vertical gusts (Johannsen, Pfendler, & Stein, 1976), and another in which helicopter pilots had to respond to unexpected wind-shear bursts while using an automated altitude-hold (Bortolussi & Vidulich, 1989).

Only a few studies have examined the effects of automation on monitoring performance with decision-making tasks. Idaszak and Hulin (1989) carried out a study using a simulated process control task. Subjects were required to monitor the system and respond to limits, alarms, and deviations in process parameter values. In the active monitoring condition subjects controlled the process and monitored it at the same time. In the passive condition, subjects only monitored the process while watching a video of nature scenes. The active group monitored better than did the passive group, being faster at detecting both out-of-limits conditions and alarms than were the passive operators. Idaszak and Hulin (1989) suggested that active participation increases the operator's workload and perceptions of task difficulty, and, therefore, active participation acts as a source of motivation and benefits monitoring performance. However, one problem with this study is that the passive group was required to answer questions about the video they watched; hence, it is likely that they were forced to allocate resources away from the monitoring task, and this diversion of resources may have been greater than in the active group doing both the monitoring and the controlling task. It is therefore questionable whether the passive condition used in this study is representative of the type of monitoring of automation in many real systems.

Liu, Fuld, and Wickens (1993) tested subjects on a scheduling task requiring assignment of incoming customers to the shortest of three parallel service lines or queues. In the manual mode, subjects had to make assignments (by pressing one of three buttons, one for each queue) as well as monitor their own assignments and press another key if they detected an error. In the automated mode, subjects were told that the queuing assignments would be made by the computer and that their task was only to detect wrong assignments. In fact, the "automation" condition was simulated by giving each subject a replay of their earlier manual performance; this was done to control for visual display differences between the automated and manual modes. In contrast to Wickens and Kessel (1979), Liu et al. (1993) found that error detection was superior in the automated rather than in the manual assignment condition. However, this effect may have been influenced by the control method that Liu et al. used. Because subjects performed the automated condition after the manual condition and received the same sequence of stimuli, they may have benefited from seeing difficult errors (that they made initially in the manual condition) again in the automated condition, whereas subjects seeing such errors for the first time may not have detected them. They interpreted the different pattern of results in the two studies in the framework of multiple-resource theory (Wickens, 1984): The greater competition between the monitoring and assignment tasks for domain-specific attentional resources would result in poorer manual monitoring, as compared to the Wickens and Kessel (1979) study in which monitoring would not compete with the response-related resources associated with tracking.

Hilburn, Jorna, and Parasuraman (1995) also found a benefit of automation on monitoring performance. Using a realistic air traffic control (ATC) simulation of the Netherlands airspace, they examined the effects of strategic ATC decision aiding (e.g., a descent advisor) on the performance of licensed en-route controllers. They found that, compared to unaided performance, decision aiding was associated with reduced reaction time to respond to secondary malfunctions (failures by pilots to respond to datalinked clearances).

Thackray and Touchstone (1989) had subjects perform a simulated ATC task either with or without the help of an automated aid that provided advisory messages concerning potential aircraft-to-aircraft conflicts. The automation failed twice, early and late during a two-hour session. Thackray and Touchstone (1989) reasoned that subjects using the automated aid would become "complacent" and thus fail to detect the failures. Although subjects were somewhat slower to respond to the first failure when using the automated aid, this was not the case for the later failure. Overall, subjects were as efficient at monitoring in the presence of automation as in its absence.

In general, these studies indicate that when operators do not actively control a process they are poorer at detecting malfunctions than when they are engaged both in control and in monitoring. This effect probably is due to the limited proprioceptive and other forms of feedback under conditions of passive monitoring. However, there are also some indications that monitoring under automated control may not always be poor. Moreover, these studies do not show that failure detection in general (e.g., for perceptual, cognitive, and motor tasks) is adversely affected by automation.

MONITORING BEHAVIOR IN AUTOMATED SYSTEMS

Among the factors that may influence efficiency of monitoring under automation are the task load imposed on the operator and the availability of feedback. Thackray and Touchstone (1989) also suggested that evidence of poor monitoring may not be forthcoming in the limited-duration sessions used in laboratory studies but might require lengthy field trials. However, Parasuraman, Molloy, and Singh (1993) suggested that a critical factor might be the overall workload level — whether monitoring is the only task or whether operators are simultaneously engaged in other manual tasks as well. They proposed that monitoring might be efficient when it is the only task (as in Thackray & Touchstone's 1989 study), with or without computer aiding, but that when operators are engaged in other simultaneous tasks, monitoring of an automated task is poorer than that of the same task under manual control.

Manual Task Load

To test this idea, Parasuraman et al. (1993) had nonpilot subjects simultaneously perform tracking and fuel-management tasks manually over four 30-minute sessions. At the same time, an automated engine-status task had to be monitored. Subjects were required to detect occasional automation "failures" by identifying engine malfunctions not detected by the automation. In one of the experimental conditions, the reliability of the automation remained constant over time for the first three sessions. Subjects detected over 70% of malfunctions on the engine-status task when they performed the task manually, while simultaneously carrying out tracking and fuel management. However, when the engine-status task was under automation control, detection of malfunctions was markedly reduced (see Fig. 5.2).

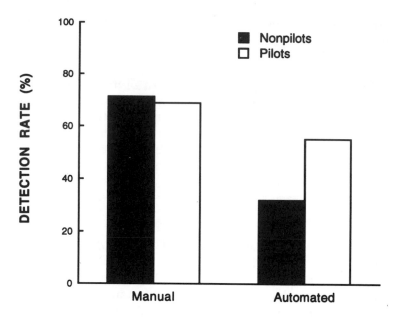

FIG. 5.2. Monitoring performance (detection rate of engine malfunctions) under manual and automated conditions by pilots and nonpilots.

This substantial reduction in failure detection rate was not accompanied by any significant change in false alarms, indicating that detection sensitivity was reduced. In a separate experiment, the same conditions were administered but subjects performed only the monitoring task, without the tracking and fuel-management tasks. Subjects were now nearly perfect (> 95%) in detecting failures in the automated control of the engine-status task, which was the only task. These results provide a clear indication of the potential cost of long-term "static" automation on system performance, and show that human operator monitoring of automation can be poor when subjects simultaneously perform other manual tasks.

Experienced pilots were found to show performance trends similar to those of nonpilots (Parasuraman, Mouloua, & Molloy, 1994). Fig. 5.2 compares the performance of pilots and nonpilots. Although the overall performance level of the pilots was higher than that of the nonpilots, the pilots showed the same pattern of performance decrement under automation as did the nonpilots.

Automation Reliability and Consistency

The monitoring performance patterns of both pilots and nonpilots in these studies is consistent with the view that reliable automation engenders operator trust (Lee & Moray, 1992). This leads to a reliance on automation that is associated with only occasional monitoring of its efficiency, suggesting that a critical factor in the development of this phenomenon might be the constant, unchanging reliability of the automation. Conversely, automation with inconsistent reliability should not induce trust and should therefore be monitored more closely. This prediction was supported in the Parasuraman et al. (1993) study. The data shown in Fig. 5.2 were obtained from conditions in which the reliability of the automation remained constant over time. Parasuraman et al. also examined monitoring performance for another group of subjects for whom the automation reliability varied from low to high every 10 minutes in each session. Monitoring performance was significantly higher in this group than in the constant-reliability group, as shown in Fig. 5.3.

The absolute level of automation reliability may also affect monitoring performance. In a subsequent study in which automation reliability was varied from very low ($\sim$ 25%) to very high (> 90%), the detection rate of automation failures varied inversely with automation reliability (May, Molloy, & Parasuraman, 1993). Interestingly, even when automation reliability was very low, which should lead to complete mistrust of the automation, monitoring performance under automation was slightly (although not significantly) poorer than performance under manual control.

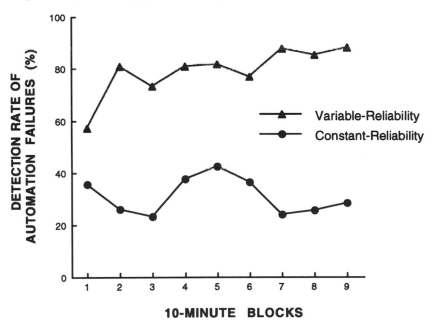

FIG. 5.3.   Effects of consistency of automation reliability on monitoring performance under automation.

The automation failure rates used in these studies were high and unrepresentative of real systems. It could therefore be argued that the results are artifactual. Two pieces of evidence mitigate against this possibility. First, infrequent failures of the automation were also simulated. The joint system (human + computer) probability of detection of system malfunctions for a "low" (for simulation purposes) failure rate of 10^{-2} was estimated from empirical operator performance data at higher failure rates. System performance was simply the weighted average of the detection probabilities of the automation and the human. As Fig. 5.4 shows, system performance was near perfect (100% detection) during the early stages of performance. Late in the simulation, however, a "catastrophic" or total failure of the automation was staged (in Block 11); when this occurred the automation failed to detect any engine malfunctions, essentially forcing the subject to a manual mode. As Fig. 5.4 indicates, under these conditions, the system probability of detection drops precipitously. These results show that high (i.e. near-perfect) automation reliability can compensate for poor operator performance except when the automation fails catastrophically, in which case system performance may be markedly compromised.

Molloy and Parasuraman (in press) extended this analysis by carrying out a similar study in which only a single failure of the automation occurred during a session. As in the Parasuraman et al. (1993) study, subjects performed tracking and fuel management tasks manually, and monitored an engine status task under automation control. In this study, however, the automation failed on only one occasion, either early or late, during the

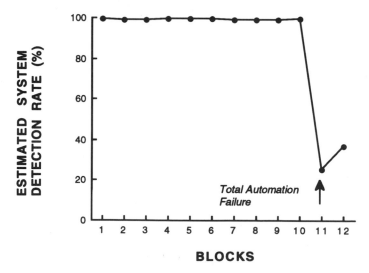

FIG. 5.4.   Estimated system monitoring performance (human + computer) for a system with a low failure probability of 10^{-2}.

course of a 30-minute sesssion. Molloy and Parasuraman (in press) found essentially the same result as did Parasuraman et al. (1993), namely that monitoring was less accurate when the task was automated than when it was under manual control (see Fig. 5.5). As in the previous study by Parasuraman et al. (1993), these effects were found only under multiple task conditions; when monitoring was the only task, it was equally efficient under manual and automation control.

Display Separation

In the studies finding poor monitoring of automated tasks under multiple-task conditions, the manual tasks performed by the subjects were presented in the central visual field, above and below the center of the display, whereas the automated engine-status task was displayed about 6° to the left of center. It could therefore be argued that poor monitoring of automation arises primarily because operators are busy with other tasks in their central visual field and fail to fixate automated tasks presented in the periphery. Many studies have shown that in accordance with sampling theory, the

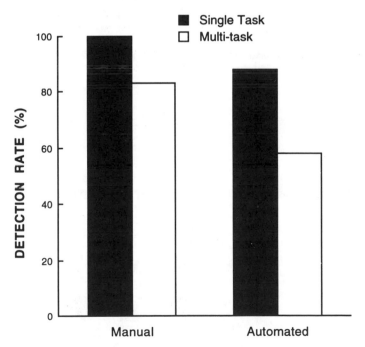

FIG. 5.5. Monitoring performance under manual and automated conditions for a single automation failure.

scanning patterns of operators reflects the frequency or bandwidth (Senders, 1964) and information content (Bohnen & Leermakers, 1991) of displays; high-frequency displays are fixated more frequently (see also Moray, 1984). This line of reasoning would suggest that inefficiency in monitoring automation does not represent solely an attentional resource problem but occurs whenever there is a heavy visual information load that commands the operator's central vision elsewhere.

To test this view, the previous experiments were repeated in a study conducted by Steve Westerman in our laboratory in which the automated monitoring task was spatially superimposed on the manual tracking task in the central visual field. The superimposition was carefully designed by interleaving display elements and using color to avoid masking of one display by the other (see Fig. 5.6). This should allow monitoring of the automated task without fixations away from manual tracking, given the continuous nature of the tracking task. It was predicted that monitoring performance would be better in this superimposed condition than in the normal situation of spatially separate automated and manual tasks. The results, however, did not confirm this hypothesis. Whereas the detection rate of automation failures was poor in the standard separate display condition (53.3%), it was equally poor in the superimposed condition (52.7%). The only significant effect was that performance in both automated conditions was poorer than when the same task was performed under

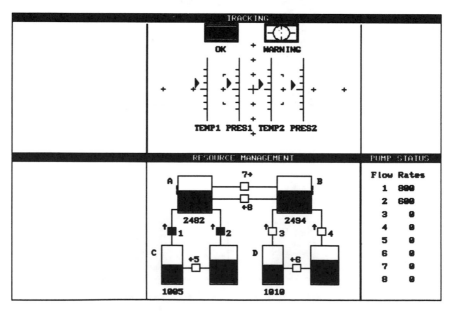

FIG. 5.6.   Spatial superimposition of an automated engine status on a manual tracking task.

manual control. Thus, no benefit of spatial superimposition was found for detection accuracy. These results provide strong evidence against a purely visual-fixation account of monitoring inefficiency, and are consistent with the view that the performance decrement from manual to automated control represents an attentional effect related to operator reliance on the automation. These findings suggest also that display redesign based on task layout cannot solve the automation monitoring problem.

COUNTERMEASURES

The standard engineering solution to human error is to automate. In the case of monitoring, this is possible and perhaps even desirable for low-level system functions. As noted previously, however, for higher-order system functions, machine monitoring may not provide a general solution to the monitoring problem. This is because high-level machine monitors will themselves have to be monitored, and designers might be tempted to develop another machine monitor to track the lower-level monitors, and so on in infinite regress. Back-ups to back-ups can also increase the peripheralization of the ultimate back-up, the human operator, and provide only an illusion of redundancy, not true redundancy (Palmer & Degani, 1991).

There are several potential ways in which efficient human monitoring of automation might be promoted. As mentioned earlier, general ergonomic principles for improving vigilance have been published (Craig, 1984; Davies & Parasuraman, 1982), but additional methods specific to the problem of monitoring automated systems also need to be considered. There is not the space here to consider all possible options. Two methods, display integration and adaptive task allocation, are discussed. Other possibilities include individual and team training (see Scerbo, chapter 3, this volume) and group management of monitoring functions (Satchell, 1993).

Integrated Displays

In the study discussed earlier that examined the effects of display separation on monitoring (Parasuraman et al., 1994), spatial superimposition of automated and primary manual tasks was found not to improve monitoring performance. This finding was taken to suggest that poor monitoring of an automated task reflects attentional as opposed to (or in addition to) visual fixation factors. This suggests that display redesign efforts should be aimed at reducing the attentional demands associated with detecting a malfunction in an automated task, rather than focus simply on display layout. Integration of elements within a display is one method for reducing attentional demands of fault detection, particularly if the integrated components combine to form an emergent feature (Bennett & Flach, 1992). Emergent features that have been examined include various object forms and shapes (e.g., a polygon), but parts of objects (e.g., the apex of a triangle) have also been used (Sanderson, Flach, Buttigieg, & Casey, 1989; Woods, Wise, & Hanes, 1981) If the emergent feature is used to index a malfunction, detection of the malfunction could occur preattentively and in parallel with other tasks. In principle, then, an automated task presented in an integrated display should not be subject to the factors described earlier that lead to poor monitoring.

Molloy and Parasuraman (1994) examined this possibility with a version of an engine-status display that is currently implemented in many cockpits, the Engine Indicator and Crew Alerting System (EICAS). The EICAS display used in this study consisted of four circular gauges showing different engine parameters. The integrated form of this display was based on one developed by Abbott (1990), the Engine Monitoring and Crew Alerting System (EMACS), in which the four engine paramaters are shown as columns on a deviation bar graph. Parameter values above or below normal were displayed as deviations from a horizontal line (the emergent feature) that represent normal operation. Molloy and Parasuraman (1994) tested nonpilots with these engine-status displays using the same monitoring paradigm developed by Parasuraman et al. (1993), that is, under manual and automated conditions, with occasional failures of the automation. Performance (detection rate) under manual conditions was initially equated for the EICAS and EMACS tasks and averaged about 70%. Under automation, however, although subjects detected only 36% of failures with the nonintegrated EICAS display, they detected 64% of automation failures with the integrated EMACS display. The results showed that display integration can significantly improve monitoring of an automated task. Note, however, that performance with the integrated display was still somewhat lower when the task was automated (64%) than when it was performed manually (72%). Similar results were obtained in a recent replication of this study in which experienced pilots served as subjects (Molloy, Deaton, & Parasuraman, 1995).

Adaptive Task Allocation

Given that display integration reduces inefficiency of monitoring of automation but does not eliminate it completely, what other countermeasures could be used? The problem was articulated by Liu et al. (1993) as one of deciding between computer control of functions (with "passive" human monitoring) and active human control of the same function. Posed in this manner, the question might imply that task allocation between humans and machines is all or none. This need not be the case. The traditional approach to automation is based on a policy of allocation of function in which either the human or the machine has full control of a task (Fitts, 1951). Although this policy may work for functions that cannot be performed by either agent, it faces difficulties for functions that can be performed by either the human or the machine. An alternative philosophy, variously termed *adaptive task allocation* or *adaptive automation*, sees function allocation between humans and machines as flexible (Hancock & Chignell, 1989; Rouse, 1988; see also Scerbo, chap. 3, this volume).

According to proponents of adaptive systems, the benefits of automation can be maximized and the costs minimized if tasks are allocated to automated subsystems or to the human operator in an adaptive, flexible manner rather than in an all-or-none fashion (Rouse, 1988). For example, the operator can actively control a process during moderate workload, allocate this function to an automated subsystem during peak workload if necessary, and retake manual control when workload diminishes, thereby minimizing some of the costs of conventional automation, (e.g., monitoring inefficiency). If these assertions are correct, than adaptive task allocation might help promote better monitoring of automation. Monitoring under conditions of manual control is often quite good. This suggests that one method to improve monitoring of automation might be to insert brief periods of manual task performance after a long period of automation. Manual task reallocation might have a beneficial impact on subsequent operator monitoring under automation.

This idea was tested in a study using the same flight-simulation paradigm developed by Parasuraman et al. (1993). Eighteen subjects performed the flight-simulation task for three 30-minute sessions. For the control group, the engine-status task was automated throughout. There were two adaptive task allocation groups, corresponding to two possible methods of adaptation (Parasuraman, Bahri, Deaton, Morrison, & Barnes, 1992; Rouse, 1988). For the model-based adaptive group, a single 10-minute block of fully manual performance on the systems-monitoring task was allocated to subjects in the middle of the second session (i.e., on Block 5). This type of function allocation is termed "model-based" because it reflects a model indicating that operator performance of that function is likely to be inefficient at that point in time (as shown by Parasuraman et al., 1993). This method, however, is insensitive to the actual performance of an individual operator. For the *performance-based* adaptive group, function allocation was changed in the middle of the second session for an individual subject only if the past history of that subject's monitoring performance did not meet a criterion. If the performance criterion was met, the function was not allocated to the subject but continued under automation control. For both adaptive groups, the change in allocation was signaled 30 seconds prior to the change. Following 10 minutes of manual performance in Block

5, a prewarned reallocation of the monitoring task to the automation was made. Subjects completed the rest of the second session and the entire third session (Blocks 6 through 9) with automation.

Adaptive allocation of a task to the operator improved monitoring performance in subsequent blocks. The detection rate of automation failures was not significantly different for the three groups during the first 40 minutes spent with automation. Detection rates were higher for the adaptive groups than for the control group in the manual block. This finding is not surprising given the previous evidence supporting superior monitoring under manual compared to automated conditions (Parasuraman et al., 1993). A novel finding, however, was that the performance benefit for both adaptive groups persisted for the next four blocks (6 through 9) when the task was returned to automated control. Overall monitoring performance for the control group was the same in the pre- (Blocks 1 through 4) and postallocation phases (Blocks 6 through 9) (see Fig. 5.7). For the two adaptive groups, however, mean detection rate was higher in the postallocation than in the preallocation phase. The important point to note is that the task conditions were identical in both these

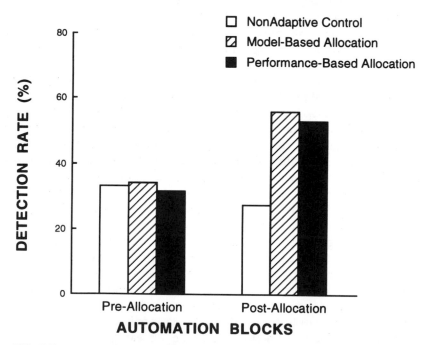

FIG. 5.7. Overall effects of adaptive task allocation on monitoring performance under automation. The nonadaptive group performed under automated control throughout all blocks. For the two adaptive groups the automated task was returned to manual control between the preallocation and postallocation blocks.

phases—the engine-status task was automated while subjects manually controlled tracking and fuel management. The performance benefit exceeded 50% for both groups. Similar results were obtained in a subsequent study in which experienced pilots served as subjects (Parasuraman et al., 1994).

For the same set of flight-simulation tasks more sustained benefits have also been obtained with multiple or repetitive manual task reallocation (Mouloua, Parasuraman, & Molloy, 1993). One cautionary note should be sounded, however. The automated engine-status task did not incorporate a "history" or memory component: When it was allocated with only minimal warning, operators did not need to refresh the previous history of malfunctions or review normal operation on the task in order to perform the task satisfactorily. This might not be the case for other systems-monitoring tasks; for example, in process control where faults develop over relatively long periods of time (Moray, 1986). Use of historical and predictive information on engine-status displays in the cockpit may also aid pilots in fault detection (Trujillo, 1994). If adaptive task allocation were to be used in these cases, the allocation procedure would have to be coordinated with additional decision aiding that updated the monitor on the prior status of the automated task.

With this caveat, the results of these studies show that adaptive task allocation provides a potentially important countermeasure against automation-related monitoring inefficiency. More generally, intermittent manual task reallocation may also provide another benefit, namely reduced degradation of manual skills that may occur following long periods of automation. However, additional studies with different monitoring tasks, other methods for adaptive control, and under different conditions of multiple-task performance need to be conducted to test the general effectiveness of this new approach to function allocation.

CONCLUSIONS

Automation has increased the importance of the need to understand and control the factors that influence human monitoring behavior. Although it is often pointed out that human monitoring can be subject to errors, in many instances operational monitoring can be quite efficient. Where human monitoring tends to be poor is in work environments that do not conform to well-established ergonomic design principles (Craig, 1984; Davies & Parasuraman, 1982) and in highly automated systems where there is little opportunity for manual experience with the automated tasks. Among the major influences on monitoring of automated systems for which direct empirical evidence is available are automation reliability, automation consistency, the level of manual task load on the operator, and display factors. Integration of display elements and the use of emergent features to signify automation malfunctions can reduce inefficiency of monitoring (Molloy & Parasuraman, 1994). Another possible countermeasure is adaptive task allocation (Parasuraman, 1993). However, these provide only partial solutions and may not be generally applicable in complex systems in which there are hundreds of automated subtasks. Given that the development of automation shows little evidence of declining, human factors professionals will be severely challenged to come up with effective methods to help those who will be required to "watch the computers."

ACKNOWLEDGMENTS

Supported by research grant NAG-1-1296 from the National Aeronautics and Space Administration, Langley Research Center, Hampton, VA (Alan Pope, technical monitor).

REFERENCES

Abbott, T. S. (1990). *A simulation evaluation of the engine monitoring and control system display* (Technical Report No. 2960). Hampton, VA: NASA Langley Research Center.

Air Transport Association. (1989). *Airport data from 10/01/88 to 5/31/89.* (ACS-100 Security Division Report). Washington, DC: Author.

Bainbridge, L. (1983). Ironies of automation. *Automatica, 19,* 775–779.

Bennett, K., & Flach, J. M. (1992). Graphical displays: Implications for divided attention, focused attention, and problem solving. *Human Factors, 34,* 513–533.

Bohnen, H. G. M., & Leermakers, M. A. M. (1991). Sampling behavior in a four-instrument monitoring task. *IEEE Transactions on Systems, Man, and Cybernetics, SMC-21,* 893–897.

Bortulossi, M. R., & Vidulich, M. A. (1989). The benefits and costs of automation in advanced helicopters: An empirical study. In *Proceedings of the 5th International Symposium on Aviation Psychology* (pp.271–275). Columbus, OH: The Ohio State University.

Bulloch, C. (1982). Cockpit automation and workload reduction. Too much of a good thing? *Interavia, 3,* 263–264.

Craig, A. (1984). Human engineering: The control of vigilance. In J. S. Warm (Ed.), *Sustained attention in human performance* (pp. 247–291). London: Wiley.

Davies, D. R., & Parasuraman, R. (1982). *The psychology of vigilance.* London: Academic Press.

Deaton, J. E., & Parasuraman, R. (1993). Sensory and cognitive vigilance: Age, event rate, and subjective workload. *Human Performance, 4,* 71–97.

Duffy, E. (1957). The psychological significance of the concept of "arousal" or "activation." *Psychological Review, 64,* 265–275.

Edwards, E. (1976). Some aspects of automation in civil transport aircraft. In T.B. Sheridan & G. Johannsen (Eds.), *Monitoring behavior and supervisory control.* New York: Plenum.

Fitts, P. M. (1951). (Ed.) *Human engineering for an effective air navigation and traffic control system.* Washington, DC: National Research Council.

Gonzalez, R. C., & Howington, L. C. (1977). Machine recognition of abnormal behavior in nuclear reactors. *IEEE Transactions on Systems, Man, and Cybernetics, SMC-7,* 717–728.

Hancock, P. A., & Chignell, M. H. (1989). (Eds.). *Intelligent interfaces: Theory, research, and design.* Amsterdam: North Holland.

Hancock, P. A., & Warm, J. S. (1989). A dynamic theory of stress and sustained attention. *Human Factors, 31,* 519–537.

Hilburn, B., Jorna, P. G. A. M., & Parasuraman, R. (1995). The effect of advanced ATC automation on mental workload and monitoring performance: An empirical investigation in Dutch airspace. In *Proceedings of the 8th International Symposium on Aviation Psychology.* Columbus, OH: The Ohio State University.

Idaszak, J. R., & Hulin, C. L. (1989). *Active participation in highly automated systems: Turning the wrong stuff into the right stuff.* Urbana–Champaign, IL: Aviation Research Laboratory, Institute of Aviation, University of Illinois at Urbana Champaign.

Johannsen, G., Pfendler, C., & Stein, W. (1976). Human performance and workload in simulated landing-approaches with autopilot failures. In T.B. Sheridan & G. Johannsen (Eds.), *Monitoring behavior and supervisory control* (pp. 83–93). New York: Plenum.

Kessel, C. J., & Wickens, C. D. (1982). The transfer of failure detection skills between monitoring and controlling dynamics. *Human Factors, 24,* 49–60.

Koshland, D. E., Jr. (1989). Low probability-high consequence accidents. *Science, 244,* 405.

Lee, J. D., & Moray, N. (1992). Trust, control strategies, and allocation of function in human-machine systems. *Ergonomics, 35,* 1243–1270.

Liu, Y. R., Fuld, R., & Wickens, C. D. (1993). Monitoring behavior in manual and automated scheduling systems. *International Journal of Man-Machine Studies, 39,* 1015–1029.

Matthews, G., Davies, D. R., & Lees, J. (1990). Arousal, extraversion, and visual sustained attention: The role of resource availability. *Personality and Individual Differences, 11,* 1159–1173.

May, P., Molloy, R., & Parasuraman, R. (1993, October). *Effects of automation reliability and failure rate on monitoring performance in a multitask environment.* Paper presented at the Annual Meeting of the Human Factors Society, Seattle, WA.

McDaniel, J. W. (1988). *Rules for fighter cockpit automation.* In *Proceedings of the IEEE National Aerospace and Electronics Conference* (pp. 831–838). New York: IEEE.

Molloy, R., Deaton, J.E., & Parasuraman, R. (1995). Monitoring performance with the EMACS display in an automated environment. In *Proceedings of the 8th International Symposium on Aviation Psychology.* Columbus, OH: The Ohio State University.

Molloy, R., & Parasuraman, R. (1994). Automation-induced monitoring inefficiency: The role of display integration and redundant color coding. In M. Mouloua & R. Parasuraman (Eds.), *Human performance in automated systems: Current research and trends* (pp. 224–228). Hillsdale, NJ: Lawrence Erlbaum Associates.

Molloy, R., & Parasuraman, R. (in press). Monitoring an automated system for a single failure: Vigilance and task complexity effects. *Human Factors.*

Moray, N. (1984). Attention to dynamic visual displays in man-machine systems. In R. Parasuraman & D. R. Davies (Eds.), *Varieties of attention* (pp. 485–513). New York: Academic Press.

Moray, N. (1986). Monitoring behavior and supervisory control. In K. Boff, L. Kaufman, & J.Thomas (Eds.), *Handbook of perception and human performance. Vol. 2. Cognitive processes and performance* (pp. 40.1–40.51). New York: Wiley.

Mosier, K. L., Skitka, L. J., & Korte, K. J. (1994). Cognitive and social psychological issues in flight crew/automation interaction. In M. Mouloua & R. Parasuraman (Eds.), *Human performance in automated systems: Current research and trends* (pp. 191–197). Hillsdale, NJ: Lawrence Erlbaum Associates.

Mouloua, M., & Parasuraman, R. (1994). (Eds.). *Human performance in automated systems: Current research and trends.* Hillsdale, NJ: Lawrence Erlbaum Associates.

Mouloua, M., Parasuraman, R., & Molloy, R. (1993). Monitoring automation failures:

Effects of single and multiadaptive function allocation. In *Proceedings of the Human Factors Society* (pp. 1-5). Seattle, WA: Human Factors and Eugonomics Society.

Muir, B. M. (1988). Trust between humans and machines, and the design of decision aids. In E. Hollnagel, G. Mancini, & D.D. Woods (Eds.), *Cognitive engineering in complex dynamic worlds* (pp. 71-83). London: Academic Press.

National Transportation Safety Board. (1973). *Eastern Airlines L-1011, Miami, Florida, December 29, 1972* (Report No. NTSB-AAR-73-14). Washington DC: Author.

Norman, S., Billings, C. E., Nagel, D., Palmer, E., Wiener, E. L., & Woods D. D. (1988). *Aircraft automation philosophy: A source document.* Moffett Field, CA: NASA Ames Research Center.

Palmer, E., & Degani, A. (1991). Electronic checklists: Evaluation of two levels of automation. In *Proceedings of the International Symposium on Aviation Psychology*, 6th Annual Conference (pp. 178-183). Columbus, OH: The Ohio State University.

Parasuraman, R. (1986). Vigilance, monitoring, and search. In K. Boff, L. Kaufman, & J. Thomas (Eds.), *Handbook of perception and human performance. Vol. 2. Cognitive processes and performance* (pp. 43.1-43.39). New York: Wiley.

Parasuraman, R. (1987). Human–computer monitoring. *Human Factors, 29,* 695-706.

Parasuraman, R. (1993). Effects of adaptive function allocation on human performance. In D. J. Garland & J. A. Wise (Eds.), *Human factors and advanced aviation technologies* (pp. 147-157). Daytona Beach, FL: Embry-Riddle Aeronautical University Press.

Parasuraman, R., Bahri, T., Deaton, J., Morrison, J., & Barnes, M. (1992). *Theory and design of adaptive automation in aviation systems* (Progress Report No. NAWCADWAR-92033-60). Warminster, PA: Naval Air Warfare Center, Aircraft Division.

Parasuraman, R., Molloy, R., & Singh, I.L. (1993). Performance consequences of automation-induced "complacency." *International Journal of Aviation Psychology, 3,* 1-23.

Parasuraman, R., Mouloua, M., & Molloy, R. (1994). Monitoring automation failures in human–machine systems. In M. Mouloua & R. Parasuraman (Eds.), *Human performance in automated systems: Current research and trends* (pp. 45-49). Hillsdale, NJ: Lawrence Erlbaum Associates.

Perrow, C. (1984). *Normal accidents.* New York: Basic Books.

Peterson, W. L. (1984, June). The rage to automate. *Air Line Pilot,* pp. 15-17.

Riley, V. (1994). A theory of operator reliance on automation. In M. Mouloua & R. Parasuraman (Eds.), *Human performance in automated systems: Current research and trends* (pp. 8-14). Hillsdale, NJ: Lawrence Erlbaum Associates.

Rouse, W.B. (1988). Adaptive aiding for human/computer control. *Human Factors, 30,* 431-438.

Sanderson, P. M., Flach, J. M., Buttigieg, M. A., & Casey, E. J. (1989). Object displays do not always support better integrated task performance. *Human Factors, 31,* 183-198.

Satchell, P. (1993). *Cockpit monitoring and alerting systems.* Aldershot, England: Ashgate.

Senders, J. (1964). The human operator as a monitor and controller of multidegree of freedom systems. *IEEE Transactions on Human Factors in Electronics, HFE-9,* 1-6.

Sheridan, T. B. (1970). On how often the supervisor should sample. *IEEE Transactions on Systems Science and Cybernetics, SSC-6,* 140-145.

Sheridan, T. B. (1992). *Telerobotics, automation, and human supervisory control.* Cambridge, MA: MIT Press.

Singh, I. L., Molloy, R., & Parasuraman, R. (1993). Automation-induced "complacency": Development of the complacency-potential rating scale. *International Journal of Aviation Psychology, 3,* 111-121.

Stanton, N. (1994). *Human factors in alarm design.* London: Taylor and Francis.

Thackray, R. I., & Touchstone, R. M. (1989). Detection efficiency on an air traffic control monitoring task with and without computer aiding. *Aviation, Space, and Environmental Medicine, 60,* 744-748.

Trujillo, A. (1994). *Effects of historical and predictive information on ability of transport pilot to predict an alert* (NASA Technical Memorandum 4547). Hampton, VA: NASA Langley Research Center.

Warm, J. S. (1984). *Sustained attention in human performance.* London: Wiley.

Weick, K. E. (1988). Enacted sensemaking in crisis situations. *Journal of Management Studies, 25,* 305–317.

Wickens, C. D. (1984). Processing resources in attention. In R. Parasuraman & D. R. Davies (Eds.), *Varieties of attention* (pp. 63–102). New York: Academic Press.

Wickens, C. D. (1992). *Engineering psychology and human performance.* New York: HarperCollins.

Wickens, C. D. (1994). Designing for situational awareness and trust in automation. In *Proceedings of the IFAC Conference on Integrated Engineering.* (pp. 171–175). Baden-Baden, Germany: IFAC.

Wickens, C.D., & Kessel, C. (1979). The effects of participatory mode and task workload on the detection of dynamic system failures. *IEEE Transactions on Systems, Man, and Cybernetics, SMC-9,* 24–34.

Wickens, C. D., & Kessel, C. (1981). Failure detection in dynamic systems. In J. Rasmussen & W. Rouse (Eds.), *Human detection and diagnosis of system failures (pp. 433–461).* New York: Plenum.

Wiener, E. L. (1988). Cockpit automation. In E. L. Wiener & D. C. Nagel (Eds.), *Human factors in aviation* (pp. 433–461). San Diego, CA: Academic Press.

Wiener, E. L., & Curry, R. E. (1980). Flight-deck automation: Promises and problems. *Ergonomics, 23,* 995–1011.

Will, R. P. (1991). True and false dependence on technology: Evaluation with an expert system. *Computers in Human Behavior, 7,* 171–183.

Woods, D. D., Wise, J., & Hanes, L. (1981). An evaluation of nuclear power plant safety parameter display systems. *Proceedings of the Human Factors Society, 25,* 110–114.

Young, L. R. (1969). On adaptive manual control. *IEEE Transactions on Man–Machine Systems, MMS-10,* 292–331.

6 Pilot Workload and Flightdeck Automation

Barry H. Kantowitz
John L. Campbell
Battelle HF Transportation Center, Seattle

INTRODUCTION

The relationship between automation and operator mental workload is an important consideration to efficiency and safety in many modern human--machine systems. Aviation is a domain where this relationship is particularly crucial. This chapter focuses on some of the potential problems that can occur with automated glass cockpits, especially as these problems relate to pilot workload. As a consequence of errors and accidents in complex human–machine systems, there has been an increase in the use of automation in many of these systems. Automation is presently used to perform a variety of tasks, from merely providing assistance to the operator to replacing the operator completely. The use of automation is particularly evident in the flight decks of modern commercial aircraft, in which functions such as flight control, navigation, and fuel management have increasingly been placed under computer control. In military cockpits, these functions, in addition to others such as sensor management, threat warning, target acquisition, and weapons delivery, are frequently performed with at least some degree of automation.

The highly automated glass cockpit offers many advantages to pilots. For example, automating fuel management procedures can save fuel compared to manual flight if weather conditions are not severe, and automating navigation functions frees the pilot from having to perform routine psychomotor tasks. In a more general sense, flightdeck automation can reduce the number of human errors by providing predictable, programmable, and reliable systems as well as reducing the involvement of people in the operation of the system. Pilots, however, regard automation as a mixed blessing (Wiener et al., 1991). From a human factors perspective, automation is not always an innate good, because it introduces the possibility of new kinds of errors that might not occur with older, nonautomated flight decks. For example, increased monitoring requirements (Lerner, 1983), overreliance on the system (Danaher, 1980), and a proliferation of flight deck components (Wickens, 1984) have all emerged as unintended consequences of automation that can increase pilot workload.

As is the case with the introduction of any new technology, designers of automated flightdecks must take into account how pilots might respond to automation, and how automation can change the fundamental nature of the flying task. With this central issue in mind, we begin this chapter with an overview of the concept of pilot workload and briefly discuss key factors that can influence pilot workload. Next, we summarize various ways in which automation can be applied, and describe workload-related problems that have been associated with flightdeck automation. In this context, we then discuss three key human factors issues—allocation of function, stimulus–response compatibility, and the internal model of the operator—which should be considered during the design and implementation of automated flightdecks.

The goal of the chapter is not to condemn automation but, instead, to emphasize potential problems with automation so that they can be avoided in the future. Designers should not eschew high levels of automation. But the implementation of automation should be human centered, with due appreciation of the limitations and expectations of the pilots who are the end users of such automation.

PILOT WORKLOAD

Pilot workload is defined (Kantowitz, 1988a) as an intervening variable, similar to attention, that modulates or indexes the tuning between the demands of the environment and the capacity of the operator. As an intervening variable, workload cannot be directly evaluated or observed. Instead, it is a conceptual, multifaceted construct that must be inferred from changes in observable data. Pilot workload is measured by evaluating subjective ratings, objective performance on both primary (flying) and secondary (incidental) tasks, and physiological indices (Kantowitz & Casper, 1988). The great amount of research on pilot workload is justified by the belief that this unobservable construct will help scientists and practitioners predict pilot performance and thus increase aviation safety. This requires that workload measurement meet reasonable scientific standards, which has not always been achieved; discussion of such measurement problems is beyond the scope of this chapter, but see Kantowitz (1992) for examples.

Factors That Influence Pilot Workload

Human factors research has demonstrated that many of the errors and accidents that occur during complex task performance are associated with levels and types of operator workload (see also Kantowitz & Sorkin, 1983). In Fig. 6.1, some of the factors that can influence pilot workload have been grouped according to how they affect workload.

Increases in the factors represented by the minus sign in the leftmost box of Fig. 6.1 will generally decrease workload. For example, increases in pilot

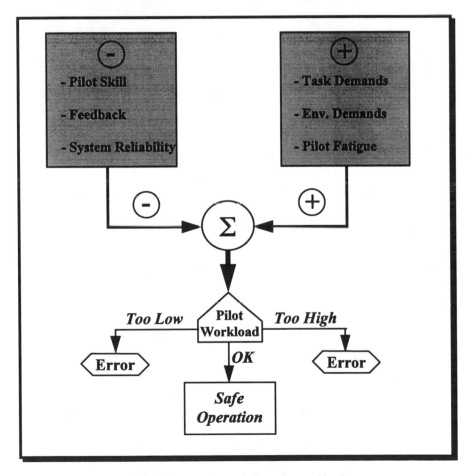

FIG. 6.1.   Key factors influencing workload.

skill, that is, improvements in operating methods (Leplat, 1978) and overall performance, will typically decrease pilot workload, so this factor is represented by the minus sign in the leftmost box of Fig. 6.1. Providing feedback helps the pilot understand what the automated system is doing and decreases workload by decreasing the number of alternatives the pilot must consider. Increases in the reliability of the automated system will reduce uncertainties associated with decision making and task performance, leading to decreases in workload.

Increases in the factors represented by the plus sign in the rightmost box of Fig. 6.1 will generally increase workload. For example, increasing environmental demands (e.g., landing at higher speeds or imposing a last minute change of runway) increases workload. Increasing task demands, such as increasing the number of displays that the pilot must monitor, also increases workload, particularly during demanding flight segments. Pilot fatigue also increases workload because it can reduce the capacity of the pilot to respond to task demands in a timely or appropriate manner.

Importantly, the factors just discussed are not independent, and will often interact with one another to influence resulting levels of workload. For example, an increase in environmental demands that results in a negligible effect on workload for highly skilled pilots (e.g., landing with heavy crosswinds) may have a much greater effect on workload for less-skilled or novice pilots. Thus, the aggregate effect of these factors on pilot workload may vary as a function of additional influences such as levels of experience, control and display design, and pilot age.

As seen in Fig. 6.1, pilot workload can change at any instant as the sum of these factors changes. When workload is too high or too low, the pilot is far more likely to make errors of omission or commission than when workload is satisfactory. The dangers of overload are widely recognized, and much of the impetus for more automation results from attempts to decrease pilot workload. However, underload is also an important danger, especially in glass cockpits where much of the work is accomplished automatically (see also Warm, Dember, & Hancock, chap. 9, this volume). In this regard, Kantowitz and Casper (1988) in a critical review of human workload in aviation concluded: "It would be tragic and ironic if our current efforts to measure pilot workload succeeded, only to be faced with a new generation of aircraft where pilot workload was so low that nobody bothered to measure it at all" (p. 184).

Errors are least likely to occur when workload is moderate and does not change suddenly or unpredictably (Kantowitz & Casper, 1988). Consistent and safe flightdeck operations call for levels of pilot workload that are neither too high nor too low. As discussed in more detail later in the chapter, the decision to automate flightdeck functions will have, a priori, neither a negative nor a positive affect on pilot workload. The effects of automation on pilot workload will vary as a function of the manner in which key human factors issues are addressed during the design of the automated system.

Automation and Pilot Workload

Results from a recent field study using the Aviation System Reporting System (ASRS) illustrate how automation can influence pilot workload. In this study, Bittner, Kantowitz, and Bramwell (1992) examined all incidents occurring between January 1989 and March 1991 involving wide-body four-engine aircraft. The search yielded 38 glass and 206 nonglass reports, of which 20 glass and 120 nonglass were selected for human factors coding and analysis. The goal of the study was to determine, from actual incident data, if there were any underlying factors that could distinguish between incidents involving aircraft equipped with the highest levels of automation technology (glass) versus older aircraft with lesser automation capabilities (nonglass).

The selected 140 reports were coded according to two independent ergonomic taxonomies (shown in Table 6.1). The Human Information Processing (HIP) taxonomy was derived from a general HIP model (Kantowitz, 1988b) that divides the operator into perceptual, cognitive, motor, and attentional subsystems. The basic model, originally designed to explain the behavior of a single operator, was extended by adding a fifth subsystem: communication and coordination. The basic model has been used successfully to guide verbal protocol analysis of rapid motor tasks (Triggs, Kantowitz, Terrill, Bittner, & Fleming, 1990) as well as maintenance activities. The air crew cockpit concerns taxonomy was derived from earlier work on crew reactions to automation (Wiener, 1989).

The coded data were analyzed by a principal factor analysis that revealed five underlying factors: crew coordination, instrument scan, procedure following, decision making unrelated to automation/workload, and unspecified motor-automation understanding. The names of these factors were determined from the factor loadings on the coding variables and consideration of the ASRS reports that showed highest correlations with the factor under evaluation. For example, the first factor, crew coordination, exhibited its largest loading (.69) with cockpit concern variable 1C.

The utility of the five factors was then evaluated for sensitivity to possible differences between Glass and Non-Glass flightdecks. For each of the 140 incidents, factor scores for the five factors were correlated with a dichotomous glass versus nonglass variable. Significant correlations were obtained for the first three factors: crew coordination ($r = .36$, $p < .0005$), instrument scan ($r = .25$, $p < .002$), and procedure following ($r = .23$, $p < .007$).

TABLE 6.1
Human Factors Codes Used in the ASRS Field Study

Human Information Processing Model	*Air Crew Cockpit Concerns*
I. Perceptual	1. Flying Challenge
A. Nonspecific (PERNON)	A. Flying Skills Deficit
B. Signal detection (PERSIG)	i. Manual (FLYMAN)
C. Pattern interpretation	ii. Instrument scan (FLYINS)
II. Cognitive	iii. Out of cockpit (FLYOUT)
A. Nonspecific (COGNON)	iv. Procedures (FLYPROC)
B. Short-term memory (COGSHT)	v. Other (FLYOTH)
C. Long-term memory (COGNLG)	B. Situational awareness (SITAW)
D. Decision making (COGCDEC)	C. Crew coordination (FLYCREW)
III. Motor	D. Fatigue (FLYFAT)
A. Nonspecific (MOTNON)	E. Complacency (FLYCOM)
B. Activation (MOTACT)	F. Transition between Aircraft (FLYTRAN)
C. Interdiction (preservation)	2. ATC
(MOTINT)	A. Communication (ATCCOM)
IV. Attention	B. Coordination (ATCCOR)
A. Nonspecific (ATTNON)	3. Automation
B. Speed (ATTSPD)	A. Understanding (AUTUND)
C. Load (ATTLOD)	B. Trust/reliability (AUTTST)
V. Interpersonal	C. Misuse (AUTMIS)
A. Communication (INCOM)	D. Workload (AUTWORK)
B. Coordination (INTCOR)	E. Update speed (AUTUPDT)
	F. Too little trust external (AUTETST)
	4. Equipment Malfunction (EQUIPMAL)
	5. Automation Malfunction (AUTMAL)

*Short codes for variables are given in parentheses.

The automation-workload coding variable (Table 1, 3D) showed a high loading on Factor 1 (.48) and a moderate loading on Factor 3 (.31); both loadings were statistically significant ($p < .01$). This implies that when incidents involving automation-workload occur, such incidents are more frequent in glass cockpit aircraft. Examination of these incident reports suggests that such incidents are also more severe, causing relatively greater disruption in glass cockpit aircraft.

Finally, it is important to note that the methodology used in this field study offers a quantitative basis for assessment of incident reports filed by crew members (see also Bittner, Morrissey, Bramwell, & Kinghorn, 1994, for a more extended discussion of this methodology). Previously, an analyst faced with a set of such reports could only rely on his or her professional experience to indicate which factors implied in the reports were most important. The present factor analysis methodology allows the incident reports to be partitioned according to their objective loadings upon factors revealed by an impartial statistical analysis. Of course, statistics does not remove the need for careful thought, and good aviation judgment will always be required to interpret even the most objective analysis of incident reports. However, having a tool to provide an objective analysis of a set of incident reports represents an improvement over current subjective methods.

EFFECTS OF AUTOMATION

Automation is defined as having equipment perform functions that could be performed by the pilot manually. Four levels of technology are relevant (Kantowitz & Sorkin, 1983). At the lowest level, the human provides power and control (e.g., someone using a shovel). At the next level, the machine provides power while the human retains control (e.g., using hydraulics to lower landing gear instead of cranking the gear down manually). Increasing technology another level has the machine providing power and information while the human retains control (e.g., a nuclear power plant). At the highest level of technology, the machine provides power, information, and control while the human monitors the operation (e.g., the Flight Management System, or FMS, in a 747−400 aircraft).

Potential Problems With Automation

The use of automation has provided many benefits to pilots. For example, automated systems can perform many calculations associated with flight navigation much faster and more accurately than can a pilot. Also, automated systems can be used to perform some of the pilot's tasks during particularly demanding flight segments, and automated sensors can be used to detect threats or failures that are beyond the pilot's ability to detect. However, during the development of automated flightdecks, these benefits need to be weighed against the possible costs associated with automation. In particular, automation has the potential to decrease or increase pilot workload, depending on how the automation is designed and implemented within the flight deck. Increases in the use of automation in modern flightdecks have led to a number of problems, some of which have implications for pilot workload levels; six of these problems are discussed next.

Increased Monitoring Requirements.   Automation has had a profound impact on the human operator, and increases in automation have led to a redefinition of the operator's role. In advanced aircraft, for example, the role of the pilot has been rapidly changing from one of continuous manual control to one of supervisory control. Although the pilot is required to directly control fewer functions in this new role, many more automated functions must now be monitored. Indeed, pilots of advanced aircraft are often called *flight managers*, indicating that a significant portion of their time is now spent monitoring a large number of automated subsystems (National Research Council, 1982). Thus, increases in automation increase vigilance requirements, leading to increases in workload. In addition to this increase in workload levels, there is considerable evidence (Kantowitz & Hanson, 1981; Mackworth, 1970) that humans are poor monitors, especially when required to attend to multiple tasks or channels. Recent studies have confirmed that human operators are relatively poor at monitoring automated subsystems when they are simultaneously engaged in other manual tasks (Parasuraman, Molloy, & Singh, 1993).

Increased Training Requirements.   Training requirements are increased by automation, with the potential for commensurate changes in pilot workload. Automation introduces unique training requirements, because pilots must become proficient in both manual and automatic flight modes. As an example, pilots must become familiar with flight procedures that accompany transitions between automatic and manual flight modes, as well as unique requirements for intercrew communications.

Furthermore, flightdeck automation is not binary, that is, either off or on. There are typically several modes or levels associated with automated systems, all of which must be learned by pilots and remembered during flight. In particular, pilots must become familiar with both the system capabilities and operator responsibilities at each level of automation. Flying with automation deprives the pilot of practice in the manual mode, which may induce a loss of proficiency that requires additional training time. Finally, as flightdeck automation is able to do more, it takes longer to learn a more complex system.

Inappropriate Levels of Pilot Trust.   As a monitor, many of the pilot's actions are determined by the information presented to him or her by the automated system. Thus, the pilot's level of trust in the system can have an important effect on workload and subsequent performance. Increases in the number of system components increase the likelihood of false alarms, leading to a reduction in the trust that pilots will place in the automated system. For example, the original TCAS (Traffic Collision Avoidance System) suffered from a high false-alarm rate that made it unusable; indeed, an improved version (TCAS II) still has high false-alarm rates in high-density traffic areas, although software improvements are intended to fix this. In general, users are most reluctant to rely on equipment they do not trust (Lee & Moray, 1992). When trust in the automated system is too low and an alarm is presented, pilots may spend additional time verifying the problem or they may simply ignore the alarm altogether. Alternatively, putting too much trust in an imperfect automated system may lead to a false sense of security, causing the operator to ignore other sources of flight data or to forego established and prudent flight procedures (Danaher, 1980). When trust in the automated system is too high, the pilot may assume that the automated system has a particular situation under control when, in fact, the situation may require direct pilot intervention. For example, a pilot facing a potential mid-air collision may try to program his or her way out of the problem instead of turning off the automation and flying manually. In sum, levels of trust in an automated system that are inappropriately low can lead to workload levels that are too high, whereas levels of trust that are inappropriately high can lead to workload levels that are too low.

Decrements in the Pilot's Familiarity With the System and Overall Skill Levels.   The increased monitoring load placed on pilots of automated aircraft may not free up the pilot's attentional resources to perform other duties, as is frequently the intended goal of automation. Rather, such monitoring may so remove the pilot from the active control loop that the pilot loses familiarity with the key system elements and processes for which he or she is responsible. For example, out-of-loop time may reduce the pilot's skills to the point that he or she is no longer effective in a system emergency or failure. In a number of experiments investigating participatory mode in automated systems, active controllers had exhibited faster response times and more accurate failure detection performance than have passive monitors (Ephrath & Young, 1981; Wickens & Kessel, 1981; Young, 1969). Thus, as the "distance" between the operator and the system under control increases, workload can be increased if the operator is suddenly required to jump back into the active control loop and directly control the system. Long-term participation as a system monitor rather than as an active controller can also lead to reductions in baseline skill levels, weaker internal models of system processes, and reduced decision-making abilities, particularly for highly automated system functions.

New Forms of Human Error.   Designers often fail to anticipate new problems created by automated systems. This failure arises both from the designer's natural tendency to concentrate on the benefits, rather than the costs, of automation, and from a lack of familiarity with human factors design analyses. Also, traditional analyses of operator error focus on industrial work environments or environments in which the operator is a direct controller. Although the use of automation can lead to a decrease in the number or severity of traditional errors, human errors are not eliminated by automation. In their new role as system monitors, pilots can become susceptible to forms of error that would have been impossible under fully manual modes of flight operation. Wickens (1984) argued that automation "merely relocates the sources of human error to a new level" (pp. 492–493). For example, incorrect entry of navigation waypoints in inertial navigation systems is a new kind of pilot error made possible by higher levels of technology. Importantly, such an error can greatly increase future workload when adjustments to the navigation system must be made in real time, concurrent with other flight tasks. Later, we briefly discuss how a model of stimulus–response compatibility might predict and reduce such forms of error.

Poor Implementation of Automation. A more general problem is the "clumsy" (Wiener, 1989) implementation of many automated systems. Functions that are easy to automate get automated; functions that are difficult to automate do not. Thus, instead of being driven by requirements analyses and human factors guidelines for allocations of functions (Kantowitz & Sorkin, 1987), automation has been machine centered. This results in generally low workload for pilots under routine conditions, followed by excessively high workload when automation can no longer do the job.

KEY HUMAN FACTORS ISSUES TO BE CONSIDERED IN THE DESIGN OF AUTOMATED SYSTEMS

Laboratory simulation research has confirmed that automation can reduce pilot workload. But this is not the whole story. When a critical incident occurs, high levels of flightdeck automation can lead to a ripple effect (Bittner et al., 1992) whereby workload is dramatically increased. This implies not that automation is necessarily harmful per se, but that a more intelligent implementation of automation is required in modern aircraft. This section discusses three key human factors issues—allocation of function, stimulus–response compatibility, and the internal model of the operator—that should be considered during the design and implementation of automated flightdecks.

Allocation of Function

Allocation of functions is a traditional human factors responsibility that determines if a particular task is better assigned to a human or to automation (Kantowitz & Sorkin, 1987). At first, this was accomplished by making lists of the functional advantages and disadvantages of people versus machines. This logical approach to allocation of functions was not successful, as noted by Kantowitz and Sorkin: "Any table that can compare human and machine, especially if numerical indexes of relative performance or equations can be listed, as any good engineer would attempt, is bound to favor the machine. Machines and people are not really comparable subsystems, even though human-factors specialists go to great length to relate the two" (1987, p. 359). Thus, allocating functions between people and automation is not at all equivalent to allocating functions among automated subsystems.

The technology now exists to create intelligent interfaces between people and automation (Kantowitz, 1989). The traditional binary interface (the automation is either off or on) does not always ensure that pilot workload decreases as more automation is engaged. The pilot must still keep track of the states of various subsystems, and humans have only a limited ability to keep track of variable states in working memory. However, if one conceptualizes intelligent control as a continuum, an optimal interface would be able to assume any state in this continuum ranging from complete machine control to complete human control of the aircraft. Furthermore, an optimal intelligent interface would not create any overhead cost associated either with the state itself or the path used to reach the state. Although the traditional binary interface can move along this continuum (albeit in discrete steps), this creates overhead for the pilot. The overhead depends not only on the number of automatic subsystems engaged, but also on the order in which the crew address these subsystems.

An intelligent flightdeck interface would take into account both the state of the pilot (e.g., workload) and the state of the environment (e.g., fuel load, weather, equipment malfunction). It would then recommend (or implement) an optimal allocation of functions for that aircraft and aircrew state. This goes beyond dynamic allocation of functions, where the pilot decides what flight automation is to be used at any moment in the flight. Although current automation is useful in low- and moderate-workload situations, it fails under high pilot workload and can even introduce ripple effects (Bittner et al., 1992) that exacerbate workload difficulties. The reason for this deficit is clear: Current flightdeck automation is machine-centered rather than human-centered, and is implemented with little resident intelligence.

Stimulus–Response (SR) Compatibility

Stimulus–response compatibility is another key concept in human factors (Kantowitz, Triggs, & Barnes, 1990). It refers to the relationship, both geometric and conceptual, between a stimulus such as a display, and a response, such as a control action. The debate about the relative merits of outside in (moving airplane) versus inside out (moving horizon) artificial horizon indicators is an argument about stimulus–response compatibility. (Actually, a better display has both moving; see Kantowitz & Sorkin, 1983, p. 221.)

Fig. 6.2 shows a recent model of stimulus–response compatibility (Kantowitz et al., 1990) based on a nested hierarchy of frames, rules, and response tendencies. Without going into fine detail, it is sufficient for present purposes to merely note that a frame is a well-developed knowledge structure based on pilot experience and training. Plans and actions that run counter to established frames (i.e., low stimulus–response compatibility) are potential flightdeck problems.

One such problem is altitude deviation in the MD-80 under control of flightdeck automation (Kantowitz & Bittner, 1992). Pilots flying manually tend to slow their rate of ascent when within 1,000 or 2,000 feet of their desired altitude. But in the MD-80 automation maintains a high rate of climb, up to eight times what a pilot might select. Thus, pilots tended to worry that the high rate implied exceeding the desired altitude. They manipulated the trim pitch wheel to slow the ascent, not realizing, under the stress of the moment, that this disarmed the altitude capture function. This created the altitude "bust" they feared by removing control from the automation without the pilot taking over manually. From a human factors perspective, this is a compatibility problem. Automation did not behave in the manner expected by the pilot, that is, as another pilot would fly. Although improved training and familiarity with the MD-80 has decreased

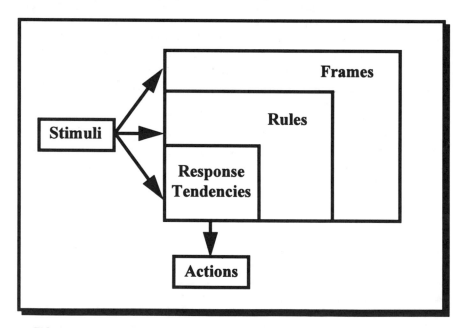

FIG. 6.2.   Nested hierarchical relations among frames, rules, and response tendencies as sources of stimulus–response compatibilities.

the frequency of this error, a better solution would have been to have the original design be more compatible with pilot frames and rules.

Another example of poor stimulus–response compatibility, suggested by Rolf Braune of Boeing Commercial Airplane Group, is the vertical navigation (VNAV) functions of the FMS. The VNAV-path mode violates compatibility expectations and thus is difficult to use correctly. Because of this problem, at least one airline has instructed its pilots not to use VNAV functions below 10,000 feet.

In summary, implementations of automation that exhibit low stimulus–response compatibility can confuse the pilot, increasing task demands unnecessarily. They create extra workload and lower trust in flightdeck automation.

Internal Model of the Operator

The internal model of the operator refers to the operator's internal representation of the system under control; that is, his or her understanding about system elements, dynamics, processes, and inputs and outputs. As a logical abstraction, the internal model will vary tremendously as a function of individual operators, tasks, and environments. The operator uses this model as an organizing schema to plan future activities, to derive hypotheses about expected relationships among system components, and to perform system tasks. Having an accurate (or at least functional) internal model is crucial to successful performance in any complex task or system.

The lack of understanding about how pilots conceptualize both their tasks and the performance characteristics of cockpit displays and controls has been a key issue in the design of the automated flightdeck. Indeed, a National Research Council (NRC, 1982) report on automation concluded that "the effectiveness of automation depends on matching the designs of automated systems to pilots' representations of their combat tasks" (p. 59). More recently, Sarter (1991) noted that current flight deck automation fails to take into account pilots' mental models of the Flight Management System. Violations of the pilots' expectations, such as the examples previously discussed under the heading stimulus-response compatibility, are common in current automation. Such violations can increase pilot workload by increasing uncertainties, response times, and errors in task performance. For example, if the system is displaying unexpected data or maneuvering the aircraft in an unexpected manner, additional time and effort may be required in order for the pilot to discover how and why the system is taking these actions.

Ideally, there should exist a "match" between the actual operating characteristics of the system, the pilot's internal model of the system, and the designer's model of the system. In the context of supervisory control system design, Sheridan and Hennessy (1984, p. 20) noted the need to "bring all of these models into harmony, since they ultimately influence the decision processes of the human supervisor and the consequences of the system's operation."

A key component in designing the system in a manner consistent with the pilot's internal model is the provision of timely and accurate feedback. Feedback helps the pilot understand what the automation is doing. The crucial importance of feedback is well documented in human factors (e.g., Kantowitz & Sorkin, 1983) and in experimental psychology (e.g., Annett, 1969). The ability to utilize feedback information, defined as a flow of information counter to the main flow of a system, is an essential characteristic of both human and machine systems. One Aviation Safety Reporting System incident reveals the high cost of minimal feedback. Fuel transfer had not been balanced in a plane under auto-pilot control. The auto-pilot tried to compensate for this. Only minimal feedback, slight rotation of the yoke, was presented to the pilot. Eventually, the auto-pilot reached its limit of compensation and returned control to the pilot, who then had to quickly diagnose the problem of imbalance. Until the incident occurred, the crew did not know what the automation was doing. Better feedback would have prevented this incident.

Developers of automated flightdecks can incorporate the concept of pilots' mental models into their design activities in a number of specific ways, including matching task demands to environmental demands, providing timely and accurate feedback to the pilot, designing control configurations and display formats that are consistent with both system performance requirements and the pilots' expectations for the system, and developing training strategies that facilitate the development of appropriate internal models.

The three design issues discussed here — allocation of function, stimulus--response compatibility, and the internal model of the operator — are clearly not the only issues that the human factors professional will consider during the design of automated systems. However, thoughtful consideration of information processing concepts and human factors methods relevant to these three issues can go a long way toward maintaining appropriate levels of pilot workload on the automated flightdeck. As an example, Fig. 6.3 shows the combined effects of these three design issues on pilot workload and trust in the system. In this figure, we have contrasted the use of a nonintelligent versus intelligent flightdeck interface, the implementation of low versus high stimulus–response (S–R) compatibility between flightdeck displays and controls, and the provision of delayed versus timely feedback. These combined effects have been illustrated within the context of a very

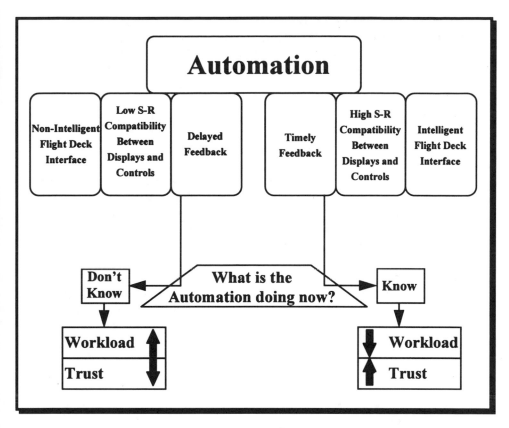

FIG. 6.3.   Effects of interface design, S–R compatibility, and feedback on automation usage.

simple but common question for a pilot new to advanced automation — "What is the automation doing now?"

CONCLUSIONS

Automation is rapidly being incorporated in modern flightdecks as a way of increasing system reliability and decreasing operator errors. Tasks that automation is presently being used for include, for example, flight control, navigation, fuel management, sensor management, threat warning, target acquisition, and weapons delivery. In many respects, automation has provided system designers with a means to increase the performance capabilities of both commercial and military aircraft.

However, despite its potential for improving overall aircraft performance, automation is not always an innate good. It has been associated
with a number of unique and unanticipated problems and has not always
been well received by pilots. If not implemented with careful consideration
of pilot requirements, automation can be accompanied by inappropriate
levels of pilot workload. Pilot workload should be neither too high nor too
low; errors are least likely to occur when workload is moderate and does not
change suddenly or unpredictably (Kantowitz & Casper, 1988).

We do not argue that designers should avoid the use of automation
during flightdeck development. As noted previously, the goal of this
chapter has been to emphasize potential problems with automation so that
they can be avoided in the future. In particular, design of automated
flightdecks should be human-centered. Designers must consider how pilots
might respond to automation, and how automation can change the pilot's
role and the nature of the flying task. The effects of automation on pilot
workload can vary, and will depend greatly on the extent to which key
human factors issues are considered during the design of the automated
system. Key human factors issues that should be addressed during the
design and implementation of automated flightdecks include allocation of
function, stimulus–response compatibility, and the internal model of the
operator.

Although we have focused on the use of automation in the design of
modern flightdecks, this discussion is relevant to the design of a wide
variety of human–machine systems. Maintaining appropriate levels of
workload during automated operating conditions is, for example, a key
issue in the design of nuclear power plants and other process control
environments. Such environments are frequently characterized by relatively
short periods of direct manual control of the system interspersed with
longer periods of automatic control and system monitoring on the part of
the operator. Thus, over- and underload continue to be critical design
issues. Closer to home for most of our population, however, are the
technological inroads being made in the area of Intelligent Transportation
Systems (ITS). Such systems include automated highways and "smart-cars,"
with increasingly sophisticated in-vehicle sensors, controls, and displays.
Such systems are being developed for eventual use by the full spectrum of
highway travelers, including the elderly and those with disabilities. Therefore, the need to consider the implications of key design decisions on
workload will become even more critical as these systems near operational
deployment (see Hancock, Parasuraman, and Byrne, chapter 16, this
volume).

Fig. 6.4 expands on some of the concepts presented earlier in Fig. 6.1,
and provides a framework for integrating and summarizing this chapter. In
the figure, we have identified critical links between the three key human

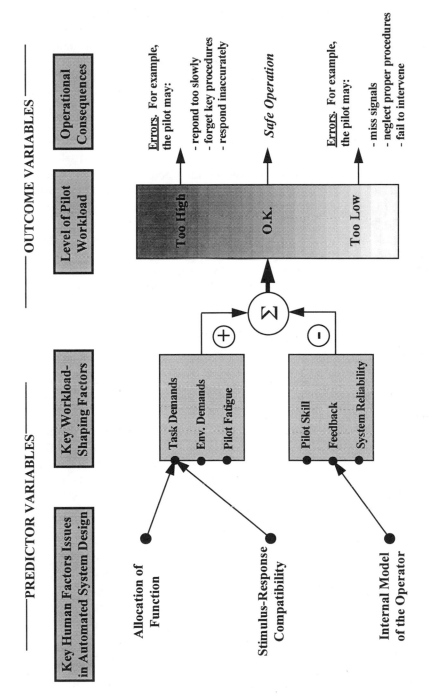

FIG. 6.4.   Summary of concepts related to automated flightdecks and pilot workload.

factors issues in automated system design and the workload-shaping factors discussed earlier in this chapter. These links are not intended to be exhaustive and can vary, depending on the nature of the system under investigation. The links simply serve to highlight the workload-shaping factor that is most frequently and most directly influenced by each of the three human factors design issues. Other relationships between these human factors issues and workload-shaping factors can occur. For example, allocation of function decisions can have an effect on system reliability; controls with poor stimulus–response compatibility can increase pilot fatigue levels.

Fig. 6.4 also shows that changes in the factors that influence pilot workload can result in either an increase or a decrease in workload (see also Fig. 6.1). In Fig. 6.4, we have broadly characterized levels of pilot workload as "too high," "O.K.," or "too low." Fig. 6.4 also depicts examples of the operational consequences associated with workload levels that are either too high or too low. Importantly, these examples are neither exhaustive nor exclusive to specific workload levels. Many other errors are possible and the same specific error (e.g., responding too slowly) might occur when workload is too low as well as when workload is too high.

Implications for the Design of Automated Flightdecks

Our discussion in this chapter suggests at least three parallel lines of human factors efforts that can improve training for, and design of, automated flightdecks. First, machine intelligence must be built into the pilot-automation interface. Some efforts in this direction are already under way, such as work to have the proper checklist automatically displayed for the pilot when a subsystem fails. Indeed, this approach can use artificial intelligence to prioritize tasks and information for the pilot so that he or she is not overwhelmed with competing inputs when several subsystems fail simultaneously. Second, flightdeck displays and controls must be designed in a manner that is compatible with pilots' interpretation of displayed information, decision-making processes, and response tendencies. In particular, we should focus our efforts on how complex rules and frames govern S–R compatibility and influence both pilots' interpretation of cockpit displays and their use of cockpit controls. Third, models of human mental states must be built, validated, and applied to the flight deck. It is poor human factors to compel pilots to conform to the Procrustean requirements of current automation. Because it is far easier to modify the automation than to modify the human (Kantowitz & Sorkin, 1983), designers must make automation serve the human rather than vice versa. The discipline of human factors has the tools to improve flight deck automation, because until pilots no longer experience difficulty in determining what the automation has done, is doing, or might do in the near future, pilot workload will not fully benefit from glass cockpits.

REFERENCES

Annett, J. (1969). *Feedback and human behavior.* Baltimore, MD: Penguin.

Bittner, A. C., Jr., Kantowitz, B. H., & Bramwell, A. (1992). *Workload assessment of automated flight decks: analysis of ASRS incident reports* (pp. 226–230). (working paper). Seattle, WA: Battelle, HARC.

Bittner, A. C., Jr., Morrissey, S. J., Bramwell, A. T., & Kinghorn, R. A. (1994). Hierarchical analysis of multiple responses (HAMR): Ergonomics applications. *Proceedings of the 3rd Pan-Pacific Conference on Occupational Ergonomics* (pp. 226–230). Pohong, Korea: Ergonomics Society of Korea.

Danaher, J. W. (1980). Human error in ATC systems operations. *Human Factors, 22,* 535–545.

Ephrath, A. R., & Young, L. R. (1981). Monitoring vs. man-in-the-loop detection of aircraft control failures. In, J. Rasmussen, & W. B. Rouse, (Eds.), *Human detection and diagnosis of system failures* (pp. 143–154). New York: Plenum.

Kantowitz, B. H. (1988a). Defining and measuring pilot mental workload. In J. R.Comstock, Jr. (Ed.), *Mental-State Estimation 1987* (pp. 179–188). Hampton, VA: National Aeronautics and Space Administration, Scientific and Technical Information Division.

Kantowitz, B. H. (1988b). Laboratory simulation of maintenance activity. *Proceedings of the 1988 IEEE 4th Conference on Human Factors and Nuclear Power Plants* (pp. 403–409). New York: IEEE.

Kantowitz, B. H. (1989). Interfacing human and machine intelligence. In P. Hancock, & M. Chignell (Eds.), *Intelligent interfaces: Theory, research, and design* (pp. 49–68). Amsterdam: North-Holland.

Kantowitz, B. H. (1992). Selecting measures for human factors research. *Human Factors, 34,* 387–398.

Kantowitz, B. H., & Bittner, A. C., Jr. (1992). Using the aviation safety reporting system database as a human factors research tool. *Proceedings of the 15th Annual Aerospace & Defense Conference* (pp. 31–39). Piscataway, NJ: IEEE.

Kantowitz, B. H., & Casper, P. A. (1988). Human workload in aviation. In E. Wiener & D. Nagel (Eds.), *Human factors in aviation* (pp. 157–185). New York: Academic Press.

Kantowitz, B. H., & Hanson, R. H. (1981). Models and experimental results concerning the detection of operator failures in display monitoring. In J. Rasmussen & W. B. Rouse (Eds.), *Human detection and diagnosis of system failures* (p. 301–315). New York: Plenum.

Kantowitz, B. H., & Sorkin, R. D. (1983). *Human factors: understanding people-system relationships.* New York: Wiley.

Kantowitz, B. H., & Sorkin, R. D. (1987). Allocation of functions. In G. Salvendy (Ed.), *Handbook of human factors* (pp. 355–369). New York: Wiley.

Kantowitz, B. H., Triggs, T. J., & Barnes, V. (1990). Stimulus–response compatibility and human factors. In R. W. Proctor & T. Reeves (Eds.), *Stimulus–response compatibility* (pp. 365–388). Amsterdam: North-Holland.

Lee, J., & Moray, N. (1992). Trust and the allocation of function in the control of automatic systems. *Ergonomics, 35,* 1243–1270.

Leplat, J. (1978). Factors determining workload. *Ergonomics, 21,* 143–149.

Lerner, E. J. (1983). The automated cockpit. *IEEE Spectrum, 20,* 57–62.

Mackworth, J. F. (1970). *Vigilance and attention*. Baltimore, MD: Penguin.

National Research Council (1982). *Automation in combat aircraft*. Washington, DC: National Academy Press.

Parasuraman, R., Molloy, R., & Singh, I. (1993). Performance consequences of automation-induced "complacency.". *International Journal of Aviation Psychology, 3*, 1-23.

Sarter, N. B. (1991). The flight management system: Pilots' interaction with cockpit automation. *Proceedings of the 1991 IEEE International Conference on Systems, Man, and Cybernetics* (pp. 1307-1310). Piscataway, NJ: IEEE.

Sheridan, T. B., & Hennessy, R. T. (1984). *Research and modeling of supervisory control behavior*. Washington, DC: National Academy Press.

Triggs, T. J., Kantowitz, B. H., Terrill, B. S., Bittner, A. C., Jr., & Fleming, T. F. (1990). The playback method of protocol analysis applied to a rapid aiming task. *Proceedings of the Human Factors Society, 34*, 1275-1279.

Wickens, C. D. (1984). *Engineering psychology and human performance*. Columbus, OH: Bell and Howell.

Wickens, C. D., & Kessel, C. (1981). Failure detection in dynamic systems. In W. B. Rouse & J. Rasmussen (Eds.), *Human detection and diagnosis of system failures* (pp. 155-170). New York: Plenum.

Wiener, E. L. (1989). *Human factors of advanced technology ("Glass Cockpit") transport aircraft* (NASA-Contractor Report No. 177 528). Moffett Field, CA: NASA-Ames Research Center.

Wiener, E. L., Chidester, T. R., Kanki, B. G., Palmer, E. A., Curry, R. E., & Gregorich, S. E. (1991). *The impact of cockpit automation on crew coordination and communication: I. overview, LOFT evaluations, error severity, and questionnaire data* (NASA Contract No. 177587). Moffettfield, CA: NASA.

Young, L. R. (1969). On adaptive manual control. *Ergonomics, 12*(4), 655-675.

7

Psychophysiological Measures of Workload: Potential Applications to Adaptively Automated Systems

Arthur F. Kramer
Leonard J. Trejo
University of Illinois

Darryl G. Humphrey
Wichita State University

INTRODUCTION

The main goal of this chapter is to provide a brief synopsis of recent research in the field of applied psychophysiology. More specifically, we describe several studies that together begin to define the techniques and situations in which psychophysiological measures may provide important insights into human information processing activities of relevance to automated systems. Our review will focus on one class of psychophysiological measures—event-related brain potentials (ERPs)—and, for the most part, the measurement of a single but multidimensional psychological construct—mental workload. We do not mean to imply from our restricted treatment of the literature that we believe that other psychophysiological techniques have limited utility in the assessment of operator state in automated systems. In fact, psychophysiological measures such as heart rate and eye movements show great promise for the assessment of human information processing activities in simulated and operational contexts. However, we have chosen the path of providing a somewhat in-depth discussion of one particular psychophysiological technique rather than a broader but more superficial treatment of the applied psychophysiology literature (for more comprehensive treatments of this literature, see Kramer, 1991; Kramer & Spinks, 1991; Wilson & Eggemeier, 1991).

As argued by a number of researchers (Sheridan, 1987; Wickens, 1992), automation has, in many cases, changed the nature of rather than diminished the processing demands imposed on human operators. Instead of manually controlling the inner loop components of systems such as aircraft, manufacturing, and chemical processes, humans are now involved in monitoring system parameters and occasionally intervening in the operation of the system to detect, diagnose, and correct system malfunctions. One important by-product of this shift in roles for human operators is that it has become more difficult to infer the operators' information processing activities and strategies. In manual and semi-automated systems, human operators were constantly engaged in making analog and discrete inputs in an effort to maintain the state of the system within an acceptable range. These control inputs and adjustments could, in turn, be used as a yardstick against which to measure the operators' performance and to infer whether the operator was under- or overloaded or had missed critical system information.

However, overt performance is often quite sparse in automated systems, because the operators only occassionally intervene to adjust system parameters or conduct tests. In such cases, it is often difficult to determine the extent of operators engagement in automated systems. In fact, there is now a sufficient body of data to suggest that operators are often slower and less accurate in detecting and diagnosing system malfunctions when they serve as system supervisors than when they are involved in actively controlling a system (Bortolussi & Vidulich, 1989; Ephrath & Young, 1981; Kessel & Wickens, 1982). Such studies clearly suggest a need for human information processing assessment techniques that do not rely solely on the occurrence of overt control actions.

One solution that has been proposed to reduce the operator's workload while still keeping him or her "in the loop" is adaptive aiding. The concept of adaptive aiding involves the use of automation only when the operator requires assistance to meet task demands. Otherwise, the operator maintains control of the system functions, often by manually controlling system parameters, and therefore remains in the loop (Rouse, 1988; Wickens, 1992). Commercial and military piloting is one environment in which adaptive aiding has been employed for many years. In this setting the pilot can offload manual control responsibilities by engaging the autopilot.

Although the use of adaptive aiding allows for more flexible distribution of tasks between computers and human operators, and therefore can potentially enhance overall system effectiveness, the concept of adaptive aiding has also generated a number of interesting and important questions. Although many of these questions are beyond the scope of this chapter (but see Scerbo, chap. 3, this volume) one important issue that could potentially benefit from use of psychophysiological measures concerns the basis for deciding whether aiding is needed. Two different approaches have been examined in previous studies. One approach involves the use of human performance models to predict how well an operator is likely to perform a task given changing task demands and human resources (Govindaraj &

Rouse, 1986; Greenstein & Ravesman, 1986). The other approach has involved online assessment of performance, which, in turn, has been used to infer operators' intentions and capabilities to successfully complete system-relevant tasks (Geddes, 1986). Given the sparsity of overt human actions in many modern-day systems as well as the imperfect mapping of performance to intentions and mental processes, it appears reasonable to ask whether psychophysiological measures can be used to improve the assessment and prediction of human performance in complex systems. In the following sections of this chapter, we discuss the potential of psychophysiological techniques for the assessment of information processing activities of human operators in adaptively automated systems.

PSYCHOPHYSIOLOGICAL MEASURES: ADVANTAGES AND DISADVANTAGES

In an effort to adhere to the truth-in-advertising dictum, we would be remiss if we did not describe both the advantages as well as the disadvantages in the use of psychophysiological measures, and more specifically ERPs, in the assessment of psychological processes of relevance to automated systems. We begin with the disadvantages. ERPs, and psychophysiological measures in general, are relatively expensive and time consuming to acquire, analyze, and interpret, at least in comparison to most performance and subjective measures that have been obtained in extra-laboratory settings. However, over the past decade the cost for the specialized equipment necessary to record ERPs (e.g., amplifiers, transducers, a/d conversion boards, large data storage media) has decreased quite substantially such that the hardware and software can now be purchased for somewhere in the neighborhood of $30,000. The interpretation of the data is another matter. The complexity of the ERP waveform as well as the substantial theoretical and empirical literature that relates ERP components to different psychological processes precludes a cookbook approach to data interpretation. Thus, it is necessary to have a knowledgeable psychophysiologist involved in any research or assessment project.

A related point concerns the complexity of signal extraction and analysis and the detection of potential artifacts. Although artifacts are certainly a concern even with the recording of reaction time and accuracy measures, the magnitude of the problem is often larger for physiological measures. For example, many ERP components can be contaminated by other electrical activity, such as that generated by eye, neck, and body movements. Artifacts also arise from inadequate electrode placement and saturation of the a/d channels. Although knowledge of signal characteristics and analytic procedures along with careful data recording protocols can eliminate or reduce the impact of many of these potentially confounding factors, a good deal of technical expertise is necessary to ensure successful data collection and signal extraction.

Another concern with many ERP recording procedures is the potential intrusiveness of the methodology. For example, although ERPs, and the P300 component of the ERP in particular, have been found to be a sensitive index of perceptual/cognitive processing demands, many of the laboratory studies that have demonstrated this relationship have done so by using a secondary task methodology. With this method subjects are asked to perform a primary task to the best of their ability and devote any spare capacity to the performance of the secondary task. ERPs are elicited by the secondary task stimuli. The underlying assumption adopted with the use of this methodology is that any processing resources that remain after the performance of the primary task will be devoted to secondary task performance. The ERPs are assumed to tap these spare resources. In fact, a number of studies have found that the amplitude of the P300 component of the ERP elicited by the secondary task stimuli systematically decreases with increases in the difficulty of the primary task (Isreal, Chesney, Wickens, & Donchin, 1980; Isreal, Wickens, Chesney, & Donchin, 1980; Kramer, Sirevaag, & Hughes, 1988; Kramer, Wickens & Donchin, 1983, 1985) and with increases in the priority of the primary task (Strayer & Kramer, 1990). Although the ERP-based secondary task technique has been quite useful in exploring theoretical issues concerning attention and resource allocation and the development of automatic processing, the requirement to perform an extraneous task renders it difficult to apply in operational contexts in which operators may already be overburdened by task demands.

Two solutions to the intrusiveness problem have been pursued. In one procedure, hereafter to be referred to as the *primary task technique*, ERPs are elicited by discrete events within the task of interest. In this context ERP components, and in particular the P300, have been found to increase in amplitude with increases in the difficulty or priority of the task, presumably reflecting the allocation of additional processing resources or attention for more difficult or high-priority tasks (Mangun & Hillyard, 1990; Sirevaag, Kramer, Coles, & Donchin, 1989; Ullsperger, Metz, & GIlle, 1988; Wickens, Kramer, Vanasse, & Donchin, 1983). The primary task method can be quite useful in settings in which it is possible to trigger ERPs on the basis of discrete task-relevant events. Such events might include the occurrence of new aircraft on an air traffic controller's console, the presentation of updated system status information on a automated manufacturing control screen, or the presentation of new navigational fixes on an aircraft pilot's CRT. Unfortunately, however, there are situations or time periods in which few such discrete events occur but an assessment of operator state is still desired (e.g., monitoring of a sonar display or the status of a process control plant). Furthermore, it is often difficult, particularly in operational settings, to modify the system hardware and software to accommodate the acquisition of ERPs. Finally, the comparison of primary task ERPs across dissimilar tasks and systems requires the tenuous assumption that all primary task events require the same variety of processing resources or attention.

One alternative to primary and secondary task methods of ERP recording has been referred to as the *irrelevant probe technique* (Papanicolaou & Johnstone, 1984). This technique involves the recording of ERPs to auditory or visual probes that accompany the task of interest. However, unlike the secondary task method, which requires that subjects actively respond or count the probes, the probes are ignored in the irrelevant probe technique. Thus, this technique has the same advantages associated with the secondary task method while minimizing the potential for disturbing the performance of the task of interest. The theoretical rationale is essentially the same for the irrelevant probe technique as it is for the secondary task method. That is, that increases in the difficulty of the primary task will result in increased resource allocation to the primary task with a concomitant decrease in the resources available for the processing of the probes. We illustrate how this method can be used to examine mental workload in simulated real-world tasks in a later part of this chapter.

One additional concern about the use of psychophysiological measures as indices of human information processing activities is the amount of data that is necessary to reliably identify changes in mental workload, alertness, or whether an operator has failed to attend to a critical signal. In laboratory situations, ERPs are elicited by a number of presentations of a stimulus, and then these single-trial ERPs are averaged in an effort to enhance the signal-to-noise ratio of the critical ERP components. Although such a procedure is reasonable in the laboratory, it may not suffice in situations in which moment-to-moment variations in operator state is of concern. Later in this chapter we describe a program of research in which we have begun to examine the degree to which ERPs can be expected to tap dynamic changes in operator state.

Thus far, we have focused on the problems, as well as some potential solutions, in the use of psychophysiological measures for the assessment of aspects of human information processing in extra-laboratory situations. However, psychophysiological measures also possess a number of strengths that make them well suited for the assessment of aspects of human cognition of relevance to adaptively automated systems. For example, mental workload has been defined as the interaction between the structure of systems and tasks on the one hand, and the capabilities, motivation, and state of the human operator on the other (Gopher & Donchin, 1986;

Kramer, 1991). More specifically, mental workload has been defined as the information processing costs a human operator incurs as tasks are performed. In recent years, processing costs have been conceptualized in terms of multiple resources with performance decrements resulting when two or more tasks exhaust the supply of a particular variety of processing resource (Wickens, 1992).

Given the multidimensional nature of mental workload and other psychological constructs (e.g., memory, attention, language processes), it is fortunate that ERP components, which are defined with respect to their polarity, scalp distribution, and latency range, have been found to be sensitive to a variety of different information processing activities. For example, the P100 component, a positive going voltage deflection that occurs within 100 milliseconds following a stimulus, is specifically sensitive to the allocation of attention to a particular region of the visual field. The mismatch negativity (MMN), which is a negative going difference wave that occurs approximately 150 to 250 milliseconds poststimulus, provides an index of the extent to which a particular stimulus matches a predefined template (e.g., Is that the musical note I heard a few seconds ago?). The P300 component appears to reflect stimulus evaluation processing, whereas the N400 component reflects the detection of semantic mismatch. Thus, one advantage of psychophysiological measures, and ERPs in particular, is that they are inherently multidimensional in nature. That is, the components that can be found in a single one-second waveform reflect a multitude of information processing activities.

A second advantage of psychophysiological measures is that they can be recorded in the absence of overt behavior. Thus, a manual or vocal action is not required for the elicitation of many ERP components. Given that control inputs are often sparse in automated systems, psychophysiological measures may be used to provide insights into human information processing activities that would otherwise be unavailable with traditional performance measures. Finally, psychophysiological measures are recorded relatively continuously and therefore offer the potential to provide a rapid assessment of changes in operator state. However, as discussed previously, an important question concerns the amount of psychophysiological data that is required to unambiguously discriminate among different operator states. In an effort to provide a partial answer to this question, we now describe a study that was designed to examine the feasibility of employing ERPs to measure dynamic changes in mental workload.

REAL-TIME ASSESSMENT OF MENTAL WORKLOAD: A FEASIBILITY STUDY

The main goal of the study that we now briefly describe was to determine the amount of ERP data that would be necessary to reliably discriminate among several different levels of single and dual-task processing load (see Humphrey & Kramer, 1994, for a detailed description of this study). To that end, 12 young adults performed two complex tasks, monitoring six constantly changing gauges and performing mental arithmetic problems, both separately and together. ERPs were recorded from discrete events in both of the tasks, the presentation of the cursors in the monitoring task and the presentation of the operators and operands in the mental arithmetic task, and a monte carlo approach was used to relate different amounts of ERP data to different levels of accuracy in discriminating among variations in mental workload.

The two tasks that were performed by the subjects are presented in Fig. 7.1. Each of the gauges in the monitoring task were divided into 12 regions. The different regions were coded numerically (i.e., the numbers 1 to 12) and with color (i.e., 1 to 4 were green, 5 through 8 were yellow, 9 through 12 were red). The subject's task was to reset each gauge as quickly as possible once its cursor had reached the critical zone which was redundantly defined by number (> 9) and color (red). Subjects reset the gauges by depressing one of the six editing keys from a standard IBM AT keyboard with their right hand. Each key corresponded to a specific gauge. The gauge-to-key mapping was spatially compatible.

In order to encourage subjects to learn the relationships among the movement of the cursors in the different gauges, and in an effort to simulate the sampling strategies required with physically displaced gauges in operational systems, subjects could only view the position of the cursors one at a time. Thus, although the gauges were always present on the screen, the cursors were not continuously visible. To sample a gauge, (i.e., to see where the cursor was located), the subjects pressed one of a set of six keys with their left hand. Once sampled the cursor remained visible for 1,200 milliseconds. Simultaneous sampling was not possible.

The difficulty of the monitoring task was manipulated by varying the degree to which the position of the cursor on one gauge could be predicted from the position of the cursor on another gauge. In the high-predictability (HP) condition, the gauge monitoring functions were equivalent for the three gauges within a row. The only difference between these gauges was a phase offset (i.e., the cursors began at different positions on the gauges). In the low-predictability (LP) condition, each of the gauges was driven by a separate forcing function.

The center of each gauge served as a display area for the operators and operands for the arithmetic task (see Fig. 7.1). All of the operators and operands were presented simultaneously, with one operand in each gauge and one operator for every two gauges. Arithmetic problems were presented every 4 to 15 seconds following the completion of a previous problem. Subjects were instructed to complete the problems as quickly and as accurately as possible. The difficulty of the mental arithmetic task was

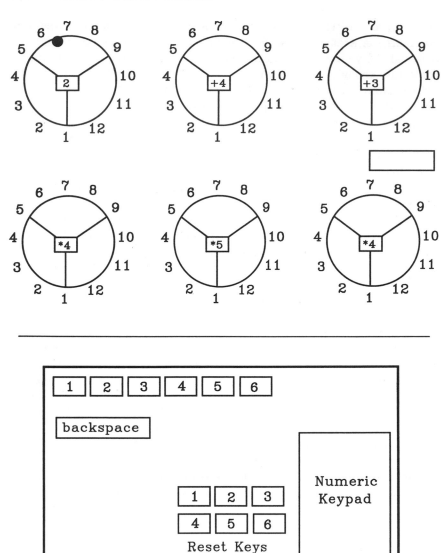

FIG. 7.1. A graphic illustration of the monitoring and mental arithmetic tasks along with the response input devices.

manipulated by varying the number of column operations necessary to complete the problem. The A2 version of the task required operations on two columns of numbers, whereas the A3 version of the task required operations on three columns of numbers. Operations included addition and multiplication. Answers were entered via the numeric keypad on the IBM AT keyboard and appeared in the window as they were typed.

Subjects performed the mental arithmetic and monitoring tasks, both separately and together, during five two-hour sessions. Electroencephalographic (EEG) activity was recorded from three midline sites — Fz, Cz, and Pz — according to the International 10–20 system (Jasper, 1958). Vertical electrooculographic (EOG) activity was recorded from electrodes placed above and below the right eye. Horizontal EOG was recorded from electrodes located lateral to each eye. The EOG and EEG were digitized every 5 miliseconds and were filtered offline (-3dB at 6.89 Hz, 0dB at 22.2 Hz). Trials that were contaminated by excessive EOG artifacts were not included in subsequent analyses of the ERP components. Fewer than 5% of the trials were rejected for excessive EOG artifacts.

RESULTS AND CONCLUSIONS

The performance data, reaction time (RT) and accuracy, were analyzed to ensure that we had successfully varied both single- and dual-task processing demands. RTs decreased from the HP to the LP condition in the monitoring task and from the two to the three column problems in the mental arithmetic task. Error rates behaved in a similar manner. Furthermore, performance was significantly poorer in the single-task than in the dual-task conditions.

After establishing that we had, in fact, successfully manipulated task difficulty as indexed by the performance measures, our next step was to determine if average ERPs differed among different task conditions. This was necessary because it would not make sense to ask how much ERP data was necessary to discriminate between different levels of mental workload if we could not show reliable ERP differences when large numbers of single trials were averaged. The grand average ERPs elicited in a subset of the task conditions for both the monitoring and mental arithmetic tasks are presented in Fig. 7.2. As can be seen from the figure, single- and dual-task ERPs are visually dissimilar for both the monitoring and mental arithmetic tasks. These differences, which were corroborated in ANOVAs, are most obvious in the region of the P300 component (i.e., 300 to 500 milliseconds poststimulus) and the later slow wave (i.e., from 750 to 1200 milliseconds poststimulus).

Given that we had now established that the different task conditions could be distinguished on the basis of averages of large numbers of single-trial ERPs, we were now able to proceed in addressing our original research question: How much ERP data is necessary to discriminate between different levels of mental workload? Of course, the answer to this question depends on the specification of a level of accuracy with which workload conditions can be discriminated. Given that the level of acceptable discrimination accuracy might vary in different situations and for

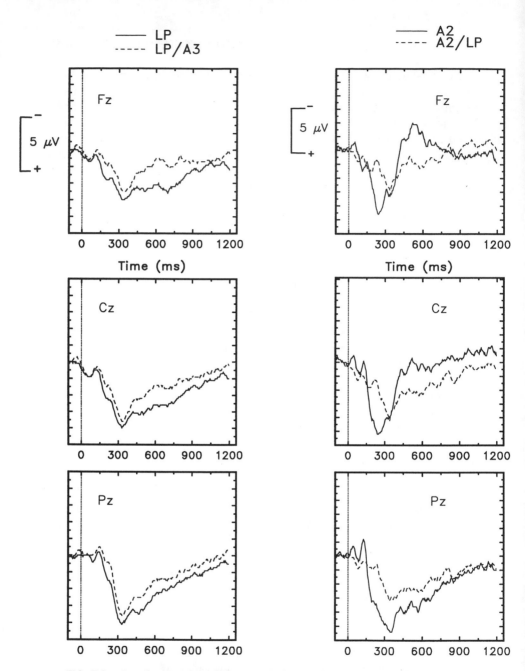

FIG. 7.2. Grand average waveforms across the 12 subjects for single and dual-task conditions in the monitoring and mental arithmetic tasks at the three electrode sites. LP refers to the single-task low predictability monitoring condition. LP/A3 refers to the ERPs elicited by the cursor updates in the monitoring task low predictability condition when performed concurrently with the three column mental arithmetic task. A2 refers to the single-task two column mental arithmetic condition. A2/LP refers to the ERPs elicited by the presentation of the operator and operands in the two column mental arithmetic task when performed concurrently with the low predictability monitoring task.

different systems, we adopted a monte carlo approach in which we systematically incremented the amount of ERP data that were averaged prior to discriminating between workload conditions. This approach has enabled us to relate, in a relatively continuous manner, different levels of accuracy of discrimination to different amounts of ERP data.

Our approach to this issue included the following steps. First, we chose two sets of experimental conditions that differed in perceived workload, performance, and average ERP measures. In an effort to evaluate the reliability of ERP measures in discriminating between levels of workload in different tasks, we also constrained our choice of conditions such that one set was chosen from the monitoring task whereas the other set of conditions was chosen from the mental arithmetic task. These conditions included the LP and LP/A3 conditions in the monitoring task and the A2 and A2/LP conditions in the mental arithmetic task.

The second step of our procedure involved the derivation of each of the ERP measures described in Table 7.1 for each of the single trials at each electrode in the selected conditions. Thus, for each experimental trial a total of 24 ERP measures were derived (i.e., the eight measures in Table 7.1 at each of the three electrode sites). The vectors of ERP measures, with one vector for each single trial, were then divided in half such that all of the even trials were placed in one pool and all of the odd trials were placed in different pool. There were approximately 75 trials in each of the two pools for each subject for the monitoring task and 35 trials in each pool for each subject for the mental arithmetic task.

The third step in our procedure involved the random selection of 1,000

TABLE 7.1
Measures Obtained From the Single-Trial Event-Related Brain Potentials in the Monitoring and Mental Arithmetic Tasks

Measure	Description
Base-peak amplitude (BPamp)	Largest positive voltage between 275 and 750 milliseconds poststimulus-
Base-peak mean voltage	Mean voltage in a 100-miliseconds window centered on the point picked as Bpamp
Base-peak root mean square	Root mean square amplitude computed in a 100-milliseconds window centered around the point picked as Bpamp
Cross-correlation mean voltage	Mean voltage in a 100-milliseconds window centered on the point of maximum cross-correlation between the ERP and a 300-milliseconds cosine template
Slow wave 1	Mean voltage between 750 and 1,250 milliseconds poststimulus
Slow wave 2	Mean voltage between 900 and 1,100 milliseconds poststimulus
Slow wave 1 rms	Root mean square amplitude computed between 750 and 1,250 milliseconds poststimulus
Slow wave 2 rms	Root mean square amplitude computed between 900 and 1,100 milliseconds poststimulus

samples, with replacement, of from 1 to n single-trial vectors of ERP measures from each of the even and odd pools of measures in each of the four conditions (i.e., the two conditions from the monitoring task and the two conditions from the mental arithmetic task). The vectors were averaged after each selection of 1,000 samples. Thus, for example, in the 1,000 samples of four trials the four vectors of measures selected in each sample were averaged to produce a single vector of measures. This averaging procedure was undertaken to increase the signal/noise ratio of the ERP measures.

The fourth step in our analysis procedure involved the classification of the sample vectors as representing one of the two workload levels for each of the two tasks. The classification algorithm that we applied was a linear stepwise discriminant analysis (LSDA). The discriminant functions were developed for each task and subject on one half of the data and cross-validated on the other half of the data set. In an effort to evaluate the utility of spatial information (e.g., the distribution of the ERPs across different scalp sites) in the discrimination between workload levels, separate discriminant functions were computed for the ERP data vectors at the Fz, Cz, and Pz electrode sites as well as for a combined vector of the measures across scalp electrodes. Thus, 8 ERP measures were submitted to the LSDA procedure for each of the individual electrode functions and 24 measures were submitted for each sample vector for the combined electrode function.

Figs. 7.3 and 7.4 provide a graphic representation of the efficiency of discriminating between two workload levels for the mental arithmetic and monitoring tasks, respectively. Plots are included for both the validation data, on which the discriminant functions were derived (left side), and the cross-validation data (right side), which was classified using the discriminant functions developed with the validation data. Separate panels are provided for the single electrode functions as well as for the functions, which included measures from the three different scalp locations (combined). The 12 functions in each of the panels represent the 12 subjects who participated in the study.

There are several noteworthy aspects of the figures. First, classification accuracy is monotonically related to the amount of ERP data. This is not particularly surprising, because the signal-to-noise ratio of the ERP components increases with the square root of the number of trials averaged to produce each sample. Second, combining information across different spatial locations (in the present case the Fz, Cz, and Pz recording sites) clearly improves the classification efficiency. This improved classification efficiency is observed in (a) reduced variability among subjects, (b) increases in the average asymptotic level of classification accuracy across subjects, and (c) a reduction in the amount of ERP data necessary for correctly discriminating between workload levels.

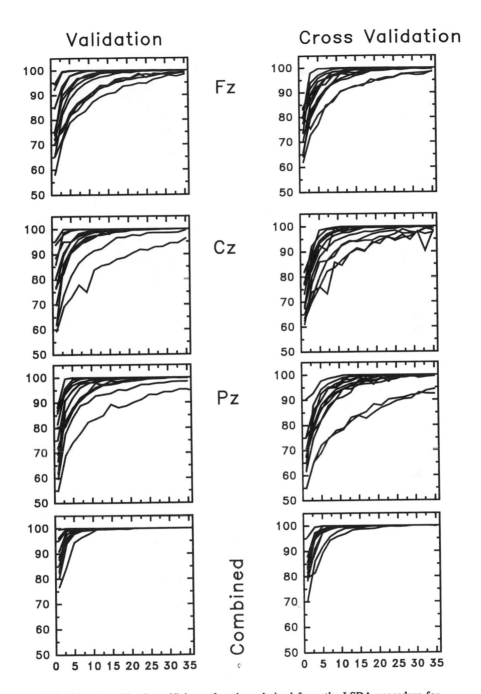

FIG. 7.3. Classification efficiency functions derived from the LSDA procedure for the mental arithmetic task for Fz, Cz, and Pz and combined electrode ERP measures. The plots on the left represent the data that was used to derive the discriminant functions while the plots on the right represent the other half of the data which was fit with the derived discriminant functions. Each of the functions in each of the panels represents a single subject. The conditions that were discriminated in these analyses were A2 and A2/LP. The y axis represents the accuracy of discriminating between the two workload conditions while the x axis represents the number of ERP samples (trials).

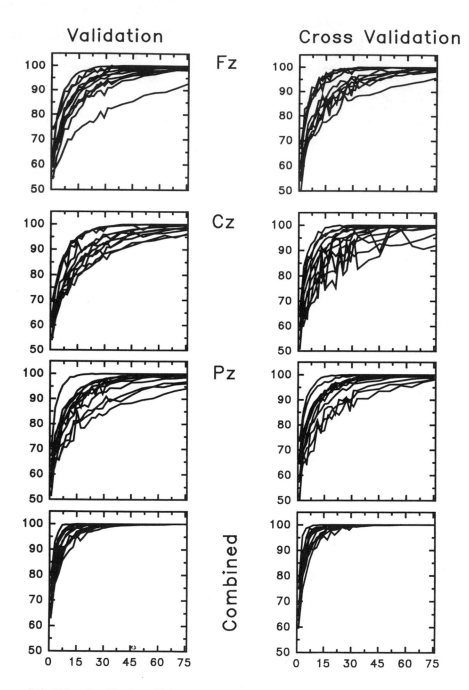

FIG. 7.4. Classification efficiency functions derived from the LSDA procedure for the monitoring task for Fz, Cz, Pz and combined electrode ERP measures. The plots on the left represent the data that was used to derive the discriminant functions while the plots on the right represent the other half of the data which was fit with the derived discriminant functions. Each of the twelve functions in each of the panels represents a single subject. The conditions that were discriminated in these analyses were LP and LP/A3. The *y* axis represents the accuracy of discriminating between the two workload conditions while the *x* axis represents the number of ERP samples (trials).

Another important aspect of the figures is the relatively small loss of classification accuracy when the discriminant functions derived on one data set (validation panels) are applied to a different set of ERP data (cross-validation panels). In fact, in the combined cross-validation sample, 90% correct classification is achieved for all of the 12 subjects within 11 and 5 trials for the monitoring and mental arithmetic tasks, respectively. These results suggest that the within-subject discriminant functions are quite reliable across similar data sets.

Another important output of the discriminant analyses is provided in Table 7.2. The table provides a summary of the ERP measures, at each electrode site and for the two tasks, that had the highest weight in the discriminant equations. Thus, the frequency of entries in different cells of the table provides an indication of the relative importance of different ERP measures in discriminating among the workload levels. There are a number of interesting aspects of the data. First, the pattern of frequencies suggests that no single measure was superior to the other measures across tasks and electrode sites. There were, however, some interesting trends. For instance, at the Pz electrode site the majority of the measures with the highest weights (i.e. the first five measures) pertained to the P300 component. On the other hand, at the Fz and Cz recording sites the measures with the highest weights were more evenly divided among measures that pertained to the P300 component and measures of different aspects of late slow wave components (i.e., the last four measures in the table). Second, the pattern of frequencies suggest that no single measure of a particular ERP component was the best discriminator for all of the subjects. Thus, in the case of the P300, both the base-peak amplitude and the cross-correlation measures proved to be good discriminators for different subjects in the sample. For the slow wave component, the best univariate measures appeared to be the mean voltage

TABLE 7.2

Number of Subjects With the Highest Weight in the Discriminant Equations for Each of the ERP Measures at Each of the Three Electrode Sites for the Monitoring and Mental Arithmetic Tasks

Measure	Monitoring Task			Arithmetic Task		
	Fz	Cz	Pz	Fz	Cz	Pz
Base-peak amplitude	2	2	3	2	1	6
Base-peak mean amplitude	1	1	1	0	2	2
Base-peak rms	0	1	1	2	1	0
Cross-correlation mean voltage	3	3	3	3	1	2
Slow wave 1	3	1	2	2	2	2
Slow wave 2	1	1	1	0	2	0
Slow wave 1 rms	2	3	1	1	1	0
Slow wave 2 rms	0	0	0	2	2	0

and root mean square error for the longer measurement interval (e.g., slow wave 1, slow wave 1 rms measures)

In summary, these data suggest that psychophysiological measures, and ERPs in particular, might have some utility as measures of momentary fluctuations in mental workload and, therefore, might serve as a trigger for adaptive aiding. Of course, this conclusion must remain tentative until our findings are validated with a larger variety of tasks, processing demands, and subjects. The issue of the generalizability of these findings to noisier settings such as high-fidelity simulators and operational systems also remains an open question (although for some promising results see Kramer, Sirevaag, & Braune, 1987; Sirevaag et al., 1993).

MENTAL WORKLOAD ASSESSMENT WITH IRRELEVANT PROBES

Our examination of the feasibility of employing ERPs to discriminate among momentary variations in processing demands was performed within the context of single- and dual-tasks in which ERPs were elicited by task-relevant discrete events. Although this ERP-eliciting procedure is advantageous because it does not require the addition of any extraneous tasks or stimuli, as does the secondary task procedure, it is often difficult, particularly in operational settings, to modify the system hardware and software to incorporate ERP recording. Additionally, it may be difficult to identify task-relevant discrete events that occur with sufficient frequency in many automated systems. Thus, it would be useful to possess other recording procedures for use in situations in which the primary task method is impractical or inappropriate. As we briefly described earlier, the irrelevant probe procedure may prove useful in situations that preclude the use of the primary task method.

In the study we describe next we assessed the utility of the irrelevant probe technique in a high-fidelity radar simulator with 10 highly experienced Navy radar operators. The radar operators performed a standard training exercise that contained periods of low- and high-processing demands. The ERPs were elicited by three different tones that differed in frequency of occurrence. One of the three tones occurred on 80% of the probe trials, whereas each of the other two tones occurred on 10% of the probe trials. Prior to the radar monitoring task, in which the tones were to be ignored, the radar operators performed a baseline oddball condition in which they pushed a response button every time one of the two low-probability tones was presented. The other low-probability tone and the high-probability tone did not require an overt response. The baseline condition was used to establish a record of each individual's ERP components in the absence of the demands of the radar monitoring task. The dual-deviant/single-standard (i.e., two different low-probability tones and one high-probability tone) tone presentation was used for two reasons. First, low-probability stimuli often elicit larger amplitude ERP components than do high-probability stimuli. Thus, the use of two low-probability tones would presumably enable us to increase the frequency of detecting changes in the amplitude of ERP components in response to variations in the processing demands in the radar monitoring task. Second, we were interested in examining ERP components such as the N100, N200 and P300 as well as the mismatch negativity (MMN; Naatanen, 1990). Although the N100, N200, and P300 components are often easiest to observe when subjects actively attend or respond to the ERP eliciting events, MMNs are difference waveform components that can most easily be dissociated from other ERP components when elicited by high- and low-probability events that do not require any overt action. Thus, the use of two deviant probe events, one that required a response and one that did not, would enable us to discern the MMN as well as the other ERP components in the baseline condition.

Our expectations with regard to the sensitivity of the different ERP components to the introduction of the radar monitoring task and an increase in its difficulty were as follows. The N100 component has long been interpreted to reflect early processes of selective attention or resource allocation (Hackley, Woldoroff,& Hillyard, 1990; Hillyard, Hink, Schwent, & Picton, 1973). The N100 has also shown a graded sensitivity to processing demands across both tasks and input locations (Parasuraman, 1985). Thus, it is conceivable that the irrelevant probe-evoked N100s will show a systematic decrease in amplitude with the introduction of the radar monitoring tasks as well as with an increase in its difficulty.

Previous evidence suggests that N200s may also be sensitive to changes in processing demands within and across tasks, at least when the N200s are elicited by task-relevant events. For example, Lindholm, Cheatham, Koriath, and Longridge (1984) reported increases in the amplitude of N200s with increases in the difficulty of a simulated flight mission (see also Horst, Ruchkin, & Munson, 1987).

There are now several reports suggesting that the MMN, which was once thought to be insensitive to attention, may be susceptible to attentional demands under some circumstances (Trejo, Ryan-Jones, & Kramer, in press; Woldoroff, Hackley, & Hillyard, 1991). Thus MMNs, derived by subtracting the standard from the ignored-deviant waveforms, may reflect changes in processing demands that are associated with the performance of the radar monitoring tasks as well as an increase in its difficulty.

The P300 component is well documented to be sensitive to changes in processing demands. However, most of these demonstrations have taken place within the context of primary or secondary task methods in which the eliciting stimulus is to be actively attended. There is little evidence suggesting that P300s will reliably reflect graded changes in processing demands when the P300s are elicited by task-irrelevant stimuli (but for exceptions see Sirevaag et al., 1993; Wilson & McCloskey, 1988). In fact, there is strong evidence dating back to the early research of Sutton and colleagues that P300s are not elicited by task irrelevant stimuli (see Sutton & Ruchkin, 1984). Thus, given the challenging nature of the radar monitoring task it is our expectation that irrelevant-probe P300s will not be observed in either the low- or high-load conditions of the radar monitoring task.

The radar monitoring task required the radar operators to detect and respond appropriately to the appearance of a variety of different targets. The targets included commercial aircraft and ships as well as friendly and hostile military aircraft and ships. Appropriate responses included the activation of countermeasures, notification of commanders, and the logging of the area, time, location, and bearing of the targets. The task lasted approximately 45 minutes and was subdivided, for analysis purposes, into periods of low- and high-processing demands. Low- and high-demand periods were operationally defined by two different radar instructors and corresponded, for the most part, to periods of low- and high-target density.

ERP-eliciting tones were presented in a baseline block of 500 trials and during the performance of the radar monitoring task. One of the rare tones was responded to in the baseline condition. The radar operators ignored the tones during the performance of the radar monitoring scenario. The tones varied in frequency and probability of occurrence as follows: 1500 Hz/0.80 (standards), 1000 Hz/0.10 (low deviants), and 2000 Hz/0.10 (high deviants). The interstimulus interval between tones was 700 milliseconds. The tones were 111-milliseconds bursts presented at 80 dB SPL.

Electroencephalographic (EEG) activity was recorded from three midline sites—Fz, Cz, and Pz—and the right mastoid. These electrodes were referenced to the left mastoid. Vertical EOG activity was recorded from electrodes placed above and below the right eye. Horizontal EOG was recorded from electrodes lateral to each eye. Offline, epochs of 1,000 milliseconds, including a 200-milliseconds prestimulus baseline, were extracted and the vertical and horizontal EOG records were then used to reduce EOG contamination.

ERP averages for each stimulus and experimental condition were created separately for each subject, arithmetically re-referenced to average mastoids, digitally low-pass filtered (0 to 12 Hz), and adjusted for zero-median prestimulus baseline voltage. A difference wave was computed by subtracting the average ERP for standards from the average ERP for nontarget deviants. A number of ERP components were measured on the average waveforms in each of the experimental conditions for each of the subjects. These components included N100 (latency range of 75 to 175 milliseconds poststimulus), N200 (latency range of 200 to 400 milliseconds poststimulus), and the P300 (latency range of 300 to 600 milliseconds poststimulus). The N100 and N200 amplitudes were defined as the average of 30 points centered on their respective maximal amplitudes. P300 was defined as the mean amplitude from 300 to 600 milliseconds poststimulus. Finally, two mismatch negativity components (MMN) were measured on the deviant-standard subtraction waveforms. MMN1 was defined as the mean amplitude from 150 to 250 milliseconds poststimulus. MMN2 was defined as the mean amplitude from 250 to 350 milliseconds poststimulus.

RESULTS AND CONCLUSIONS

Several important findings were obtained in our study. First, and perhaps most important, a number of ERP components were found to be sensitive to both the introduction of the radar monitoring task as well as an increase in its difficulty. The N100s and N200s elicited by the deviant tones and recorded at the Fz site systematically decreased in amplitude from the baseline condition to the low-load radar monitoring condition and from this condition to the high-load radar monitoring condition. The MMNs, illustrated in the difference waveforms presented in Fig. 7.5, behaved in a manner similar to the N100 and N200s with respect to their sensitivity to processing demands. That is, the MMN recorded at the Fz site decreased in amplitude with the introduction of the radar monitoring task as well as with an increase in its difficulty.

An interesting and unanswered question is why attention would be allocated to task-irrelevant probes. After all, the voluntary allocation of attention to irrelevant aspects of the environment would be harmless at best and costly (in terms of a reduced processing of relevant information) at worst. Thus, why do the radar operators adopt this seemingly suboptimal strategy? The tentative answer, provided by recent research on visual attention, is that subjects do not adopt such a strategy but instead attention is captured by some features of the environment.

Research by Yantis and colleagues (Yantis & Jonides, 1984, 1990) and others (Theeuwes, 1992; Todd & Van Gelder, 1979) has demonstrated that stimuli that suddenly appear in the environment capture attention regardless, for the most part, of subject's intentions. Many of these experiments are run in the following way. Subjects are asked to search displays for a particular target. On a small proportion of the search trials the target is a sudden-onset item, whereas on most of the trials the target is revealed by removing portions of one of many premasks that appear on the display.

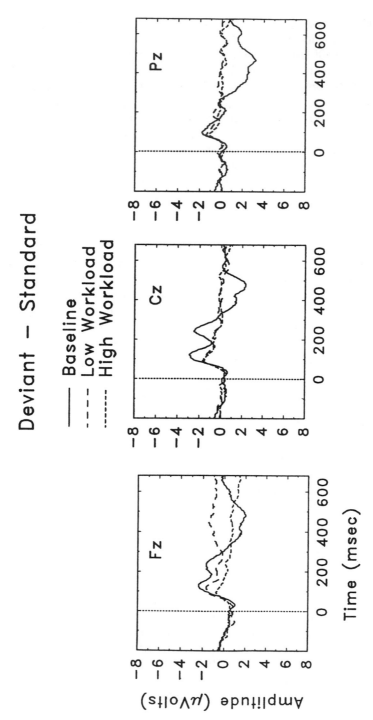

FIG. 7.5. Grand average waveforms derived by subtracting the ERPs elicited by the standard tones (e.g. the tones which occurred on 80% of the tone trials) from the ERPs elicited by the deviant tones.

These latter items are referred to as *offset-stimuli*. It is important to note that the individual onset styles of the letters within a display are independent of the target's presence or absence, and, if present, its location: The target letter is equally likely to be a sudden-onset object as any one of the remaining offset objects. Thus, there is no reason for subjects to attend to the onset object first, because it is not any more likely to be the target than any of the other (offset) objects. However, subjects do appear to attend to the sudden onset items first. This results in a search slope that is independent of the number of items on the display when the target is an onset item. Yantis (1993) suggested that the sudden appearance (sudden-onset) of a new object has a special status in that it automatically captures attention, in a bottom-up or stimulus-driven fashion, as long as attention is not tightly focused on another object in the visual field. Thus, although new objects can be said to automatically capture attention because individuals do not need to intend to process the object for the capture to occur, capture is not strongly automatic because it will fail to occur if attention is already focused elsewhere in the visual field.

We speculate that the notion of attention capture may provide an answer to why it appears that subjects attend the irrelevant probes in our study. That is, the deviant irrelevant probes capture attention because they constitute the occurrence of a new object in the environment. Attention that is not focused elsewhere (e.g., on the radar monitoring task) is captured by the deviant probes and is reflected in the amplitude of the N100, N200, and MMN components. Within this type of theoretical framework subjects do not voluntarily allocate attention to the probes but instead attentional resources that are not being used in the service of the primary task are redirected, in a stimulus-driven or bottom-up fashion, to the processing of the probes.

The P300 results can also be accommodated within this framework. It has been argued previously that P300 reflects the operation of a selection process that capitalizes on higher-order or semantic properties of stimuli. For example, whereas N100s are elicited by any stimulus that appears in an attended portion of the visual field, P300s are only elicited by target objects that occur in the attended field (Mangun & Hillyard, 1990). The fact that P300s were not elicited by the irrelevant probes in the radar monitoring task suggests that capture of attention by the probes was quite transitory. Thus, although the graded workload effect indicated by the N100, N200, and MMN components does suggest that the probes were processed to some degree during the performance of the radar monitoring task, the lack of a P300 effects suggests that this processing was aborted prior to a full evaluation of the stimulus.

We believe that our results are encouraging in that they suggest that ERPs elicited by task irrelevant probes can provide a nonintrusive method for the assessment of variations in mental workload (see also Ullsperger, Freude, & Erdmann, 1994). It is important to note, however, that the N100, N200, and MMN components are rather small relative to the ongoing electroencephalographic activity and, therefore, it is unlikely that these measures will be able to provide a real-time assessment of mental workload. On the other hand, there are many situations in which an offline assessment of mental workload is important (e.g., prototype and system evaluation, operator training assessment, fitness for duty evaluation), and it is these situations in which the ERP-based irrelevant probe technique might be effectively employed.

A BRIEF POSTSCRIPT

Our brief review of the applied psychophysiological literature suggests that psychophysiological measures have the potential to provide useful information about human information processing activities in automated systems. Our literature review was confined to one psychological construct, mental workload, and one class of psychophysiological measures, ERPs. However, other ongoing programs of research are actively exploring the utility of a variety of other psychophysiological measures for the assessment of mental workload as well as other psychological constructs of relevance to human performance in automated systems. For example, a group of Navy researchers led by Scott Makeig (Makeig & Inlow, 1993; Mullane, Makeig, & Trejo, in press) has been developing a system to predict lapses in vigilance, and the resulting decrements in performance, on the basis of an ongoing frequency domain analysis of electroencephalographic (EEG) activity. Thus far, relatively high success rates have been achieved in laboratory simulations of a sonar monitoring task using only a single recording site and a single psychophysiological measure, EEG. Potential improvements in such a system may be achieved by increasing the spatial resolution of the EEG recording (i.e., recording from more than a single electrode site) as well as by incorporating other psychophysiological measures that vary with levels of alertness such as respiration, heart rate, and eye movements.

Farwell and Donchin (1988) reported the development of an ERP-based communication device in which the P300 component of the ERP is used to index operators attention to particular objects in a 6 x 6 matrix of letters and numbers. One interesting aspect of this system is that communication can take place in the absence of eye movements to the attended objects. Thus, such a system could potentially detect the allocation of attention to a location in the periphery of the visual field while an operator is fixated, and possibly also attending, to a location in the central visual field. Gehring et al. (1993) recently reported the discovery of an ERP component, which they dubbed the *error related negativity* (ERN), and which appears to provide a manifestation of a neural system associated with the detection of and compensation for errors of responding. Such a component has the potential to provide an index of whether operators were aware of errors that they may have made in responding to system events.

In addition to the potential applicability of ERP and EEG measures described previously, a number of other psychophysiological measures (for a detailed review see Kramer, 1991) have the potential to contribute to the assessment of operator state in adaptively automated systems. One of the most promising classes of measures is heart rate and heart rate variability. Measures of heart rate have been successfully used to assess changes in mental workload in simulated and operational flight tasks in both commercial and military environments (Jorna, 1993; Roscoe, 1992; Veltman & Gaillard, 1993). One advantage of heart rate measures, compared to ERP measures, is that they can be recorded in the absence of probe stimuli. Thus, heart rate measures can be recorded in a variety of simulator and operational settings without the need to introduce extraneous and potentially disruptive stimuli. Given the continuous nature of heart rate measures it is also relatively easy to collect these data in a timely fashion. However, one concern with heart rate measures concerns the relative sensitivity of this class of measures to physical as compared to mental demands. Thus, heart rate is much more responsive to changes in metabolic demands engendered by physical than by mental demands. Fortunately, there are techniques that can be used to deconfound physical and mental influences on the heart rate measure (Jorna, 1992; Mulder, Veldman, Ruddel, Robbe, & Mulder, 1991). Heart rate measures are also less diagnostic than are ERPs with respect to changes in particular aspects of human information processing. Thus, although ERP components are sensitive to a narrow set of mental functions (e.g. N100/P100—spatial attention, N200—physical mismatch, ERN—error monitoring), heart rate measures are sensitive to changes in a wider variety of emotional and cognitive phenomena. However, although this lack of diagnosticity might be problematic in some circumstances (e.g., when trying to decide whether the perceptual or motor demands imposed on the pilot by a prototype cockpit for an advanced helicopter are excessive), there are other situations in which it is sufficient to know that the operator is overloaded and that computer-based aiding is needed. It is with this latter case that heart rate measures might be most profitably employed as an index of operator workload.

In summary, a number of laboratories are currently developing and validating psychophysiological measures of human information processing activities that have the potential to provide important insights into human cognition and performance in adaptively automated systems. Whether such potential will be realized is dependent on the demonstration of the efficacy of such psychophysiological techniques in extra-laboratory environments as well as the acceptance of these techniques by the user communities.

REFERENCES

Bortolussi, M., & Vidulich, M. (1989). The benefits and costs of automation in advanced helicopters: An empirical study. In R. Jensen (Ed.), *Proceedings of the 5th International Symposium on Aviation Psychology* (pp. 594–599). Columbus, OH: Ohio State University, Department of Aviation.

Ephrath, A., & Young, L. (1981). Monitoring versus man in the loop detection of aircraft control failures. In J. Rasmussen & W. Rouse (Eds.), *Human detection and diagnosis of system failures* (pp. 143–154). New York: Plenum.

Farwell, L., & Donchin, E. (1988). Talking off the top of your head: Toward a mental prosthesis utilizing event-related brain potentials. *Electroencephalography and Clinical Neurophysiology, 70,* 510–523.

Geddes, N. (1986). *Opportunities for intelligent aiding in naval air–sea warfare: An A-18 war at sea study* (Technical report 8502-1). Norcross, GA: Search Technology, Inc.

Gopher, D., & Donchin, E. (1986). Workload — An examination of the concept. In K. Boff, L. Kaufman, & J. Thomas (Eds.), *Handbook of perception and Human Performance, Vol. II.* (pp. 41.–41.49). New York: Wiley.

Govindaraj, T., & Rouse, W. (1986). Modeling the human controller in environments that include continuous and discrete tasks. *IEEE Transactions on Systems, Man and Cybernetics, SMC-11,* 410–417.

Greenstein, J., & Ravesman, M. (1986). Development and validation of a mathematical model of human decision making for human–computer communication. *IEEE Transactions on Systems, Man and Cybernetics, SMC-16,* 148–154.

Hackley, S., Woldoroff, M., & Hillyard, S. (1990). Cross-modal selective attention effects on retinal, myogenic, brainstem and cerebral evoked potentials. *Psychophysiology, 27,* 195–208.

Hillyard, S., Hink, R., Schwent, V., & Picton, T. (1973). Electrical signs of selective attention in the human brain. *Science, 182,* 177–180.

Horst, R., Ruchkin, D., & Munson, R. (1987). Event-related potential processing negativities related to workload. In R. Johnson, J. Rohrbaugh, & R. Parasuraman (Eds.), *Current trends in event-related potential research* (pp. 186–197). Amsterdam: Elsevier.

Humphrey, D., & Kramer, A. (1994). Towards a psychophysiological assessment of dynamic changes in mental workload. *Human Factors, 36,* 3–26.

Isreal, J., Chesney, G., Wickens, C., & Donchin, E. (1980). P300 and tracking difficulty: Evidence for multiple resources in dual-task performance. *Psychophysiology, 17,* 259–273.

Isreal, J., Wickens, C., Chesney, G., & Donchin, E. (1980). The event-related brain potential as an index of display monitoring workload. *Human Factors, 22,* 211–224.

Jasper, H. (1958). The ten-twenty electrode system of the International Federation. *Electroencephalography and Clinical Neurophysiology, 10,* 371–375.

Jorna, P. (1992). Spectral analysis of heart rate and psychological state: A review of its validity as a workload index. *Biological Psychology, 34,* 237–257.

Jorna, P. (1993). Heart rate and workload variations in actual and simulated flight. *Ergonomics, 36,* 1043–1054.

Kessel, C., & Wickens, C. (1982). The transfer of failure detection skills between monitoring and controlling dynamic systems. *Human Factors, 24,* 49–60.

Kramer, A. F. (1991) Physiological measures of mental workload: A review of recent progress. In D. Damos (Ed.), *Multiple task performance* (pp. 279–328). London: Taylor and Francis.

Kramer, A. F., Sirevaag, E. J., & Braune, R. (1987). A psychophysiological assessment of operator workload during simulated flight missions. *Human Factors, 29*, 145–160.

Kramer, A. F., Sirevaag, E., & Hughes, P. (1988). Effects of foveal task load on visual spatial attention: Event-related brain potentials and performance. *Psychophysiology, 25*, 512–531.

Kramer, A. F., & Spinks, J. (1991). Capacity views of information processing. In R. Jennings & M. Coles (Eds.), *Psychophysiology of human information processing: An integration of central and autonomic nervous system approaches* (pp. 179–250). New York: Wiley.

Kramer, A. F., Wickens, C. D., & Donchin, E. (1983). An analysis of the processing demands of a complex perceptual-motor task. *Human Factors, 25*, 597–622.

Kramer, A. F., Wickens, C. D., & Donchin, E. (1985). The processing of stimulus attributes: Evidence for dual-task integrality. *Journal of Experimental Psychology: Human Perception and Performance, 11*, 393–408.

Lindholm, E., Cheatham, C., Koriath, J., & Longridge, T. (1984). *Physiological assessment of aircraft pilot workload in simulated landing and hostile threat environments* (Tech. Report AFHRL-TR-83-49). Williams Air Force Base, AZ: Air Force Systems Command.

Makeig, S., & Inlow, M. (1993). Lapses in alertness: Coherence of fluctuations in performance and in the EEG spectrum. *Electroencephalography and Clinical Neurophysiology, 86*, 23–35.

Mangun, R., & Hillyard, S. (1990). Allocation of visual attention to spatial locations: Tradeoff functions for event-related brain potentials and detection performance. *Perception and Psychophysics, 47*, 532–550.

Mulder, L., Veldman, J., Ruddel, H., Robbe, W., & Mulder, G. (1991). On the usefulness of finger blood pressure measurements for studies on mental workload. *Homeostatis in Health and Disease, 33*, 47–60.

Mullane, M., Makeig, S., & Trejo, L. (in press). *Electrophysiological monitoring and management of operator alertness I: Experimental design* (NPRDC Technical Note). San Diego: Navy Personnel Research and Development Center.

Naatanen, R. (1990). The role of attention in auditory information processing as revealed by event related brain potentials and other brain measures of cognitive function. *Behavioral and Brain Sciences, 13*, 201–288.

Papanicolaou, A., & Johnstone, J. (1984). Probe evoked potentials: Theory, method and applications, *International Journal of Neuroscience, 24*, 107–131.

Parasuraman, R. (1985). Event-related brain potentials and intermodal divided attention. *Proceedings of the Human Factors Society, 29th Annual Meeting*. Santa Monica, CA: Human Factors Society.

Roscoe, A. (1992). Assessing pilot workload: Why measure heart rate, HRV and respiration? *Biological Psychology, 34*, 259–287.

Rouse, W. (1988). Adaptive aiding for human/computer control. *Human Factors, 30*, 431–443.

Sheridan, T. (1987). Supervisory control. In G. Salvendy (Ed.), *Handbook of human factors* (pp. 1243–1268). New York: Wiley.

Sirevaag, E., Kramer, A. F., Coles, M., & Donchin, E. (1989). Resource reciprocity: An event related brain potentials analysis. *Acta Psychologica, 70*, 77–97.

Sirevaag, E., Kramer, A., Wickens, C., Reisweber, M., Strayer, D., & Grenell, J. (1993). Assessment of pilot performance and workload in rotary wing helicopters. *Ergonomics, 9*, 1121–1140.

Strayer, D. L., & Kramer, A. F. (1990). Attentional requirements of automatic and controlled processing. *Journal of Experimental Psychology: Learning, Memory and Cognition, 16*, 67–82.

Sutton, S., & Ruchkin, D. (1984). The late positive complex: Advances and new problems. In R. Karrer, J. Cohen, & P. Tueting (Eds.), *Brain and information: Event-related potentials* (pp. 1–23) New York: Annals of the New York Academy of Science.

Theeuwes, J. (1992). Perceptual selectivity for color and form. *Perception and Psychophysics*, *51*, 599–606.

Todd, J., & Van Gelder, P. (1979). Implications of a sustained-transient dichotomy for the measurement of human performance. *Journal of Experimental Psychology: Human Perception and Performance*, *5*, 625–638.

Trejo, L. J., Kramer, A. F., & Arnold, J. A. (1995). Event-related potentials as indices as display monitoring performance. *Biological Psychology, 40,* 33–72.

Trejo, L., Ryan-Jones, D., & Kramer, A. (1995). Attentional modulation of the mismatch negativity elicited by frequency differences between binaurally presented tone bursts. *Psychophysiology, 32,* 319–328.

Ullsperger, P., Freude, G., & Erdmann, U. (1994, October). *Novelty P3 as an index of resource allocation during mental workload.* Paper presented at Thirty-Fourth Annual Meeting of the Society for Psychophysiological Research, Atlanta, GA.

Ullsperger, P., Metz, A., & Gille, H. (1988). The P300 component of the event-related brain potential and mental effort. *Ergonomics*, *31*, 1127–1137.

Veltman, J., & Gaillard, A. (1993). Indices of mental workload in a complex task environment. *Neuropsychobiology*, *28*, 72–75.

Wickens, C. D. (1992). *Engineering psychology and human performance* (2nd ed.). New York: HarperCollins.

Wickens, C. D., Kramer, A. F., Vanasse, L., & Donchin, E. (1983). The performance of concurrent tasks: A psychophysiological analysis of the reciprocity of information processing resources. *Science, 221,* 1080–1082.

Wilson, G., & Eggemeier, F. T. (1991). Psychophysiological assessment of workload in multi-task environments. In D. Damos (Ed.), *Multiple task performance* (pp. 329–360). London: Taylor and Francis.

Wilson, G., & McCloskey, K. (1988). Using probe evoked potentials to determine information processing demands. *Proceedings of the Human Factors Society* (pp. 1400–1403). Santa Monica, CA: Human Factors Society.

Woldoroff, M., Hackley, S., & Hillyard, S. (1991). The effects of channel selective attention on the mismatch negativity wave elicited by deviant tones. *Psychophysiology*, *28*, 30–42.

Yantis, S. (1993). Stimulus-driven attention capture. *Current Directions in Psychological Science, 2,* 156–161.

Yantis, S., & Jonides, J. (1984). Abrupt visual onsets and selective attention: Evidence from visual search. *Journal of Experimental Psychology: Human Perception and Performance*, *10*, 601–620.

Yantis, S., & Jonides, J. (1990). Abrupt visual onsets and selective attention: Voluntary versus automatic allocation. *Journal of Experimental Psychology: Human Perception and Performance, 16,* 121–134.

8 Automation and Situation Awareness

Mica R. Endsley
Texas Tech University

INTRODUCTION

Automation represents one of the major trends of the 20th century. The drive to provide increased levels of control to electromechanical systems, and with it a corresponding distancing of the human from direct system control, has grown out of the belief that automated systems provide superior reliability, improved performance, and reduced costs for the performance of many functions. Through the auspices of the technological imperative, automation has steadily advanced as means have been found for automating physical, perceptual, and, more recently, cognitive tasks in all kinds of systems.

In many cases automation has provided the desired benefits and has extended system functionality well beyond existing human capabilities. Along with these benefits, however, a certain price has been extracted. The role of the human operator has changed dramatically. Instead of performing tasks, the human's job has become that of monitor over an automated system—a role to which people are not ideally suited.

Contrary to the implication of the term *automated*, humans have remained a critical part of most automated systems. They must monitor for failures of the automated system and the presence of conditions the system is not designed to handle. Furthermore, as most automation has been piecemeal, covering certain functions but not others, humans have remained in the system as integrators—monitoring the automation for some functions and performing others themselves.

Because the systems to be monitored continue to increase in complexity

163

with the addition of automation, an increased trend toward large catastrophic failures often accompanies the incorporation of automation
(Wickens, 1992; Wiener, 1985). When things go wrong, they go wrong in a
big way. In examining these failures, it becomes apparent that the coupling
of human and machine in the form of observer and performer is far from
perfect in terms of optimizing the overall functioning of the joint human–
machine system.

A central shortcoming associated with the advent of automated systems
has been dubbed the *out-of-the-loop performance problem*. When acting as
monitor of an automated system, people are frequently slow in detecting
that a problem has occurred that necessitates their intervention. Once
detected, additional time is also needed to determine the state of the system
and sufficiently understand what is happening in order to be able to act in
an appropriate manner. The extra time associated with performing these
steps can be critical, prohibiting performance of the very activities the
human is present to handle. The result ranges from a slight delay in human
performance to catastrophic failures with major consequences.

In 1987, a Northwest Airlines MD-80 crashed on takeoff at Detroit
Airport due to an improper configuration of the flaps and slats, killing all
but one passenger (National Transportation Safety Board, 1988). A major
factor in the crash was the failure of an automated takeoff configuration
warning system, on which the crew had become reliant. They did not realize
the aircraft was improperly configured for takeoff and had neglected to
check manually (due to other contributing factors). When the automation
failed, they were not aware of the state of the automated system or the
critical flight parameters they counted on the automation to monitor.

In 1989, a USAir B-737 failed to takeoff at New York's LaGuardia
Airport, landing in the nearby river (National Transportation Safety Board,
1990). The precipitating cause was an accidental disarming of the auto-
throttle. Neither the captain nor the first officer monitored the critical flight
parameters in order to detect and correct the problem, and thus the takeoff
was not aborted in a timely manner, resulting in the loss of the aircraft and
two passengers.

In 1983, a Korean Airlines flight was shot down over what was then the
USSR, with no survivors. The aircraft was interpreted as hostile when it
traveled into Soviet airspace without authorization or radio contact.
Although critical equipment was never recovered, it is believed that an
erroneous entry was made into the flight navigation system early in the
flight (Stein, 1983). The crew unknowingly flew to the wrong coordinates,
reliant on the automated system and unaware of the error.

In each of these cases, the human operators overseeing the automated
systems were unaware of critical features of the systems they were operating. They were unaware of the state of the automated system and of the
aircraft parameters for which the automation was responsible.

In addition to difficulties in detecting these types of automation errors, it is also frequently difficult for human operators to correctly understand what the problem is once they have detected that something is amiss. In the USAir accident, for instance, both crew members tried to gain control of the aircraft, but were unable due to the mistrimmed rudder. The lost time associated with trying to overcome the aircraft control problem without understanding what was causing it fatally delayed aborting the takeoff until it was too late.

With many automated systems, partially due to their complexity, understanding the meaning of displayed information, once attended to, can represent a significant difficulty. For instance, in aircraft systems, pilots have reported significant difficulties in understanding what their automated flight management systems are doing and why (Sarter & Woods, 1992; Wiener, 1989). Similarly, the accident at the Three Mile Island nuclear power plant was attributed to an erroneous override of the automated emergency handling system by the human operators. They had misdiagnosed the situation based on displayed information and believed an excessive coolant level was causing the problem rather than too little coolant (Wickens, 1992). They did not correctly understand the meaning of the information that was displayed to them.

Each of these problems can be directly linked to a lower level of situation awareness that exists when people operate as monitors of automated systems. Situation awareness (SA), a person's mental model of the world around him or her, is central to effective decision making and control in dynamic systems. This construct can be severely impacted by the implementation of automation.

SITUATION AWARENESS

Originally a term used in the aircraft pilot community, situation awareness has developed as a major concern in many other domains where people operate complex, dynamic systems, including the nuclear power industry, automobiles, air traffic control, medical systems, teleoperations, maintenance, and advanced manufacturing systems. Achieving situation awareness is one of the most challenging aspects of these operators' jobs and is central to good decision making and performance. Hartel, Smith, and Prince (1991) found poor situation awareness to be the leading causal factor in military aviation mishaps. In a recent review of commercial aviation accidents, 88% of those with human error involved a problem with situation awareness (Endsley, 1994a). Situation awareness clearly is critical to performance in these environments.

Situation awareness is formally defined as "the perception of the elements in the environment within a volume of time and space, the comprehension of their meaning and the projection of their status in the near future" (Endsley, 1988a, p. 97). Situation awareness involves perceiving critical factors in the environment (Level 1 SA), understanding what those factors mean, particularly when integrated together in relation to the person's goals (Level 2 SA), and, at the highest level, an understanding of what will happen with the system in the near future (Level 3 SA). These higher levels of situation awareness are critical for allowing decision makers to function in a timely and effective manner.

For instance, in an aircraft environment, operators must be aware of critical flight parameters, the state of their on-board systems, their own location and the location of important reference points and terrain, and the location of other aircraft along with relevant flight parameters and characteristics. This information forms the "elements" they need to perceive to have good Level 1 SA. But a great deal has to do with how the operators interpret the information they take in. They need to comprehend that a certain pattern of flight parameters indicates that they are near stall point, or that the displayed altitude is below their assigned altitude. This understanding forms their Level 2 SA. At the highest level, Level 3 SA, their understanding of the state of the system and its dynamics can allow them to be able to predict its state in the near future. A group of enemy aircraft flying in a particular formation will thus be projected to attack in a given manner. With accurate and complete situation awareness, operators can act to bring their systems into conformance with their goals.

IMPACT OF AUTOMATION ON SITUATION AWARENESS

Automation can be seen to directly impact situation awareness through three major mechanisms: changes in vigilance and complacency associated with monitoring, assumption of a passive role instead of an active role in controlling the system, and changes in the quality or form of feedback provided to the human operator (Endsley & Kiris, 1995) . Each of these factors can contribute to the out-of-the-loop performance problem. In addition, automated systems, by nature of their complexity, also challenge the higher levels of situation awareness (comprehension and projection) during ongoing system operations.

Vigilance, Complacency, and Monitoring

There is a long history of cases in which operators are reportedly unaware of automation failures and do not detect critical system state changes when acting as monitors of automated systems (Ephrath & Young, 1981; Kessel & Wickens, 1982; Wickens & Kessel, 1979; Young, 1969). Although monitoring failures have typically been associated with simple, low-event tasks, Parasuraman (1987) concluded that "vigilance effects can be found in complex monitoring and that humans may be poor passive monitors of an automated system, irrespective of the complexity of events being monitored" (p. 703). There are many cases of problems in monitoring aircraft automation. Billings (1991) reported that the probability of human failure in monitoring automation increases when devices behave reasonably but incorrectly, and when operators are simply not alert to the state of automation.

Complacency—overreliance on automation—is one major factor associated with a lack of vigilance in monitoring automation. Complacency has been attributed to the tendency of human operators to place too much trust in automated systems (Danaher, 1980; Parasuraman, Molloy, & Singh, 1993; Wiener, 1985). Singh, Molloy, and Parasuraman (1993) found that complacency was a function of a person's trust in, reliance on, and confidence in automation. Trust in the automated system is a critical factor necessary for it to be employed by operators (Lee & Moray, 1992; Riley, 1994). Associated with this trust, however, operators may elect to neglect the automated system and the system parameters overseen by the automation in favor of other tasks through a shifting of attention (Parasuraman, Mouloua, & Molloy, 1994), resulting in low situation awareness on these factors. The demands of other tasks in complex, multitask environments have also been directly linked to complacency effects (Parasuraman et al., 1993). Because an operator's attention is limited, this is an effective coping strategy for dealing with excess demands. The result, however, can be a lack of situation awareness on the state of the automated system and the system parameters it governs.

Monitoring problems have also been found with systems that have a high incidence of false alarms, leading to a lack of trust in the automation. Wiener and Curry (1980) and Billings (1991) reported on numerous failures by aircrews to heed automatic alarms, leading to serious accidents. Even though the system provides a noticeable visual or auditory signal, the alarms are ignored or disabled by flight crew who have no faith in the system due to its high false alarm rate.

Thus, significant reductions in situation awareness can be found with automated systems, as people may neglect to monitor the automation and its parameters, attempt to monitor them, but fail due to vigilance problems, or be aware of problems via system alerts but not comprehend their significance due to high false alarm rates.

Active versus Passive

In addition to vigilance problems, the fact that operators are passive observers of automation instead of active processors of information may add to their problems in detecting the need for manual intervention and in reorienting themselves to the state of the system in order to do so. Evidence suggests that the very act of becoming passive in the processing of information may be inferior to active processing (Cowan, 1988; Slamecka & Graf, 1978). This factor could make a dynamic update of system information and integration of that information in active working memory more difficult.

In a recent study, Endsley and Kiris (1995) found that subjects' situation awareness was lower under fully automated and semi-automated conditions than under manual performance in an automobile navigation task. Only level 2 SA, understanding and comprehension, was negatively impacted, however; Level 1 SA was unaffected. Thus, although they were aware of low-level data (effectively monitoring the system), they had less comprehension of what the data meant in relation to operational goals. The out-of-the-loop performance problem associated with automation was also observed in the automated conditions.

This finding was specifically attributed to the fact that operators in the automated conditions were more passive in their decision-making processes, drawing on the automated expert system's recommendations. Under the conditions of the experiment, there was no change in information displayed to the operators, and vigilance and monitoring effects were insufficient to the explain the situation awareness decrement. Turning a human operator from a performer into an observer can, in and of itself, negatively affect situation awareness, even if the operator is able to function as an effective monitor, and this can lead to significant problems in taking over during automation failure.

Feedback

A change in the type of system feedback or a complete loss of feedback has also been cited as a problem associated with automation (Norman, 1989). "Without appropriate feedback people are indeed out-of-the-loop. They may not know if their requests have been received, if the actions are being performed properly, or if problems are occurring" (Norman, 1989, p. 6). He attributed this problem largely to an erroneous belief by system designers that information on certain parameters is no longer needed by system operators once relevant functions are assumed by automation.

In some cases, critical cues may be eliminated with automation and replaced by other cues that do not provide for the same level of performance. In many systems, important cues may be received through auditory, tactile, or the olfactory senses. When processes are automated, new forms of feedback are created, frequently incorporating more accurate visual displays; yet, the fact that information is in a different format may make it harder to assimilate with other information or less salient in a complex environment. Young (1969) and Kessel and Wickens (1982) found that proprioceptive feedback received during manual control was important to performance and denied in automated tracking tasks in which information was only presented visually. The development of electronic fly-by-wire flight controls in the F-16 led to problems in determining airspeed and maintaining proper flight control, because the vibration information that usually came through the flight stick was suddenly missing (even though the needed information was clearly indicated on traditional visual displays) (Kuipers, Kappers, van Holten, van Bergen, & Oosterveld, 1989). Artificial stick-shakers are now routinely added to fly-by-wire systems to put back in the feedback to which operators are accustomed (Kantowitz & Sorkin, 1983).

In some cases, the design of the automated system intentionally conceals information from the operator. Some autofeathering systems, for instance, have neglected to notify pilots of their actions in shutting down engines, leading to accidents (Billings, 1991). In some notable accidents, the fact that the automated system had failed was not clearly indicated to the operator, as in the Northwest Airlines accident in Detroit (National Transportation Safety Board, 1988). In addition, there is a tendency for some displays to eliminate raw system data in favor of processed, integrated information. Important information as to the source of information, its reliability, or the value of constituent data underlying the integrated information may be unavailable. Billings (1991) noted that the clarity of the integrated displays may be seductive, yet highly misleading if such underlying information is not known.

A noted problem in many systems is the lack of information salience that may accompany automation. Frequently, displays associated with complex automated systems involve computerized CRT screens with information imbedded in hierarchical displays that may be associated with various system modes. Problems with getting lost in menus, not finding the desired display screen, and interpreting cluttered displays have been noted. The increased display complexity and computerized display format reduces the perceptual salience of information, even if it is available. In a complex environment with many activities going on, it is easy for operators to lose track of such information.

Either intentionally or inadvertently, the design of many systems poses a considerable challenge to situation awareness through the elimination of or change in the type of feedback provided to operators regarding the system's status. Unless very careful attention is paid to the format and content of information displays, these issues can easily sabotage situation awareness when operators are working with automated systems.

Lack of Understanding of Automation

One of the major impediments to the successful implementation of auto-mation is the difficulty many operators have in understanding automated systems, even when they are attending to them and the automation is working as designed. This may be partially attributed to the inherent complexity associated with many of these systems, to poor interface design, and to inadequate training.

The development and maintenance of situation awareness involves keeping up with a large quantity of rapidly changing system parameters, and then integrating them with others parameters, active goals, and one's mental model of the system to understand what is happening and project what the system is going to do. This allows operators to behave proactively to optimize system performance and take actions to forestall possible future problems. As complexity increases (as it is apt to do with automation), this task becomes even more challenging. The number of parameters increases, and they change and interact according to complex underlying functions. Achieving an accurate mental model of the system can be very difficult, and this taxes the ability of the operator to attain the higher levels of situation awareness (comprehension and projection) from information that is per-ceived. By adding to system complexity, therefore, automated systems may make achieving good situation awareness more difficult.

Wiener (1989) documented many problems with a lack of understanding of automated systems in aircraft by the pilots who fly with them. Mc-Clumpha and James (1994) conducted an extensive study of nearly 1,000 pilots from across varying nationalities and aircraft types. They found that the primary factor explaining variance in pilots' attitudes toward advanced technology aircraft was their self-reported understanding of the system. Although understanding tended to increase with number of hours in the aircraft, this also was related to a tendency to report that the quality and quantity of information provided was less appropriate and more excessive. Although pilots are eventually developing a better understanding of auto-mated aircraft with experience, the systems do not appear to be well designed to meet their information needs. Rudisill (1994) reported from the same study that "What's it doing now?", "I wonder why its doing that?", and "Well, I've never seen that before" are widely heard comments in advanced cockpits, echoing similar concerns by Wiener (1989).

Many of these problems clearly can be attributed to standard human factors short-comings in interface design. For instance, transitions from one system mode to another may not be salient, designated by small changes in the displayed interface yet creating very different system behavior. Although the systems are operating properly, operators may be confused as they misinterpret observed system behavior in light of their mental model of a different system mode.

With increased complexity, proving information clearly to operators so that they understand the system state and state transitions becomes much more challenging. Operators may rarely see certain modes or combinations of circumstances that lead to certain kinds of system behavior. Thus, their mental models of the systems may be incomplete. This leaves operators unable to properly interpret observed system actions and predict future system behavior, constituting a significant situation awareness problem.

Although problems with complexity and interface design are somewhat peripheral to automation per se (i.e., these problems also exist in many systems quite apart from any automation considerations), these issues often plague automated systems, and can significantly undermine operator situation awareness in working with automated systems.

Benefits to Situation Awareness

It should be noted that automation does not always result in these types of problems. Wiener (1985) pointed out that automation has, for the most part, worked quite well, and has accompanied a dramatic reduction in many types of human errors. Furthermore, he believed that it may improve situation awareness by reducing the display clutter and complexity associated with manual task performance, and through improved integrated displays (Wiener, 1992, 1993).

It has been suggested that automation may also improve situation awareness by reducing excessive workload (Billings, 1991). Curry and Ephrath (1977) found that monitors of an automatic system actually performed better than did manual controllers in a flight task. As monitors, the authors argued, subjects may have been able to distribute their excess attention to other displays and tasks.

Recent research, however, demonstrates a certain degree of independence between situation awareness and workload (Endsley, 1993a). Workload may only negatively impact situation awareness at very high levels of workload. Low situation awareness can also accompany low levels of workload. If workload is reduced through automation, therefore, this may not translate into higher situation awareness.

Furthermore, whether automation actually results in lower workload remains questionable. Wiener's (1985) studies showed that pilots report automation does not reduce their workload, but actually may increase it during critical portions of the flight. Many recent studies are beginning to confirm this. Harris, Goernert, Hancock, and Arthur (1994) and Parasuraman, Mouloua, and Molloy (1994) showed that operator initiation of automation under high workload may increase workload even more. Riley (1994) augmented this with his finding that a subject's choice to use automation for a task is not related to the workload level of the task. Grubb, Miller, Nelson, Warm, and Dember (1994) showed that workload actually was fairly high in tasks where humans must act as monitors over a period of time, as they do with automated systems. Automation in many ways may serve to increase workload, particularly when workload is already high. Bainbridge (1983) called it the irony of automation that when workload is highest automation is of the least assistance. Despite this, however, workload remains the fundamental human factors consideration in many automation decisions.

DESIGN AND EVALUATION OF AUTOMATED SYSTEMS

The effect of automated systems on situation awareness and the out-of-the-loop performance problem has been established as a critical issue that can undermine the effectiveness of human–machine performance in advanced systems. Many of the factors that can lead to situation awareness problems — monitoring, passive decision making, poor feedback, poor mental models — can be directly traced to the way that the automated systems are designed. As such, it is possible, and essential, to minimize these problems during system design, thus allowing the potential benefits of automation to be realized without depriving the human operator of the situation awareness needed for good performance.

Interface Design

At a minimum, the design process should include steps to ensure that needed information is always present regarding the state of the automation and the state of the parameters being monitored in a clear, easily interpreted format. As subtle changes in the form of information can impact situation awareness, it is critical that proposed designs be tested thoroughly (within the context of the operators' problem domain and in conjunction with multiple task demands) for possible insidious effects on situation awareness. Careful consideration needs to be given to providing interpretable and comprehensible information (that maps to the operator's goals), as opposed to volumes of low-level data, in order to better meet operator needs.

New Approaches to Automation

Many of the issues surrounding the negative impact of automation on situation awareness and human performance may be attributable not to automation itself, but the way that automation has traditionally been implemented in many systems. In most cases, the performance of some task has been given over to the automation and the human's job has become that of monitor. New approaches are currently being explored that challenge this division. These approaches redefine the assignment of functions to people and automation in terms of a more integrated team approach. Two orthogonal and possibly complementary approaches can be defined along the axes of Fig. 8.1. One approach seeks to optimize the assignment of control between the human and automated system by keeping both involved in system operation. The other recognizes that control must pass back and forth between the human and the automation over time, and seeks to find ways of using this to increase human performance.

Level of Control.   One way to minimize the negative effects of automation is to devise implementation schemes that keep the human actively involved in the decision making loop while simultaneously reducing the load associated with doing everything manually. This can be accomplished by determining a level of automation that minimizes negative impacts on operator situation awareness (Endsley, 1987b, 1993b).

Wiener and Curry (1980) and Billings (1991) both discussed the fact that automation does not exist in an all-or-none fashion, but can be implemented at various levels. With regard to the automation of cognitive tasks (through artificial intelligence or expert systems), a task may be accomplished (a) manually with no assistance from the system; (b) by the operator with input in the form of recommendations provided by the system; (c) by the system, with the consent of the operator required to carry out the action; (d) by the system, to be automatically implemented unless vetoed by

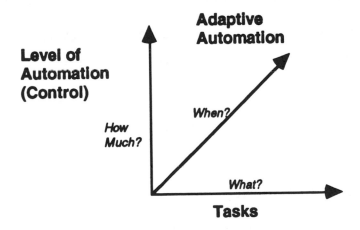

FIG. 8.1.   Automation design considerations.

the operator; or (e) fully automatically, with no operator interaction (Endsley, 1987b, 1993b). This can be viewed as five possible levels of automation from none to full, as depicted in Fig. 8.2.

Endsley and Kiris (1995) implemented automation of an automobile navigation task at each of these five levels. They found that the out-of-the-loop performance problem was significantly greater under full automation than under intermediate levels of automation. This corresponded with a greater decrement in situation awareness under full automation than under intermediate levels, as compared to manual control. By implementing functions at a lower level of automation, leaving the operator involved in the active decision making loop, situation awareness remained at a higher level and subjects were more able to assume manual control when needed.

Thus, even though full automation of a task may be technically possible, it may not be desirable if the performance of the joint human–machine system is to be optimized. Intermediate levels of automation may be preferable for certain tasks, in order to keep human operators' situation awareness at a higher level and allow them to perform critical functions.

Adaptive Automation.   Recent work on adaptive automation has also been found to aid in overcoming the out-of-the-loop performance problem (Parasuraman, 1993). Adaptive automation recognizes that over the course of time, control of tasks may need to pass back and forth between an operator and an automated system in response to changing demands. Adaptive automation attempts to optimize this dynamic allocation of tasks by creating a mechanism for determining in real time when tasks need to become automated (or manually controlled) (Morrison, Cohen, & Gluck-

		Roles	
Level of Automation		**Human**	**System**
None	1	Decide, Act	———
Decison Support	2	Decide, Act	Suggest
Consentual AI	3	Concur	Decide, Act
Monitored AI	4	Veto	Decide, Act
Full Automation	5	———	Decide, Act

FIG. 8.2.   Levels of control and automation (adapted from Endsley and Kiris, 1995).

man, 1993). In direct contrast to historical efforts that have featured fixed task allocation assignments, adaptive automation provides the potential for improving operator performance with automated systems by continually adjusting to operator needs and keeping operators in the loop. In recent research, Carmody and Gluckman (1993) found Level 2 SA to be impacted by adaptive automation of certain tasks.

Issues.   Many questions remain in exploring the problem space set forth by the approach in Fig. 8.1. Most notably, the characteristics of tasks that determine their optimal level of control and suitability for adaptive automation need to be investigated. Gluckman, Carmody, Morrison, Hitchcock, and Warm (1993), for instance, found different effects on workload and performance for adaptive automation involving a static (system monitoring) task versus a dynamic (resource management) task. Carmody and Gluckman (1993) also found situation awareness to be more affected by adaptive automation of a dynamic task than a static task. Lewandowski, Durso, and Grounlund (1994) proposed that if parts of an integrated task are automated, more performance decrements will occur than if the whole task is automated. Endsley and Kiris (1995) proposed that psychomotor, perceptual, and cognitive tasks may be differentially affected by automation level. Considerable work is needed to determine the critical dimensions of tasks for successful implementation of adaptive automation and specifying an optimal level of control for a given task.

Second, when adaptive automation should be invoked needs to be determined. In a first look at this issue, Parasuraman (1993) examined whether manual control implemented at a preset periodic interval differed in effect from manual control implemented on the basis of poor monitoring performance (indicating a loss of situation awareness). He found no differences between the two conditions in terms of their effect on subsequent human monitoring performance under automation. The insertion of a period of manual control was equally beneficial in both cases. This type of work needs to be extended to address questions of periodic insertion of automation into manual tasks. Research is also needed to explore the interaction between adaptive automation and level of control—how much automation needs to be employed may be a function of when it is employed.

How adaptive automation should be implemented is also a question. Various schemes have been proposed, from purely manual control for turning the automation on and off to system control for invoking automation based on real-time monitoring of human performance, physiology, or flight conditions. This is not a simple matter. Many systems have left it up to operators to invoke automation at their discretion. In critical situations, however, the operator may be so overloaded as to make this an extra burden, incapacitated or otherwise unable to do so, unaware that the situation calls for automated assistance, or a poor decision maker. Leaving the system with the ability to turn itself on and off may be even more problematic, because this taxes the operator with the task of keeping up with what the system is doing. A major question lies in determining how adaptive automation should be implemented so as to provide the potential benefits without leading to new problems of loss of system awareness.

Thus, an increased level of human control and adaptive automation may provide means for keeping operators sufficiently in the loop to avoid the situation awareness decrements that can occur with traditional all-or-nothing function allocations. Significant research issues still need to be resolved to determine how to best implement these concepts within operational settings.

Evaluation of Automated Systems

A direct consideration of the operator interface, the proper level of control for a task, and means of transitioning between automated and nonautomated states needs to made during the design process in order to address fundamental problems associated with automation. Although these concepts have shown merit for improving situation awareness, research to establish precise guidelines for such prescriptions is, to date, just beginning. Lacking such guidance, and because the effects of automation can be quite insidious, careful testing during the development of automated systems is imperative.

Several methods have been established for the measurement of situation awareness. (For a complete review see Endsley, 1994b.) Most involve the creation of a simulation of the system under consideration. Impact of a particular design concept on situation awareness can be measured directly through either objective or subjective means, or it can be inferred through less direct performance measures.

Objective Measurement. The most commonly used means of objectively evaluating a design concept's impact on situation awareness involves directly questioning operators as to their perceptions of critical aspects of the system they are operating. The Situation Awareness Global Assessment Technique (SAGAT) (Endsley, 1987a, 1988b) is a technique wherein the simulation is frozen at randomly selected times, the system displays blanked, and the simulation suspended while subjects quickly answer questions about their current perceptions of the situation. Subject perceptions are then compared to the real situation based on simulation computer databases to provide an objective measure of situation awareness.

SAGAT includes queries about all operator situation awareness requirements, including Level 1 (perception of data), Level 2 (comprehension of meaning), and Level 3 (projection of the near future) components. This includes a consideration of system functioning and status as well as relevant features of the external environment. This approach minimizes possible biasing of attention, as subjects cannot prepare for the queries in advance because they could be queried over almost every aspect of the situation to which they would normally attend.

SAGAT provides an objective, unbiased assessment of operator situation awareness that overcomes memory problems incurred when collecting data after the fact, yet minimizes biasing of subject situation awareness due to secondary task loading or artificially cueing the subject's attention. Empirical, predictive, and content validity has been demonstrated for this technique (Endsley, 1989, 1990a, 1990b).

Subjective Measurement. Subjective measures of situation awareness are easier and less expensive to administer than are objective measures, but may lack the same degree of accuracy and diagnosticity. The most commonly used method is to have operators provide ratings of their situation awareness with system concepts along a designated scale.

Taylor (1989) developed the Situational Awareness Rating Technique (SART), which has operators rate system designs on the amount of demand on attentional resources, supply of attentional resources, and understanding of the situation provided. As such, it considers operators' perceived workload (supply and demand on attentional resources) in addition to their perceived understanding of the situation. Although SART has been shown to be correlated with performance measures (Selcon & Taylor, 1989), it is unclear whether this is due to the workload or the understanding components.

Performance Measurement. In general, performance measures provide the advantage of being objective and are usually nonintrusive. Simulation computers can be programmed to record specified performance data automatically, making the required data relatively easy to collect. Hansman et al. (1992), for example, used detection of clearance amendment errors as a measure of aircrew situation awareness in evaluating the use of an automated datalink system for updating the onboard flight management computer. Because many other factors can act to influence subject performance measures (e.g., decision making, workload impacts, actions, or individual strategy differences), these are also limited for inferring subject situation awareness by themselves.

General Measurement Considerations. One of the biggest difficulties associated with automation is its insidious effect on situation awareness and performance. An increased tendency for out-of-the-loop performance problems may be difficult to detect during testing. For this reason, it is preferable to measure situation awareness directly, in addition to evaluating operator performance with an automation concept.

It also needs to be recognized that automation can affect situation awareness in a global, not readily predicted manner. A new system may provide more situation awareness on one factor, but simultaneously reduce situation awareness on others. Assessment of the impact of automation on situation awareness needs to take into account operator situation awareness across the range of requirements.

Finally, it can take quite a bit of training for operators to feel comfortable with an automation concept and be able to adequately perform with it. Assessments of any automation concept need to be made after operators have become proficient with the system, if a fair evaluation is to be made. Of course, if that proficiency takes an inordinate amount of time (years in the case of previously mentioned aircraft systems), system deficiencies are certainly indicated.

CONCLUSION

In conclusion, the successful implementation of automation is a complex issue. The traditional form of automation that places humans in the role of monitor has been shown to negatively impact situation awareness and thus their ability to effectively perform that function. Losses in situation awareness can be attributed to the unsuitability of humans to perform a monitoring role, assumption of a passive role in decision making, and inadequate feedback associated with automation. As a result many automated systems have been suboptimized, with infrequent but major errors attributed to a failure of the human component. This unsatisfactory state of affairs currently plagues many automation efforts. New approaches to automation design that seek to fundamentally alter the role of the human operator in interacting with the automated system provide a great deal of promise for surmounting this problem. Careful test and evaluation of proposed automation concepts is imperative for establishing that adequate situation awareness is maintained to keep human operators in the loop and able to fulfill their functions.

REFERENCES

Bainbridge, L. (1983). Ironies of automation. *Automatica, 19,* 775–779.
Billings, C. E. (1991). *Human-centered aircraft automation: A concept and guidelines* (NASA Technical Memorandum 103885). Moffet Field, CA: NASA-Ames Research Center.
Carmody, M. A., & Gluckman, J. P. (1993). Task specific effects of automation and

automation failure on performance, workload and situational awareness. In R. S. Jensen & D. Neumeister (Eds.), *Proceedings of the Seventh International Symposium on Aviation Psychology* (pp. 167-171). Columbus, OH: Department of Aviation, The Ohio State University.

Cowan, N. (1988). Evolving conceptions of memory storage, selective attention, and their mutual constraints within the human information processing system. *Psychological Bulletin, 104*(2), 163-191.

Curry, R. E., & Ephrath, A. R. (1977). Monitoring and control of unreliable systems. In T. B. Sheridan & G. Johannsen (Eds.), *Monitoring behavior and supervisory control* (pp. 193-203). New York: Plenum.

Danaher, J. W. (1980). Human error in ATC system operations. *Human Factors, 22*(5), 535-545.

Endsley, M. R. (1987a). *SAGAT: A methodology for the measurement of situation awareness* (NOR DOC 87-83). Hawthorne, CA: Northrop Corporation.

Endsley, M. (1987b). The application of human factors to the development of expert systems for advanced cockpits. In *Proceedings of the Human Factors Society 31st Annual Meeting* (pp. 1388-1392). Santa Monica, CA: Human Factors Society.

Endsley, M. R. (1988a). Design and evaluation for situation awareness enhancement. In *Proceedings of the Human Factors Society 32nd Annual Meeting* (pp. 97-101). Santa Monica, CA: Human Factors Society.

Endsley, M. R. (1988b). Situation awareness global assessment technique (SAGAT). In *Proceedings of the National Aerospace and Electronics Conference (NAECON)* (pp. 789-795). New York: IEEE.

Endsley, M. R. (1989). A methodology for the objective measurement of situation awareness. In *Situational Awareness in Aerospace Operations (AGARD-CP-478)* (pp. 1/1-1/9). Neuilly Sur Seine, France: NATO-AGARD.

Endsley, M. R. (1990a). Predictive utility of an objective measure of situation awareness. In *Proceedings of the Human Factors Society 34th Annual Meeting* (pp. 41-45). Santa Monica, CA: Human Factors Society.

Endsley, M. R. (1990b). *Situation awareness in dynamic human decision making: Theory and measurement*. Unpublished doctoral dissertation, University of Southern California, Los Angeles.

Endsley, M. R. (1993a). Situation awareness and workload: Flip sides of the same coin. In R. S. Jensen & D. Neumeister (Eds.), *Proceedings of the Seventh International Symposium on Aviation Psychology* (pp. 906-911). Columbus, OH: Department of Aviation, The Ohio State University.

Endsley, M. R. (1993b). Situation awareness: A fundamental factor underlying the successful implementation of AI in the air traffic control system. In D. J. Garland & J. A. Wise (Eds.), *Human factors and advanced automation technologies* (pp. 117-122). Daytona Beach, FL: Embry-Riddle Aeronautical University Press.

Endsley, M. R. (1994a, March). *A taxonomy of situation awareness errors*. Paper presented at the Western European Association of Aviation Psychology 21st Conference, Dublin, Ireland.

Endsley, M. R. (1994b). Situation awareness in dynamic human decision making: Measurement. In R. D. Gilson, D. J. Garland, & J. M. Koonce (Eds.), *Situational awareness in complex systems* (pp. 79-97). Daytona Beach, FL: Embry-Riddle Aeronautical University Press.

Endsley, M. R., & Kiris, E. O. (1995). The out-of-the-loop performance problem and level of control in automation. *Human Factors, 37*(2), 381-394.

Ephrath, A. R., & Young, L. R. (1981). Monitoring vs. man-in-the-loop detection of aircraft control failures. In J. Rasmussen & W. B. Rouse (Eds.), *Human detection and diagnosis of system failures* (pp. 143-154). New York: Plenum.

Gluckman, J. P., Carmody, M. A., Morrison, J. G., Hitchcock, E. M., & Warm, J. S. (1993). Effects of allocation and partitioning strategies of adaptive automation on task performance and perceived workload in aviation relevant tasks. In R. S. Jensen & D. Neumeister (Eds.), *Proceedings of the Seventh International Symposium on Aviation Psychology* (pp. 150–155). Columbus, OH: Department of Aviation, The Ohio State University.

Grubb, P. L., Miller, L. C., Nelson, W. T., Warm, J. S., & Dember, W. N. (1994). Cognitive failure and perceived workload in vigilance performance. In M. Mouloua & R. Parasuraman (Eds.), *Human performance in automated systems: Current research and trends* (pp. 115–121). Hillsdale, NJ: Lawrence Erlbaum Associates.

Hansman, R. J., Wanke, C., Kuchar, J., Mykityshyn, M., Hahn, E., & Midkiff, A. (1992, September). *Hazard alerting and situational awareness in advanced air transport cockpits.* Paper presented at the 18th ICAS Congress, Beijing, China.

Harris, W. C., Goernert, P. N., Hancock, P. A., & Arthur, E. (1994). The comparative effectiveness of adaptive automation and operator initiated automation during anticipated and unanticipated taskload increases. In M. Mouloua & R. Parasuraman (Eds.), *Human performance in automated systems: Current research and trends* (pp. 40–44). Hillsdale, NJ: Lawrence Erlbaum Associates.

Hartel, C. E., Smith, K., & Prince, C. (1991, April). *Defining aircrew coordination: Searching mishaps for meaning.* Paper presented at the Sixth International Symposium on Aviation Psychology, Columbus, OH.

Kantowitz, B. H., & Sorkin, R. D. (1983). *Human factors: Understanding people-system relationships.* New York: Wiley.

Kessel, C. J., & Wickens, C. D. (1982). The transfer of failure-detection skills between monitoring and controlling dynamic systems. *Human Factors, 24*(1), 49–60.

Kuipers, A., Kappers, A., van Holten, C. R., van Bergen, J. H. W., & Oosterveld, W. J. (1989). Spatial disorientation incidents in the R.N.L.A.F. F16 and F5 aircraft and suggestions for prevention. In *Situational awareness in aerospace operations (AGARD-CP-478)* (pp. OV/E/1–OV/E/16). Neuilly Sur Seine, France: NATO–AGARD.

Lee, J., & Moray, N. (1992). Trust, control strategies and allocation of function in human–machine systems. *Ergonomics, 35*(10), 1243–1270.

Lewandowski, S., Durso, F. T., & Grounlund, S. D. (1994). Modular automation: Automating sub-tasks without disrupting task flow. In M. Mouloua & R. Parasuraman (Eds.), *Human performance in automated systems: Current research and trends* (pp. 326–331). Hillsdale, NJ: Lawrence Erlbaum Associates.

McClumpha, A., & James, M. (1994). Understanding automated aircraft. In M. Mouloua & R. Parasuraman (Eds.), *Human performance in automated systems: Current research and trends* (pp. 183–190). Hillsdale, NJ: Lawrence Erlbaum Associates.

Morrison, J., Cohen, D., & Gluckman, J. P. (1993). Prospective principles and guidelines for the design of adaptively automated crewstations. In R. S. Jensen & D. Neumeister (Eds.), *Proceedings of the Seventh International Symposium on Aviation Psychology* (pp. 172–177). Columbus, OH: Department of Aviation, The Ohio State University.

National Transportation Safety Board (1988). *Aircraft accident report: Northwest Airlines, Inc., McDonnell-Douglas DC-9-82, N312RC, Detroit Metropolitan Wayne County Airport, August, 16, 1987* (NTSB/AAR-99-05). Washington, DC: Author.

National Transportation Safety Board (1990). *Aircraft accident report: USAir, Inc., Boeing 737-400, LaGuardia Airport, Flushing New York, September 20, 1989* (NTSB/AAR-90-03). Washington, DC: Author.

Norman, D. A. (1989). *The problem of automation: Inappropriate feedback and interaction not overautomation* (ICS Report 8904). La Jolla, CA: Institute for Cognitive Science, University of California–San Diego.

Parasuraman, R. (1987). Human–computer monitoring. *Human Factors, 29*(6), 695–706.

Parasuraman, R. (1993). Effects of adaptive function allocation on human performance. In D.

J. Garland & J. A. Wise (Eds.), *Human factors and advanced aviation technologies* (pp. 147–158). Daytona Beach, FL: Embry-Riddle Aeronautical University Press.

Parasuraman, R., Molloy, R., & Singh, I. L. (1993). Performance consequences of automation-induced complacency. *International Journal of Aviation Psychology, 3*(1), 1–23.

Parasuraman, R., Mouloua, M., & Molloy, R. (1994). Monitoring automation failures in human–machine systems. In M. Mouloua & R. Parasuraman (Eds.), *Human performance in automated systems: Current research and trends* (pp. 45–49). Hillsdale, NJ: Lawrence Erlbaum Associates.

Riley, V. (1994). A theory of operator reliance on automation. In M. Mouloua & R. Parasuraman (Eds.), *Human performance in automated systems: Current research and trends* (pp. 8–14). Hillsdale, NJ: Lawrence Erlbaum Associates.

Rudisill, M. (1994). Flight crew experience with automation technologies on commercial transport flight decks. In M. Mouloua & R. Parasuraman (Eds.), *Human performance in automated systems: Current research and trends* (pp. 203–211). Hillsdale, NJ: Lawrence Erlbaum Associates.

Sarter, N. B., & Woods, D. D. (1992). Pilot interaction with cockpit automation: Operational experiences with the flight management system. *The International Journal of Aviation Psychology, 2*(4), 303–321.

Selcon, S. J., & Taylor, R. M. (1989). Evaluation of the situational awareness rating technique (SART) as a tool for aircrew systems design. In *Situational awareness in aerospace operations (AGARD-CP-478)* (pp. 5/1–5/8). Neuilly Sur Seine, France: NATO-AGARD.

Singh, I. L., Molloy, R., & Parasuraman, R. (1993). Automation-induced complacency: Development of the complacency-potential rating scale. *International Journal of Aviation Psychology, 3*(2), 111–122.

Slamecka, N. J., & Graf, P. (1978). The generation effect: Delineation of a phenomenon. *Journal of Experimental Psychology: Human Learning and Memory, 4*(6), 592–604.

Stein, K. J. (1983, October 3). Human factors analyzed in 007 navigation error. *Aviation Week & Space Technology,* pp. 165–167.

Taylor, R. M. (1989). Situational awareness rating technique (SART): The development of a tool for aircrew systems design. In *Situational awareness in aerospace operations (AGARD-CP-478)* (pp. 3/1–3/17). Neuilly Sur Seine, France: NATO-AGARD.

Wickens, C. D. (1992). *Engineering psychology and human performance* (2nd ed.). New York: HarperCollins.

Wickens, C. D., & Kessel, C. (1979). The effect of participatory mode and task workload on the detection of dynamic system failures. *IEEE Transactions on Systems, Man and Cybernetics, SMC-9*(1), 24–34.

Wiener, E. L. (1985). Cockpit automation: In need of a philosophy. In *Proceedings of the 1985 Behavioral Engineering Conference* (pp. 369–375). Warrendale, PA: Society of Automotive Engineers.

Wiener, E. L. (1989). *Human factors of advanced technology ("glass cockpit") transport aircraft* (NASA Contractor Report No. 177528). Moffett Field, CA: NASA-Ames Research Center.

Wiener, E. L. (1992, June). *The impact of automation on aviation human factors.* Paper presented at the NASA/FAA Workshop on Artificial Intelligence and Human Factors in Air Traffic Control and Aviation Maintenance, Daytona Beach, FL.

Wiener, E. L. (1993). Life in the second decade of the glass cockpit. In R. S. Jensen & D. Neumeister (Eds.), *Proceedings of the Seventh International Symposium on Aviation Psychology* (pp. 1–11). Columbus, OH: Department of Aviation, The Ohio State University.

Wiener, E. L., & Curry, R. E. (1980). Flight deck automation: Promises and problems. *Ergonomics, 23*(10), 995–1011.

Young, L. R. A. (1969). On adaptive manual control. *Ergonomics, 12*(4), 635–657.

9 Vigilance and Workload in Automated Systems

Joel S. Warm
William N. Dember
University of Cincinnati

Peter A. Hancock
University of Minnesota

INTRODUCTION

Active vs. Supervisory Control

Vigilance or sustained attention refers to the ability of observers to maintain their focus of attention and to remain alert to stimuli for prolonged periods of time (Davies & Parasuraman, 1982; Parasuraman, Warm, & Dember, 1987; Warm, 1984, 1993). This aspect of human performance is of considerable interest to human factors/ergonomics specialists because of its vital role in automated human–machine systems (Howell, 1993; Warm, 1984). As Sheridan (1970, 1987) noted, the development and utilization of automatic control and computing devices for the acquisition, storage, and processing of information has altered the role of the human operator in many work settings from that of active controller to that of executive or supervisor. Thus, in settings such as military surveillance, industrial quality control, robotic manufacturing, seaboard navigation, nuclear power plant operations, long-distance driving, and prodromal symptom monitoring in intensive-care units, observers must attend to a wide variety of displays for untoward events and take effective action when they are noted. As Parasuraman (1986) said, the responsibility for target detection in today's highly automated systems may be allotted to instruments and controls, but human operators are still needed when systems malfunction or unusual events occur. In some cases, such as aircraft and nuclear power plant accidents, vigilance failures can be disastrous.

Historical Background

The systematic study of vigilance began in World War II when the Royal Air Force commissioned Norman Mackworth to investigate an unexpected and perilous finding. After only about 30 minutes on watch, airborne radar observers began failing to notice the "blips" on their radar screens that signified potential targets—enemy submarines—in the sea below. As a result, the undetected U-boats were free to prey on Allied shipping. Thus, a serious problem existed with the human component of a then newly developed automation system for military surveillance.

In an effort to attack this problem experimentally, Mackworth (1948, 1950/1961) devised a simulated radar display called the "Clock Test," in which observers were asked, over a two-hour period, to view the movements of a pointer along the circumference of a blank-faced clock that was otherwise devoid of scale markings or reference points. Once every second, the pointer would move 0.3 inches to a new position. From time to time, it executed a "double jump" of 0.6 inches. This was the critical signal for detection to which observers were to respond by pressing a key. In Mackworth's pioneering experiments, as in most of those that have followed, observers were tested alone; the task was prolonged and continuous; the signals to be detected were clearly perceptible when observers were alerted to them, but were not compelling changes in the operating environment; the signals appeared in a temporally unpredictable manner with low probability of occurrence; and the observers' responses had no bearing on the probability of occurrence of critical signals.

Using the Clock Test in this way, Mackworth was able to chart the course of performance over time and to confirm the suspicions arising in the field that the efficacy of sustained attention is fragile, waning quickly over time. He found the initial level of signal detections in his experiment to be quite good—signals were detected about 85% of the time. However, performance efficiency fell off by 10% after only 30 minutes of watch and continued to decline gradually for the remainder of session.

Since Mackworth's pioneering work, his discovery that the quality of sustained attention deteriorates over time has been confirmed repeatedly (See, Howe, Warm, & Dember, 1995). This "decrement function" or "vigilance decrement" is the most ubiquitous finding in vigilance experiments. Lest one be tempted to argue that the vigilance decrement is an artificial laboratory phenomenon and that performance decrements are rarely found in operational settings (cf. Mackie, 1984), it is important to note that potentially serious decrements have been found to occur with both experienced monitors, as in Mackworth's early experiments, as well as inexperienced operators, and it has been noted in actual operational environments or in settings that closely simulate such environments (Baker, 1962; Colquhoun, 1967, 1977; Schmidke; 1976).

INTRODUCTION TO WORKLOAD

Vigilance as Underarousal

Given the pervasiveness of the vigilance decrement and of operators' needs for vigilance in the control of automated systems, it is important to understand why the quality of sustained attention is fragile. One possibility comes from the traditional view that vigilance tasks represent tedious and understimulating situations, in which the level of workload is exceptionally low, a view that might can easily be inferred from the description of conditions in the typical vigilance experiment indicated previously. Such a view has been the foundation for the well-known arousal or activation model of vigilance, which accounts for the decrement function in terms of the lack of stimulation necessary to support alertness. According to that model, the repetitious and monotonous aspects of sustained attention tasks reduce the level of stimulation needed by elements of the central nervous system (e.g., the ascending reticular formation and the diffuse thalamic projection system) to maintain wakefulness and alertness. As a result, the brain becomes less responsive to external stimulation, leading to a decline in the efficiency of signal detection (Aston-Jones, 1988; Frankmann & Adams, 1962; Heilman, 1995). Thus, the arousal model reflects what might be termed the paradox of automation: Although automation is designed to reduce the workload of operators, it may place them at a functional disadvantage through understimulation.

It is worth noting that the belief that vigilance tasks are understimulating stems principally from a rather superficial task analysis. It is not based on any quantitative evidence regarding the degree of underload inherent in the task. Recent studies from our laboratory have attempted to provide such evidence through measurements of the perceived mental workload of vigilance tasks—the information processing load or resource demands imposed by a task (cf. Eggemeier, 1988; O'Donnell & Eggemeier, 1986).

Our experiments have featured workload measurements obtained by means of the NASA Task Load Index (NASA-TLX; Hart & Staveland, 1988), an instrument that provides a reliable index of overall workload (test–retest correlation = .83) on a scale from 0 to 100, and also identifies the relative contributions of six sources of workload. Three of those sources reflect the demands that tasks place on operators (mental, physical, and temporal demand), whereas three others characterize the interaction between the operator and the task (performance, effort, and frustration). The NASA-TLX is considered to be one of the most effective measures of perceived workload currently available (Hill, Iavecchia, Byers, Zaklad, & Christ, 1992; Nygren, 1991).

Although the NASA-TLX is a convenient and reliable instrument for measuring mental workload, it is essentially a subjective scale, and as Natsoulas (1967) pointed out, there is always some question as to whether any form of self-report accurately reflects respondents' true perceptual experiences. He argued that this problem might be overcome by linking perceptual reports to psychophysical factors known to influence task difficulty. Following Natsoulas' lead, one way to establish the validity of subjective ratings of perceived workload in vigilance experiments would be to bring such ratings under experimental control by demonstrating that factors that degrade performance efficiency increase workload ratings, whereas factors that enhance performance diminish perceived workload. This strategy, which we call the *Natsoulas imperative*, has successfully guided the investigations described later in this chapter. These experiments have revealed, contrary to previous belief, that vigilance tasks are *not quintessential examples of task underload*. Instead, *the cost of mental operations in vigilance is substantial*, and mental demand and frustration tend to be the primary contributors to workload. These surprising results strike to the heart of an arousal model of vigilance performance. On the other hand, they seem compatible with a resource model of attention that maintains that a limited-capacity information processing system allocates processing resources or processing space to cope with tasks that confront it (Wickens, 1984). Within this view, the vigilance decrement reflects depletion of information processing resources over time as a result of the need to make continuous signal/nonsignal decisions under conditions of great uncertainty and little opportunity for situational control (Davies & Parasuraman, 1982; Warm, 1993).

THE WORKLOAD STUDIES

Signal Salience

As in many perceptual tasks, the quality of sustained attention varies directly with the salience or the discriminability of the signals to be detected (Loeb & Binford, 1963; Parasuraman & Mouloua, 1987). In our initial experiment (Gluckman, Warm, Dember, Thiemann, & Hancock, 1988), we explored the effects of signal salience on perceived workload using signals that were either highly salient or more difficult to detect. As can be seen in the left panel of Fig. 9.1, perceptual sensitivity varied directly with signal

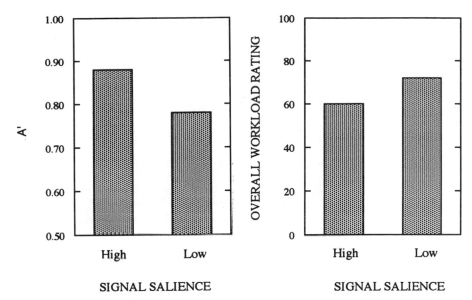

FIG. 9.1.   Perceptual sensitivity (left panel) and perceived workload (right panel) under conditions of high- and low-signal salience (data from Gluckman et al., 1988).

salience: A′ scores, a measure of perceptual sensitivity, were greater in the high- than the low-salience condition. As shown in the right panel of the figure, a mirror image of this effect was evident in the overall workload rating: Workload was lower in the high- as compared to the low-salience condition. Note especially that the workload ratings in the experiment were generally quite elevated, falling within the upper levels of the NASA-TLX scale. Such ratings are greater than those typically obtained in several other types of tasks, including time estimation, grammatical reasoning, and simple tracking (Hancock, 1988; Hart & Staveland, 1988; Liu & Wickens, 1987; Sanderson & Woods, 1987).

Scores for the several components of workload were similar for the two signal salience conditions. As shown Fig. 9.2, the major components of workload, collapsed across salience, were mental demand (MD) and frustration (F). Findings similar to these have also been reported in other studies from our laboratory (Becker, Warm, Dember, & Hancock, 1991; Dittmar, Warm, Dember, & Ricks, 1993; Warm, Dember, & Parasuraman, 1991) and also by Deaton and Parasuraman (1993) and Scerbo, Greenwald, and Sawin (1993).

Our initial study and the others mentioned previously focused on the workload of the vigil as reported at its termination. In order to track the growth of workload over time, we repeated the salience experiment (Dember, et al., 1993) under conditions in which workload was assessed, in independent groups, after either one, two, three, four, or five 10-minute

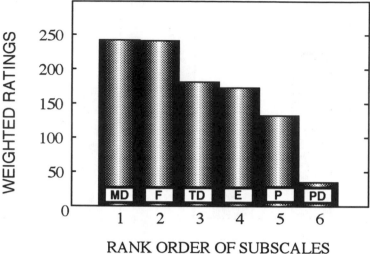

FIG. 9.2.   Relative contribution of six workload dimensions to the vigilance task (MD = mental demand, F = frustration, TD = temporal demand, E = effort, P = performance, PD = physical demand; data from Gluckman et al., 1998).

periods of watch. As in the first experiment, performance efficiency varied inversely with signal salience, and overall workload was greater in the low- as compared to the high-salience condition. The workload ratings increased over time on watch in a manner that was similar for both conditions of signal salience. Fig. 9.3 shows the overall rate of gain in workload over time.

The very low ratings given to a simple card sorting task served as a control, assuring us that the participants' workload ratings were not spuriously high. Once again, the workload scores for the vigilance task fell at the upper end of the NASA-TLX scale. Moreover, they increased linearly over the watch at the rate of 3.2 units of workload/10 minute period. These findings suggest that there are at least two determinants of the workload of vigilance: (a) the salience of the signals to be detected and factors that increase the workload over time such, such as asthenopia (visual fatigue) that comes from monitoring video displays such as those used in our experiments, and (b) musculoskeletal fatigue and restlessness associated with the constrained posture required for optimal monitoring of the display. Indeed, a recent study by Galinsky, Rosa, Warm, and Dember (1993) found that fatigue symptoms and restlessness increase considerably over a 50-minute vigil.

Event Rate

Most vigilance tasks make use of dynamic displays in which critical signals appear within an ensemble of recurrent nonsignal events. For example, in

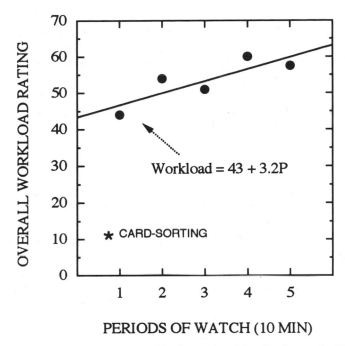

FIG. 9.3.   The rate of gain of workload over time (after Dember et al., 1993).

the Clock Test described earlier, the small 0.3 inch movements of the pointer constituted the background of neutral events in which critical signals for detection, the larger 0.6 inch movements, were embedded. Although the background events may be neutral in the sense that they typically require no response from observers, they are not at all neutral in their effects on performance efficiency. Many investigations have demonstrated that detection efficiency varies inversely with the rate of repetition of neutral background events, or the background event rate (Lanzetta, Dember, Warm, & Berch, 1987; Parasuraman, 1979). Moreover, using a dual-task procedure in which observers were instructed that a vigilance task was their primary responsibility and a probe detection task secondary, Parasuraman (1985) demonstrated that response times to the secondary probes were greatly elevated when they occurred in the presence of a fast as compared to a slow primary task event rate. Thus, in terms of performance measures, the Parasuraman (1985) study indicated that event rate has a considerable impact on the resource demands of a vigilance task. A study by Galinsky, Dember, and Warm (1989) confirmed this conclusion utilizing workload measures. In that study, an average of 80% of the signals was detected in the context of a slow event rate of 5 events/minute; this value dropped to 60% when the event rate was increased to 40/minute. The change in signal detectability was accompanied by a 138% increase in overall workload on the NASA-TLX scale, from a score of 45 at the slow event rate to 62 at the fast event rate. As before, the overall subscale profile indicated that mental demand and frustration were the primary determinants of workload. In addition, temporal demand was rated as a critical component of workload in the fast event-rate condition, a result that might be anticipated given the increased pace at which monitors must attend to the vigilance display in that condition. We replicated these effects in a later investigation (Grubb et al., 1994).

Spatial Uncertainty

Together with variations in signal salience and event rate, the difficulty of vigilance tasks can also be influenced by variations in the monitor's spatial uncertainty as to the location of the events to be observed. Stimulus events can be displayed in a fixed or certain location, or they can occur unpredictably in different positions of the monitor's display. In the former case, monitors need only fix their gaze on a limited area of the display in order to detect signals. In the latter, or spatial uncertainty case, they are confronted with a scanning requirement, and performance efficiency in this condition is poorer than that in the certainty condition (Adams & Boulter, 1964; Milôsević, 1974). As shown by one of our students, Tom Sullivan (1991), the drop in performance efficiency under conditions of spatial uncertainty is accompanied by an increase in the perceived workload of the vigilance task, a result that is consistent with previous findings indicating that attention to visual space is capacity demanding (Jonides, 1981). In his experiment, Sullivan asked observers to monitor a video screen under one of three conditions: (a) static certainty — stimuli (neutral events and critical signals) always appeared center screen, (b) dynamic certainty — stimuli appeared at different locations but in a predictable sequence, and (c) uncertainty — stimuli appeared at random in different screen locations. Using the d' index of perceptual sensitivity, Sullivan found that performance in both the dynamic certainty ($M = 1.75$) and the uncertainty ($M = 1.60$) conditions was degraded in comparison to the static certainty condition ($M = 2.44$), and that this effect was accompanied by changes in the overall workload scores, which were lowest in the certainty condition ($M = 54$) and highest in the uncertainty condition ($M = 73$). As in the studies that preceded his, Sullivan found that mental demand and frustration were the major components of workload in all conditions of his experiment.

Display Uncertainty

In all of the studies described previously, monitors were asked to attend to only one display. However, in operational environments, as, for example, those in aviation (Satchell, 1993), monitors often must attend to multiple displays. In a recent effort to examine the effects of display uncertainty on perceived workload in a vigilance situation, another of our students, Paula Grubb (1995), performed an experiment in which observers were asked to monitor one, two, or four displays resulting in 0, 1, or 2 bits of display uncertainty, respectively. Her results provide another demonstration of the manner in which perceived workload can be brought under experimental control and of the elevated scores associated with vigilance performance. As is evident in Fig. 9.4, detection probability declined with increments in display uncertainty, consistent with an earlier report by Craig and Colquhoun (1977). The decline in signal detections as display uncertainty increased was accompanied by a rise in workload.

Noise

All of our experiments described to this point have manipulated perceived workload through changes in task parameters. Vigilance performance, however, is also sensitive to the effects of several environmental determinants, one of the most important of which is noise. According to Loeb (1986), noise is perhaps the most ubiquitous pollutant in our industrialized society, and it has been extensively investigated in the context of vigilance

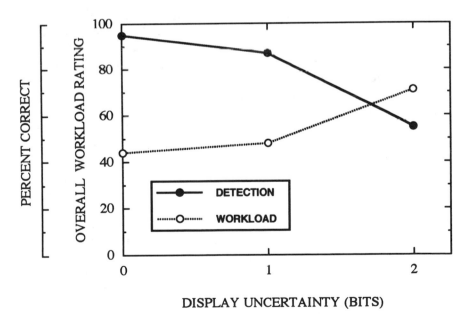

FIG. 9.4.   Percentage of correct detections (outer scale) and perceived workload (inner scale) as a function of display uncertainty (data from Grubb, 1995).

performance. Although there is debate regarding the lawfulness of the effects of noise on vigilance (Koelega & Brinkman, 1986), there is some evidence to indicate that the quality of vigilant behavior is degraded in tasks that impose high information processing demands in the presence of high levels of intermittent noise (Hancock, 1984). Under such conditions, it seemed likely that noise would also elevate perceived workload, especially in light of Cohen's (1980) argument that operators must expend processing resources in order to compensate for the distracting effects of noise. To investigate this possibility, we asked observers to perform a monitoring task in the context of quiet or two levels of jet engine noise, a low-intensity noise level of 70 dB(A) or a high-intensity level of 95 db(A) at the monitor's ear (Becker, Warm, Dember, & Hancock, 1995). The noise employed had a Doppler-like character in that planes seemed to approach from the observer's left and move away to the right. Perceptual sensitivity was degraded by noise: mean d' values for the quiet, low-noise, and high-noise conditions were 1.97, 1.80, and 1.50, respectively, and this effect was accompanied by an elevation in workload in the noise conditions. The mean NASA-TLX scores increased from 65 in quiet to 69 and 75 in the low- and high-noise conditions.

Cueing

A primary assumption in our research on the workload of sustained attention is that the elevated workload scores arise from costs associated directly with vigilance tasks themselves. However, this may not be the case. Vigilance tasks have typically been found to be quite tedious and boring, and it is conceivable that the high workload we have uncovered is not directly task induced. Instead, it might originate in efforts to combat the tedium associated with having to perform a dull task (Sawin & Scerbo, 1994; Scerbo, Greenwald, & Sawin, 1992; Thackray, 1981). To test that possibility, we, together with our students, Ted Hitchcock, Brian Moroney, and Judi See, have recently completed a study using cueing to forewarn monitors of the impending arrival of critical signals. In this investigation, observers monitored a simulated air traffic control display for cases in which planes were on a "collision course" over the center of a city. The event rate used in the experiment was high (30 events/minute) and the signal probability low (.03) so that monitors in a control condition were confronted with the need to observe frequently in order to detect low-probability events. In the cueing condition, monitors were given a verbal warning that a critical signal for detection would occur within the next five displays of aircraft bearings. They were also instructed that the cueing was perfectly reliable and that critical signals would only appear within cued intervals. If fighting through the "cloud of boredom" was the root of the elevated workload scores in vigilance, monitors in the cueing condition would have uncommonly little to do in comparison to controls who were not forewarned; hence, their workload scores might be expected to exceed those of the controls. In contrast, if the workload of sustained attention were directly task related, cueing, by reducing monitors' need for continuous observing/decision making, should reduce perceived workload. The results of the study supported the latter alternative in a dramatic way. As can be seen in Fig. 9.5, cueing was associated with almost-perfect performance and a greatly reduced level of perceived workload. The high level of performance in the cued condition is consistent with previous research using this technique (Aiken & Lau, 1967).

THE PERFORMANCE-WORKLOAD ISSUE

At the outset of this chapter, we indicated that the "Natsoulas imperative" was a guiding theme of our work, and that in addition to measuring the workload of sustained attention, one of our goals was to validate the measures provided by the NASA-TLX by bringing them under experimental control. The results described in the preceding section suggest that we have achieved that aim. On the other hand, our results could lead to the impression that the NASA-TLX does nothing more than parallel performance. Such an impression would be unfortunate, because a fundamental concern in the design of workload measures is to uncover factors that might

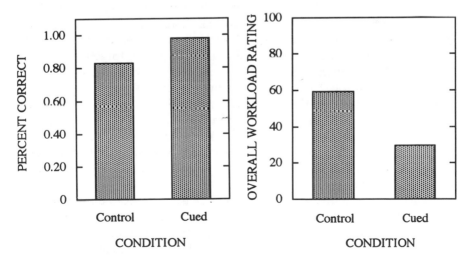

FIG. 9.5. Percentage of correct detections (left panel) and perceived workload (right panel) in noncued (control) and cued conditions.

not be evident in performance itself (O'Donnell & Eggemeier, 1986). Fortunately, our research has also provided evidence that the NASA-TLX can indeed meet that requirement in regard to vigilance.

Workload Signature

A demonstration of the added information to be secured through the use of the NASA-TLX is the consistent subscale profile obtained in our studies, in which mental demand and frustration are the primary components of the workload associated with vigilance tasks. As Becker et al. (1991) noted, the similarity in the pattern of subscale scores across experiments suggests that there may be a typical workload profile that reflects the particular demands imposed by vigilance tasks. Although the absolute magnitude of overall workload may vary as a result of experimental conditions, the workload signature seems to be relatively fixed. This signature, of course, could not have been discovered through the use of performance measures alone.

Cognitive Failure

As noted by Davies and Parasuraman (1982), and by Berch and Kanter (1984), a common finding in vigilance research is that there are substantial individual differences in performance efficiency. Accordingly, considerable effort has been dedicated toward identifying people who might be poorly suited for monitoring tasks and/or might find such tasks to be especially demanding (Craig, 1984; Koelega, 1992). Toward that end, we (Grubb et al., 1994) examined the vigilance performance and perceived workload of monitors who could be classified as high in cognitive failure (HCF). Such individuals tend to be more absent-minded, error-prone, forgetful, and less able to allocate attentional resources adequately than those who are classified as low on the cognitive failure dimension (LCF; Broadbent, Cooper, Fitzgerald, & Parkes, 1982). Accordingly, we expected HCF monitors to perform more poorly than LCF monitors on a sustained attention task and also to give the task higher workload ratings. Our expectation regarding performance efficiency was not confirmed. The detection means for the two groups of monitors were not statistically different over the course of a 30-minute vigil. The workload scores, on the other hand, told quite a different story. The average score on the NASA-TLX for the HCF monitors (56) was considerably greater than that for their LCF (40) counterparts. Thus, although the HCF monitors performed as well as the LCF monitors on the vigilance task, they did so at a higher cost in terms of resources expended. To the extent that high workload may be related to task-induced stress (Hancock & Warm, 1989; Warm, 1993), the results of this study suggest that workload measures might be a useful adjunct to performance measures in any monitoring selection process.

Chronotype Effects

As is true of many tasks (Folkard & Monk, 1985), the quality of vigilance performance is closely linked to circadian factors. Performance efficiency in vigilance tasks generally follows the course of body temperature, rising from a low point early in the morning (0800 hours) to an apogee late in the day (1700/2100 hours), and then declining (Colquhoun, 1977). To explore the role of individual differences in this effect, we (Hamilton, Warm, Dember, Rosa, & Hancock, 1994) studied evening-type individuals, who report feeling most alert in the evening and able to do their best work at that time (Horne & Ostberg, 1977). Because the preferred work times of evening types is concordant with the circadian course of vigilance performance, we anticipated that the morning-to-afternoon gain in performance among evening types would exceed that of control observers who had neither evening nor morning preferences. We also anticipated that given the preference of evening types for working late, they would find the task more demanding in the morning than in the evening, whereas the reverse might be true for the controls. Once again, our prediction about performance efficiency was not supported. Evening types did more poorly on the vigilance task in comparison to the controls regardless of time of day. In contrast, as can be seen in Fig. 9.6, the evening types did, as expected, find the workload of the task to be more demanding in the morning than in the evening, whereas just the opposite was the case for the controls. It is clear from this example that the workload scores are capable of providing information about monitors' reactions to vigilance tasks that are not evident in performance itself.

CONCLUSIONS AND IMPLICATIONS

The extensive series of experiments that we have described was prompted by the need to provide a quantitative measure of the purported work underload inherent in vigilance tasks. Our research indicates that such underload is a myth: Rather than being understimulating, vigilance tasks are exacting, capacity-draining assignments that are associated with a considerable degree of mental demand and frustration. It is counterintuitive to suppose that these characteristics can be attributed to the neurological consequences of understimulation as the classic arousal model would demand. Nevertheless, it is important to note that more recent conceptualizations of arousal, such as that by Humphreys and Revelle (1984), have merged arousal with

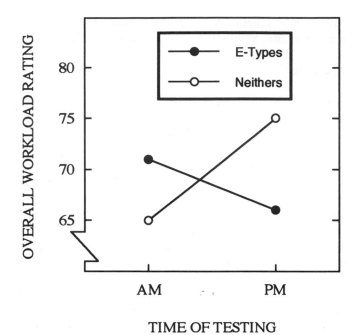

FIG. 9.6.   Perceived workload during morning and evening testing for evening-type individuals (E-types) and those with no time of day preference (Neithers; data from Hamilton et al., 1994).

resource theory. According to their view, arousal is the agent responsible for the production of resources. From that perspective, our results suggest that vigilance tasks not only deplete resources, they also drain the wellspring from which they come.

Regardless of one's theoretical perspective, our results have important ergonomic implications. Conventional wisdom about automation is that it leads to simplicity. However, as Woods (1994) pointed out, automation often complicates operators' cognitive activities and leads to errors. This is certainly true where monitoring is concerned. Evidently, human operators are not well suited for the detection of low-probability events during prolonged vigils. One solution to this problem might be to devise systems that can monitor themselves. However, in the likely case that humans will still retain system control, this solution may not be a good one because, as Parasuraman and his colleagues (Parasuraman, Mouloua, & Molloy, 1994; Singh, Molloy, & Parasuraman, 1993) showed, complacency often leads operators to do poorly in monitoring automation failures in human–machine systems. A better solution to the problem of monitoring inefficiency in the design of complex systems might be to take advantage of psychophysical factors, such as those described in this chapter, which are known to enhance vigilance performance and to lower operator workload. Given our findings regarding the workload of sustained attention, and those of Parasuraman and his colleagues in the monitoring of automation failures, it might also be advisable to seek a middle ground between active and supervisory control in system design in order to maximize operators' monitoring efficiency.

ACKNOWLEDGEMENTS

Preparation of this manuscript was supported in part by NASA Grant NAG-1-1118 to P.A. Hancock, J.S. Warm, and W.N. Dember, principal investigators. The views expressed here are those of the authors and do not necessarily represent those of NASA. Sandra Hart of the NASA-Ames Research Center, Moffett Field, CA, was the technical monitor for the grant.

REFERENCES

Adams, J. A., & Boulter, L. R. (1964). Spatial and temporal uncertainty as determinants of vigilance performance. *Journal of Experimental Psychology, 52,* 204-208.

Aiken, E. G., & Lau, A. W. (1967). Response prompting and response confirmation: A review of recent literature. *Psychological Bulletin, 68,* 330-341.

Aston-Jones, G. (1988). Cellular attributes of locus coeruleus: Implications for attentional processes. In M. Sandler, A. Dahlstrom, & R. Belmaker (Eds.), *Progress in catecholamine research: Part B. Clinical aspects* (pp. 133-142). New York: A.R. Liss.

Baker, C. H. (1962). *Man and radar displays.* New York: Macmillan.

Becker, A. B., Warm, J. S., Dember, W. N., & Hancock, P. A. (1991). Effects of feedback on perceived workload in vigilance performance. *Proceedings of the Human Factors Society 35th Annual Meeting* (pp. 1991-1994). Santa Monica, CA: Human Factors Society.

Becker, A. B., Warm, J. S., Dember, W. N., & Hancock, P. A. (1995). Effects of jet engine noise and performance feedback on perceived workload in a monitoring task. *International Journal of Aviation Psychology, 5*(1), 49-62.

Berch, D. B., & Kanter, D. R. (1984). Individual differences. In J. S. Warm (Ed.), *Sustained attention in human performance* (pp. 143-178). Chichester, England: Wiley.

Broadbent, D. A., Cooper, P., Fitzgerald, P., & Parkes, K. (1982). The cognitive failures questionnaire (CFQ) and its correlates. *British Journal of Clinical Psychology, 21,* 1-16.

Cohen, S. (1980). Aftereffects of stress on human performance and social behavior: A review of research and theory. *Psychological Bulletin, 88,* 82-108.

Colquhoun, W. P. (1967) Sonar target detection as a decision process. *Journal of Applied Psychology, 51,* 187-190.

Colquhoun, W. P. (1977). Simultaneous monitoring of a number of auditory sonar outputs. In R. R. Mackie (Ed.), *Vigilance: Theory, operational performance and physiological correlates* (pp. 163-188). New York: Plenum.

Craig, A. (1984). Human engineering: The control of vigilance. In J. S. Warm (Ed.), *Sustained attention in human performance* (pp. 247-291). Chichester, England: Wiley.

Craig, A., & Colquhoun, W.P. (1977). Vigilance effects in complex inspection. In R. R. Mackie (Ed.), *Vigilance: Theory. operational performance and physiological correlates* (pp. 239–262). New York: Plenum.

Davies, D. R., & Parasuraman, R. (1982). *The psychology of vigilance*. London: Academic Press.

Deaton, J. E., & Parasuraman, R. (1993). Sensory and cognitive vigilance: Effects of age on performance and mental workload. *Human Performance, 6,* 71–97.

Dember, W. N., Warm, J. S., Nelson, W. T., Simons, K. G., Hancock, P. A., & Gluckman, J. P. (1993). The rate of gain of perceived workload in sustained attention. *Proceedings of the Human Factors Society 37th Annual Meeting* (pp. 1388–1392). Santa Monica, CA: Human Factors and Ergonomics Society.

Dittmar, M. L., Warm, J. S., Dember, W. N., & Ricks, D. F. (1993). Sex differences in vigilance performance and perceived workload. *Journal of General Psychology, 120,* 309–322.

Eggemeier, F. T. (1988). Properties of workload assessment techniques. In P. A. Hancock & N. Meshkati (Eds.), *Human mental workload* (pp. 41–62). Amsterdam: North-Holland.

Folkard, S., & Monk, T. H. (1985). Circadian performance rhythms. In S. Folkard & T. H. Monk (Eds.), *Hours of work: Temporal factors in work-scheduling* (pp. 31–52). Chichester, England: Wiley.

Frankmann, J. P., & Adams, J. A. (1962). Theories of vigilance. *Psychological Bulletin, 59,* 257–272.

Galinsky, T. L., Dember, W. N., & Warm, J. S. (1989, March). *Effects of event rate on subjective workload in vigilance performance.* Paper presented at the meeting of the Southern Society for Philosophy and Psychology, New Orleans, LA.

Galinsky, T. L., Rosa, R. R., Warm, J. S., & Dember, W. N. (1993). Psychophysical determinants of stress in sustained attention. *Human Factors, 35,* 603–614.

Gluckman, J. P., Warm, J. S., Dember, W. N., Thiemann, J. A., & Hancock, P. A. (1988, November). *Subjective workload in simultaneous and successive vigilance tasks.* Poster session presented at the annual meeting of the Psychonomic Society, Chicago, IL.

Grubb, P. L. (1995). *Sustained attention and multiple-signal discrimination.* Unpublished doctoral dissertation, University of Cincinnati, Cincinnati, OH.

Grubb, P. L., Miller, L. C., Nelson, W. T., Warm, J. S., & Dember, W. N., & Davies, D. R. (1994). Cognitive failure and perceived workload in vigilance performance. In M. Mouloua & R. Parasuraman (Eds.), *Human performance in automated systems: Current research and trends* (pp. 115–121). Hillsdale, NJ: Lawrence Erlbaum Associates.

Hamilton, S. L., Warm, J.S., Dember, W. N., & Rosa, R. R. & Hancock, P. A. (1994, April). *Vigilance performance and stress in evening-type individuals.* Paper presented at the meeting of the Southern Society for Philosophy and Psychology, Atlanta, GA.

Hancock, P. A. (1984). Environmental stressors. In J. S. Warm (Ed.), *Sustained attention in human performance* (pp. 103–142). Chichester, England: Wiley.

Hancock, P. A. (1988). The effect of gender and time of day upon the subjective estimate of mental workload during the performance of a simple task. In P. A. Hancock & N. Meshkati (Eds.), *Human mental workload* (pp. 239–250). Amsterdam: North-Holland.

Hancock, P. A., & Warm, J. S. (1989). A dynamic model of stress and sustained attention. *Human Factors, 31,* 519–537.

Hart, S. G., & Staveland, L. E. (1988). Development of NASA-TLX (Task Load Index): Results of empirical and theoretical research. In P.A. Hancock & N. Meshkati (Eds.), *Human mental workload* (pp. 139–183). Amsterdam: North-Holland.

Heilman, K. M. (1995). Attentional asymmetries. In R. J. Davidson & K. Hugdahl (Eds.), *Brain asymmetry* (pp. 217–234). Cambridge, MA: MIT Press.

Hill, S. G., Iavecchia, H. P., Byers, A. C., Zaklad, A. L., & Christ, R. E. (1992). Comparison of four subjective workload rating scales. *Human Factors, 34,* 429–439.

Horne, J. A., & Ostberg, O. (1977). Individual differences in circadian rhythms. *Biological Psychology, 5,* 179–190.

Howell, W. C. (1993). Engineering psychology in a changing world. *Annual Review of Psychology, 44,* 231–263.

Humphreys, M. S., & Revelle, W. (1984). Personality, motivation, and performance: A theory of the relationship between individual differences and information processing. *Psychological Review, 91,* 153–184.

Jonides, J. (1981). Voluntary vs. automatic control over the mind's eye's movement. In J. B. Long & A. D. Baddeley (Eds.), *Attention and performance IX* (pp. 187–203). Hillsdale, NJ: Lawrence Erlbaum Associates.

Koelega, H. S. (1992). Extraversion and vigilance performance: 30 years of inconsistencies. *Psychological Bulletin, 112,* 239–258.

Koelega, H. S., & Brinkman, J. A. (1986). Noise and vigilance: An evaluative review. *Human Factors, 28,* 465–481.

Lanzetta, T. M., Dember, W. N., Warm, J. S., & Berch, D. B. (1987). Effects of task type and stimulus homogeneity on the event rate function in sustained attention. *Human Factors, 29,* 625–633.

Liu, Y., & Wickens, C. D. (1987). *Mental workload and cognitive task automation: An evaluation of subjective and time estimation metrics.* (Report No. EPL-87-02). Urbana, IL: Engineering Psychology Research Laboratory, Department of Mechanical and Industrial Engineering, University of Illinois.

Loeb, M. (1986). *Noise and human efficiency.* Chichester, England: Wiley.

Loeb, M., & Binford, J. R. (1963). Some factors influencing the effective auditory intensive difference limen. *Journal of the Acoustical Society of America, 35,* 884–891.

Mackie, R. R. (1984). Research relevance and the information glut. In F. Muckler (Ed.), *Human factors review* (pp. 1–11). Santa Monica, CA: Human Factors Society.

Mackworth, N. H. (1948). The breakdown of vigilance during prolonged visual search. *Quarterly Journal of Experimental Psychology, 1,* 6–21.

Mackworth, N. H. (1961). Researches on the measurement of human performance. In H. W. Sinaiko (Ed.), *Selected papers on human factors in the design and use of control systems* (pp. 174–331). (Reprinted from Medical Research Council Special Report Series 268, London, H.M. Stationary Office, 1950)

Milǒsević, S. (1974). Effect of time and space uncertainty on a vigilance task. *Perception & Psychophysics, 15,* 331–334.

Natsoulas, T. (1967). What are perceptual reports all about? *Psychological Bulletin, 67,* 249–272.

Nygren, T. E. (1991) Psychometric properties of subjective workload measurement techniques: Implications for their use in the assessment of perceived mental workload. *Human Factors, 33,* 17–33.

O'Donnell, R. D., & Eggemeier, F. T. (1986). Workload assessment methodology. In J. R. Boff, L. Kaufman, & J. P. Thomas (Eds.), *Handbook of human perception and performance. Vol. II. Cognitive processes and performance* (pp. 41-1–42-49). New York: Wiley.

Parasuraman, R. (1979). Memory load and event rate control sensitivity decrements in sustained attention. *Science, 205,* 924–927.

Parasuraman, R. (1985). Sustained attention: A multifactorial approach. In M. I. Posner & O. S. Marin (Eds.), *Attention and performance I* (pp. 593–511). Hillsdale, NJ: Lawrence Erlbaum Associates.

Parasuraman, R. (1986). Vigilance, monitoring and search. In K. R. Boff, L. Kaufman, & J. P. Thomas (Eds.), *Handbook of perception and human factors: Vol. II. Cognitive processes and performance* (pp. 43-1 - 43-39). New York: Wiley.

Parasuraman, R., & Mouloua, M. (1987). Interaction of signal discriminability and task type in vigilance decrement. *Perception & Psychophysics, 41,* 17–22.

Parasuraman, R., Mouloua, M., & Molloy, R. (1994). Monitoring automation failures in human–machine systems. In M. Mouloua & R. Parasuraman (Eds.), *Human performance in automated systems: Current research and trends* (pp. 45–49). Hillsdale, NJ: Lawrence Erlbaum Associates.

Parasuraman, R., Warm, J. S., & Dember, W. N. (1987). Vigilance: Taxonomy and utility. In L. S. Mark, J. S. Warm, & R. L. Huston (Eds.), *Ergonomics and human factors: Recent research* (pp. 11–32). New York: Springer-Verlag.

Sanderson, P. M., & Woods, M. D. (1987). *Subjective mental workload and locus of control.* (Report No. EPL-87-03). Urbana, IL: Engineering Psychology Research Laboratory, Department of Mechanical and Industrial Engineering, University of Illinois.

Satchell, P. M. (1993). *Cockpit monitoring.* Brookfield, VT: Ashgate.

Sawin, D. A., & Scerbo, M. W. (1994). Vigilance: How to do it and who should do it. *Proceedings of the Human Factors and Ergonomics Society 38th Annual Meeting* (pp. 1312–1316). Santa Monica, CA: Human Factors Society.

Scerbo, M. W., Greenwald, C. Q., & Sawin, D. A. (1992). Vigilance: It's boring, it's difficult, and I can't do anything about it. *Proceedings of the Human Factors and Ergonomics Society 36th annual meeting* (pp. 1508–1511). Santa Monica, CA: Human Factors and Ergonomics Society.

Scerbo, M. W., Greenwald, C. Q., & Sawin, D. A. (1993). The effects of subject controlled pacing and task-type on sustained attention and subjective workload. *Journal of General Psychology, 120,* 293–307.

Schmidke, H. (1976). Vigilance. In E. Simonson & P. C. Weiser (Eds.), *Psychological and physiological correlates of work and fatigue* (pp. 126–138). Springfield, IL: Thomas.

See, J. E., Howe, S. R., Warm, J. S., & Dember, W. N. (1995). A meta-analysis of the sensitivity decrement in vigilance. *Psychological Bulletin, 117,* 230–249.

Sheridan, T. (1970). On how often the supervisor should sample. *IEEE Transactions on System Science and Cybernetics SSC-6,* 140–145.

Sheridan, T. (1987). Supervisory control. In G. Salvendy, (Ed.), *Handbook of human factors* (pp. 1243–1268). New York: Wiley.

Singh, I. L., Molloy, R., & Parasuraman, R. (1993). Individual differences in monitoring failures of automation. *Journal of General Psychology, 120,* 357–373.

Sullivan, T. E. (1991). *Effects of spatial uncertainty and perceived workload in the vigilance performance of high and low resourceful individuals.* Unpublished master's thesis, University of Cincinnati, Cincinnati, OH.

Thackray, R. I. (1981). The stress of boredom and monotony: A consideration of the evidence. *Psychosomatic Medicine, 43,* 165–176.

Warm, J. S. (1984). An introduction to vigilance. In J. S. Warm (Ed.), *Sustained attention in human performance* (pp. 1–14). Chichester, England: Wiley.

Warm, J. S. (1993). Vigilance and target detection. In B. M. Huey & C. D. Wickens (Eds.), *Workload transition: Implications for individual and team performance* (pp. 139–170). Washington, DC: National Academy Press.

Warm, J. S., Dember, W. N., & Parasuraman, R. (1991). Effects of olfactory stimulation on performance and stress in a visual sustained attention task. *Journal of the Society of Cosmetic Chemists, 42,* 199–210.

Wickens, C. D. (1984). Processing resources in attention. In R. Parasuraman & D. R. Davies (Eds.), *Varieties of attention* (pp. 63–102). Orlando, FL: Academic Press.

Woods, D. D. (1994). Automation: Apparent simplicity, real complexity. In M. Mouloua & R. Parasuraman (Eds.), *Human performance in automated systems: Current research and trends* (pp. 1–7). Hillsdale, NJ: Lawrence Erlbaum Associates.

10 Human Decision Makers and Automated Decision Aids: Made for Each Other?

Kathleen L. Mosier
San Jose State University Foundation
at NASA Ames Research Center

Linda J. Skitka
University of Illinois at Chicago

INTRODUCTION

Automated decision aids and expert systems are currently being implemented in real-world, high-risk domains such as medical diagnosis, nuclear energy plant operations, engineering design, and, in the aerospace domain, in glass cockpits and in air traffic management. Human decision makers are being exposed to and required to utilize these aids as they do their jobs and make necessary decisions. As decision-making environments become more complex and data intensive, the use of automated decision aids is likely to become even more commonplace and more critical to the process.

The advantages of automated decision aids and expert systems in terms of increased efficiency and data monitoring and analysis capabilities are fairly obvious. Computers can process more and faster than humans. Ideally, the combination of human decision maker + automated decision aid should result in a high-performing team, maximizing the advantages of additional cognitive and observational power in the decision-making process. Thus far, however, the union between human decision makers and automated systems has been less than idyllic. For example:

> During surgery, an automated blood pressure cuff, which is designed to inflate periodically to measure and record a patient's blood pressure, was not properly restarted after reset to a new measurement interval. The device continued to show elevated blood pressure, and the anesthetists on the surgical team did not realize that it was still displaying the same reading it had taken prior to reset. "So, for 45 minutes (sad to say) they aggressively treated

the hypertension with powerful drugs that dilate blood vessels. . . . Finally it was discovered that there had not been a recent measurement of blood pressure, which was in fact found to be very low . . .fortunately there was no lasting harm to the patient." (Gaba, 1994, p. 59)

A China Airlines B747-SP, flying at 41,000 ft, lost power in its #4 engine. The autopilot, which was set for pitch guidance and altitude hold, attempted to correct for the loss by holding the left wing down, masking the approaching loss of control of the airplane. When the Captain disengaged the autopilot, the airplane rolled to the right, yawed, then entered a steep descent in cloud. Extensive damage occurred during descent and recovery. (NTSB Report AAR-86–03, in Billings, 1991)

A minor blockage in the secondary cooling system at Three Mile Island nuclear power plant went undiagnosed for approximately eight minutes. The annunciator lights for the closed valves were located amidst thousands of indicators and lights on the Integrated Control System panel; between 50 and 100 of the lights were flashing red or green. During those eight minutes, operators wrestled with incomprehensible and uninterpretable indicators from other parts of the system (some of which were incorrect), tried every procedure they could think of, manually overrode the automatic operation of the Emergency Core Cooling System because of contradictory indications on other indicators of pressurizer water level, and finally discovered the original stuck valve light hidden underneath the yellow caution tag for another valve. (Martin, 1980)

Many of the problems that emerge out of the interaction of human decision makers and automated decision aids arise from inappropriate expectations of these systems and their human operators, and are rooted in misperceptions or incomplete understanding of each of the cognitive agents in the human–machine decision-making system. Automated aids, for example, are based on incomplete or incorrect models of human decision makers—for example, models that do not take into account decision-making heuristics and biases, including what we have termed "automation bias," the difficulty humans have with vigilant monitoring, and the fact that human experts typically make domain-related decisions in nonanalytical, sometimes intuitive ways. Human decision makers, on the other hand, often have faulty models of how decision aids "reason," of their reliability and their effects on communication and team decision processes, and of what automated systems can and cannot do.

In reality, the facets of the human–machine decision-making system are as complex as the environments in which they function. If we are to create decision-aiding systems that are made for human decision makers and will enhance human decision making, it is essential that the models of each of these cognitive agents be accurate and complete. In the following sections, we address some of the issues involved in automating decision-aiding functions, including the fallibility of human decision makers, as well as some of the common misperceptions of each of the entities in the interaction between human decision makers and automated decisions aids.

HUMAN DECISION-MAKING HEURISTICS AND BIASES

Considerable research from social psychology indicates that people generally prefer the path of least resistance — an idea that applies to cognitive work as well as physical work. All things being equal, people typically will try to engage in the least amount of cognitive work with which they can get away (Fiske & Taylor, 1991). Decision makers often adopt various strategies designed to reduce cognitive effort and to decrease the possibility of information overload, that is, having to deal with more information than they can handle. Social psychologists have documented the widespread use of different heuristics in judgment and in everyday decision making. Heuristics are simple decision-making rules we often use to make inferences or to draw conclusions quickly and simply. To be successful, such heuristics must have two properties: They must provide a quick and simple way of dealing with a great deal of information, and they must be reasonably accurate most of the time. In sum, these mental strategies try to strike a balance between speed (or simplicity) and accuracy (reliability). Heuristics are a reasonable alternative to more analytical decision-making strategies. However, a great number of cognitive biases have been documented as the result of the use of heuristics. Biases refer to the errors in judgment that can (but do not necessarily) result from the use of heuristics to guide decision making.

The availability of automated aids in decision-making contexts such as the aircraft cockpit or the nuclear power plant control room feeds into the general tendency to take the road of least cognitive effort. These aids are readily available, widely believed to be accurate, and are a highly salient source of information. Results of research on salience effects on decision making indicate that in diagnostic situations, the brightest flashing light or the most focally located gauge — that is, the most salient cue — will bias the operator toward processing its diagnostic content over that of other stimuli (Wickens, 1984). Designers of automated warning devices have responded to this tendency through attention-getting techniques. Moreover, the implementation of automated aids may be accompanied by diminished access to traditional cues.

Focusing on salient cues may result in a misinterpretation of what is really going on, and subsequent action will be based on this faulty assessment. Or, the concentration on salient cues may result in a lack of attention to less obvious but equally important information. A classic example of salience effects is the case of the Lockheed 1011 that crashed into the Florida Everglades. This accident occurred because all three crewmembers were so fixated on a brightly illuminated *Landing Gear Inoperative* light (indicating possible malfunction of the gear or, as in this case, malfunction of the light) that they failed to notice that the autopilot had been inadvertently switched off, and was no longer maintaining altitude.

Much of the decision making in real-world contexts for which decision aids are being implemented, such as the aircraft cockpit, is heuristically guided (Orasanu, 1993). Because these decision heuristics are often grounded in training, expert knowledge, and experience, they are reasonably accurate most of the time. However, Orasanu pointed out that reliance on cognitive heuristics in decision making is likely to yield decisions that are simply "good enough" rather than "optimal," and by no means safeguards even expert decision makers from bad decisions or cognitive biases.

Automated decision aids, by virtue of their simple heuristic value, act as very salient decisional cues, and diminish the likelihood that decision makers will process information in cognitively complex ways. Automated aids are, by design, very "bright lights," and ones that may engulf or overwhelm other diagnostic information. Moreover, humans may be disinclined to make the cognitive effort to seek out other diagnostic information to the extent that automated feedback is believed to be generally reliable. They may become focused on the information provided by the decision aid, cutting off situation assessment prematurely, or, to the extent that they notice additional information, they may show either assimilation or discounting biases. Specifically, other information may be interpreted as being more consistent with the automated decision aid than it actually is—a bias called assimilation. This is a likely effect if other cues in the environment are ambiguous rather than completely incongruent with people's initial judgment, or with the automated decision aid recommendations (e.g., Darley & Gross, 1983; Glick, Zion, & Nelson, 1988; Sagar & Schofield, 1980). On the other hand, cues completely inconsistent with automated feedback may be discounted.

Confirmatory biases lead information processors to overattend to consistent information and ignore other data. When information is completely inconsistent with an initial impression or judgment, it often is attended to and processed extensively. However, the inconsistent information is often explained away so that the initial impression or judgment can remain intact (e.g., Lord, Ross, & Lepper, 1979). Exposure to inconsistent information often has the seemingly paradoxical effect of making perceivers even more convinced and wedded to their first impression or judgment (Echabe & Rivera, 1989; O'Sullivan & Durso, 1984). To the extent that decision makers are prompted toward a particular problem or diagnosis when automated decision aids bring it to their attention, other information is less likely to be used to create a new impression or to modify the impression of the problem created by the automated decision aid. In combination, normal information processing heuristics and often attending cognitive biases may prime people to overattend to automated directives and, in turn, to underattend to other sources of information that might disconfirm the automation. The tendency for automation to engulf the decision-making field is likely to be exacerbated by at least three factors: the extent to which organizations encourage the use of automated systems (i.e., the "acceptability heuristic," or doing what is sanctioned by management); the extent to which reliance on automated decision aids is mandated (e.g., certain automated aircraft traffic advisories require immediate compliance), and the extent to which automation is seen as an "expert," and one that is perceived to yield more accurate or reliable outcomes than people's own judgment.

In sum, because it is cognitively easier to delegate tasks to automation, people are likely to do it when automated aids are available. Because automated aids, when used correctly, are generally accurate, this heuristic will generally be effective. Diagnoses are made correctly, power plants function efficiently, and airplanes fly safely using automated aids. However, indiscriminate or inappropriate reliance on automation will result in errors, just as inappropriate use of other decision-making heuristics results in errors. Delegating to automation has the additional consequence of making human decision makers less attentive, and unlikely to notice aberrant events that are not brought to their explicit attention by the system. Similarly, when automated decision aids indicate problems or make recommendations, the path of least resistance is to accept these judgments at face value. Other information, to the extent that it is processed at all, is likely to be processed in such a manner that it is seen as consistent with the automated decision aid.

AUTOMATION BIAS: ERRORS OF OMISSION AND COMMISSION

The natural human tendency toward "satisficing" in judgment and decision making (i.e., doing just enough cognitive work to make reasonably acceptable decisions; Simon, 1955), creates fertile ground for what we have termed "automation bias" in the use of automated decision aids. Automation bias, that is, the tendency to use automated cues as a heuristic replacement for vigilant information seeking and processing, results in errors when decision makers fail to notice problems because an automated aid fails to detect them (an omission error) or when people inappropriately follow an automated decision aid directive or announcement (a commission error).

Omission Errors

A number of researchers have commented on and documented that overautomated systems foster what has been called "automation-induced complacency" (e.g. Parasuraman, 1987). Highly automated environments, such as glass cockpits or nuclear power plants, create complacency, boredom, and poor monitoring behavior (see Billings, 1991, Chambers & Nagel, 1985; Parasuraman, 1987).

A number of incidents have been reported when people have become dependent on automated decision aids to let them know when certain problems exist. In the absence of automated feedback to the contrary, decision makers assume that operations are as they should be, despite traditional sources of information (e.g., gauges or instruments on the traditional control panel) indicating otherwise. For example, Mosier, Skitka, and Korte (1994) examined a nonrandom sample of Aviation Safety Reporting System (ASRS) reports of incidents that included some reference to automated aids. Breaking down the 166 events investigated by flight phase revealed that the most likely flight phase for automation bias to occur was cruise, and that most of these errors were reported to be the result of crewmembers missing events or discrepancies because of complacency or nonvigilance. In 81% of the cruise events, crews mentioned either complacency or nonvigilance as a factor contributing to the incident, and in 51% of the events, the automated system was performing exactly as it had been programmed to do. What seemed to be happening in these incidents was that the crews set up a system to perform a task, and then trusted the system to do it. Once the crews delegated a task to the automated system, human monitoring was not adequate to catch inconsistencies or mistakes in task performance. One crew, for example, misprogrammed the flight path, and wandered 70 miles off course. They were not aware they were off course until notified by air traffic control (ATC).

Errors of omission often occur in conjunction with automated monitoring or warning systems, and are common in National Transportation Safety Board accident reports. A classic example involves the failure of an onboard configuration warning system to alert a Delta Airlines crew that the airplane was not properly configured for takeoff. The crew apparently relied on the warning system to the extent that they did not manually verify the positions of the flaps and slats, resulting in a crash shortly after takeoff (NTSB Report AAR-80-10, in Billings, 1991). Errors of omission may also result when automation is compensating for some abnormality, but doesn't let the operator know, as in the China Airlines incident described at the beginning of this chapter.

Commission Errors

Errors of commission, resulting from following the directives of an automated aid, have begun surfacing more recently as by-products of automated systems. Incident reports on TCAS II (Traffic Alert and Collision Avoidance System) have documented cases of airborne incursions and conflicts because "ghost" or "phantom" radar images led to erroneous action advisories (NASA ASRS Report, 1992). Preliminary evidence of commission errors has also been found in a NASA study of automated, electronic checklists. Over half of the crews that were using an automated checklist followed its recommendations and erroneously shut down one of their two engines – even though other system indicators revealed the danger in this course of action (Mosier, Palmer, & Degani, 1991). Comission errors are also occurring in other automated contexts, as evidenced by the operating room incident reported by Gaba (1994; see the beginning of this chapter).

Other research indicates that there is a broad tendency to make either omission or commission errors in highly automated decision-making contexts. Specifically, Skitka and Mosier (1994) found that people are more likely to make decisions inconsistent with objective data when a computer recommends the contradictory decision response than when a person recommends the contradictory decision response. Moreover, automation bias appears to be more likely in domains that are most likely to be highly automated – such as nuclear power plants and aircraft cockpits – than in decision domains that are less "machine" oriented, such as dating services or stock market analysis and advice.

It is important to note that automation-rooted errors have many underlying causes. Responses to TCAS alerts, for example, are in some instances mandated, and pilots are given neither the time nor the means to verify their accuracy. In other cases, human operators may make mistakes in the use of automated systems. The blood pressure cuff, for example, had never been restarted. Errors may also occur when the automation is functioning correctly. The electronic checklist was giving correct information based on its sensors. The China Airlines autopilot was performing exactly as it was set to perform. However, as we discuss later, misinterpretations of what the automated system is doing (e.g., the anesthetists thought the blood pressure cuff was still taking readings), unawareness of the limitations in automated capabilities (e.g., the electronic checklist could not take into account the condition of the other engine), or poor feedback about the system activities or status (e.g., the China Airlines crew did not realize with what the autopilot was coping), combined with automation bias, may result in errors despite perfectly functioning automated aids.

HUMAN VIGILANCE AND SYSTEM OBSERVABILITY

Cognitive processes underlying vigilance decrements may exacerbate automation bias as more and more tasks are offloaded to automated control. Many studies of vigilance have established the fact that the ability to monitor for the occurrence of infrequent, unpredictable events typically declines over time. Complacency, resulting in lack of vigilance, has often been cited as a factor in aviation incidents and accidents (e.g., Hurst & Hurst, 1982), and automated systems, especially those that are perceived to be highly reliable, may increase the complacency-inducing potential of a decision-making environment (Parasuraman, Molloy, & Singh, 1993; Wiener, 1981; see also Parasuraman, Mouloua, Molloy, & Hillburn, chap. 5, this volume).

Human inefficiency in vigilant monitoring tasks may interfere with the detection of automated errors. Most of the studies of vigilance have focused on detrimental effects on the rate of correct detection of signals (i.e., hits; Parasuraman, 1986), or what we would interpret as increases in omission errors. Vigilance decrements may also result in commission errors, as crew members who have lapsed in their attention to other system indicators follow without question the directives of automated aids.

Even vigilant decision makers, however, may be unable to detect errors in automated systems. One of the biggest obstacles to human recognition of system errors is the poor quality of feedback the systems provide (Woods, 1994). Operators often have incomplete or fuzzy mental models of how various modes of automation work (Norman, 1990; Sarter & Woods, 1992), a problem exacerbated by the fact that automation's opaque interface provides only limited information about actual status and criteria for conclusions. The operators at Three Mile Island, for example, had no integrated feedback concerning the flow of water, steam, and pressure in the unit's cooling system (Martin, 1980). In the aviation domain, it is often difficult or impossible to trace the processes or predict the activities of automated flight systems. In fact, three of the most common queries glass cockpit pilots make about automated flight systems are: What is it doing? Why is it doing that? What is it going to do next? (Wiener, 1989). Additionally, automated systems do not convey a sense of their limitations to the user, and may not make obvious what they cannot do (see also Woods, chap. 1, this volume).

The combination of these factors may result in a human–computer system that is not conducive to automated error detection, even by domain experts. Additionally, a long-range hazard of extensive use of automated decision-aiding systems is what Hart (1992) termed the "deskilling" of experts. "If people come to rely on a computer system for advice and guidance then there is the chance that human expertise will become rare. This could result in deskilling and a *regression towards the mean* of knowledge . . . [and a tendency to] trust the computer even when it is not authoritative" (italics added, p. 30).

HUMAN EXPERTISE VERSUS AUTOMATED EXPERTISE

Automated systems are often advertised as functioning just like expert human decision makers. Although this may be true in isolated, structured, combinatorial domains, or in real-world systems for which our models are accurate and complete (Dreyfus, 1989), it is not necessarily true in dynamic, naturalistic domains. Sophisticated decision-aiding systems solve problems analytically, constructing solutions selectively and efficiently from a predefined space of alternatives (Waterman, 1985). Their approach corresponds to classical theories of decision making in that they focus on analytical problem solving (i.e., alternative selection) rather than on how the problem was formulated in the first place (Winograd & Flores, 1986). They have enormous data storage and computational capabilities, and can do the kinds of prescriptive analysis that humans have neither the time nor the ability, in most cases, to do.

In contrast, when we observe human experts making decisions in dynamic real world settings, such as the operating room or the aircraft flightdeck, we see that they act and react on the basis of their prior experience. Human experts assess the environment through discriminant cue utilization (e.g., Brunswik, 1952; Hammond, Stewart, Brehmer, & Steinmann, 1975). In naturalistic settings, they tend to deliberate more during the diagnosis phase of decision making than during response selection, as it is out of this process that the generation of a workable response option evolves. In many cases, once the situation is understood, the appropriate course of action is obvious (Kaempf & Klein, 1994).

In most real-world settings, contextual factors are critical to situation diagnosis and decision making. Human experts can incorporate contextual information, and also have access to information outside of their domain of expertise that may impact the problems being solved by them. Human experts can look at the "big picture"—examine all aspects of a problem and see how they relate to the central issue. Automated and expert systems, on the other hand, can focus only on the problem itself, and do not take into account factors that are relevant to, but separate from, the basic problem (Waterman, 1985).

In terms of emulating human experts, then, there may be a vast difference between what automated systems can do and what they are advertised or perceived as being able to do. Human expertise is characterized by creativity, adaptability, the presence of a broad focus, the ability to incorporate sensory experience and contextual factors, analogical reasoning, and commonsense knowledge. Automated systems lack these qualities. The intelligence that even the most sophisticated systems possess consists of preprogrammed rule-based reasoning, based on if . . . then protocols. The process may be as simple as "IF altitude $< x$, and landing gear is not down, THEN illuminate gear warning light," or as sophisticated as some of the medical diagnostic aids, such as RECONSIDER, which acts as an encyclopedia of diseases, and lists, in probability-ranked order, the diagnoses (THENs) that are associated with a particular (IF . . .) list of symptoms (Dreyfus & Dreyfus, 1986). Decision-aiding systems can perform these protocols competently and consistently, but their processes are complementary rather than parallel to those of human experts.

CAN AUTOMATED DECISION AIDS REPLACE HUMAN EXPERTISE?

Perhaps one of the most troublesome misconceptions of automated decision aids is that they are able to make experts out of novices. The novice, armed with an expert system or decision aid, is expected to be able to perform as an expert, and to solve problems beyond the realm of his or her knowledge or experience (Will, 1991). In the domain of aviation, for example, decision aids are being proposed to augment the inexperience of general aviation pilots, and even to enable relatively inexperienced pilots to fly commercial glass cockpits in standard operations. The theory underlying this proposal is that the decision-aiding system can compensate for a lack of experiential knowledge and expertise in the human.

This proposal has several potentially dangerous pitfalls. The notion that inexperienced operator + automated aid = expert performance implies that the automated or expert systems can somehow replace the missing human expertise. This is a misleading notion. As discussed previously, computers are limited to specific technical knowledge. They are not as versatile as human experts, and the artificial expertise they offer is different from human expertise. However, we often expect automated aids to provide precisely the capabilities they cannot, for example, commonsense reasoning, interpretation of inconsistent information, or integration of contextual factors (Hart, 1992; Waterman, 1985).

This notion of automated expertise may have dangerous implications when novices, aided by decision-aiding systems, believe they are as capable or skilled as experts, and enter into situations that are at the edge of or beyond their capacity to handle. The irony of this is that automated systems are best able to aid the novice when situations are routine, or covered by standard procedures. It is the unanticipated, unprogrammable anomalies that most tax the limits of both the inexperienced operator and the automated system. Moreover, most automated aids cannot recognize or convey to the user the limits of their abilities. "When pushed beyond their limits or given problems different from those for which they were designed, expert systems can fail in surprising ways" (Waterman, 1985, p. 182). Inexperienced operators will be even less likely than experts to recognize these limitations, and may be unable to detect even blatant errors.

An additional danger associated with the combination of inexperienced operators and automated systems is that, in the long run, the novice may not get the experience or exposure requisite to the development of domain expertise. For example, if usage of automated aids is universally employed as a substitute for experience in diagnosing and handling a wide variety of events, novices may never learn the meanings associated with constellations of cues and changes in cue clusters, or develop the skills associated with situation assessment and response generation. Ultimately, this will render human decision makers less efficient and less equipped to deal with events — especially those involving any kind of automated system failure — than they might have become without the aid.

BUT THE SYSTEM IS SMARTER ...

Despite the limitations described previously, a kind of mystique of omniscience surrounds automated aids, and exacerbates the tendency toward overreliance on them. Their enormous data storage and computational capabilities evoke the perception that they can take into account far more information than can a human expert. Often, automated systems are represented as "intelligent," "knowledge intensive," "learning," or "understanding," leading even the expert user to believe the systems are smarter or more capable than they are themselves. Will (1991) commented on this when he found that experts were no more likely than novices to detect flaws in a defective decision-aiding system. In fact, both experts and novices expressed confidence in their "wrong" answer to the problem, displaying reliance on a false technology.

Wiener and Curry (1980) foreshadowed the tendency of human decision makers to overtrust automated decision aids when they discussed a phenomenon they termed "primary-backup inversion," a reference to the tendency of flight crews to utilize warning systems as primary indicators of problems, rather than as secondary checks. The implications of this reversal become more critical as automated devices become more sophisticated ("smarter"), and get involved in higher-level tasks, such as making decisions. Operators may view decision aids as having skills superior to their own, and may rely on them as the most efficient way to get decisions made, particularly in times of high workload. It is quite feasible, then, that the decision aid will be utilized as a decision maker rather than a supplementary decision aid, with the human as a backup system.

... AND MORE RELIABLE

In addition to misconceptions of automated expertise, many human decision makers have faulty perceptions of the reliability of automated aids. Signal detection theory posits that the power of an automated decision aid will depend on the strength and reliability of the signal it generates (e.g., Swets, 1986; Swets, Tanner, & Birdsall, 1961). To the extent that the system is perceived to be reliable, and the costs associated with not attending to true signals are high (as is the case with many currently existing diagnostic or warning systems), the criterion for following its directives will be lenient. Operationally, this means that the automated directives will be heavily depended on, and that the presence — or absence — of a signal will be treated as a reliable indicator of when events occur.

The advertised reliability of an automated system, however, is typically defined solely in terms of internal consistency, and thus may be somewhat inflated. Although automated systems have to meet stringent certification requirements for accuracy in detecting system problems (based on "hits" and "misses"), performing if . . . then protocols, or prescribing actions, correct performance does not necessarily lead to optimal advisories. Many contextual factors, for example, that may provide the critical nuances of situation assessment, are unavailable to automated aids. Moreover, unless all of the automated systems are integrated with each other and with all other systems, the advisories they issue may be inappropriate based on other system states or constraints, or even counterproductive to overall safety. Unless the decision domain is structured and isolated from outside elements, internal consistency is a necessary but insufficient criterion for judging system reliability.

COMMUNICATION IN TEAM CONTEXTS

The introduction of highly sophisticated automated decision aids into team contexts may fundamentally and qualitatively alter the communication aspects of the decision-making process. For example, as noted by Wiener (1993a), mounting evidence exists that crew coordination and interaction in the glass cockpit are qualitatively different than in the traditional cockpit. In the glass cockpit, crew members may "interact" with the automation rather than with the other crew member. In many cases, the automation (rather than a person) has the information needed to assess the situation and make a decision. Because of this, information may be gathered with no vocal communication between crew members. Additionally, the quality and value of nonverbal communication inherent in observing the movements of another crew member while interacting with traditional cockpit devices is diminished in the automated cockpit (Segal, 1990). Physically, due to flightdeck layout, it is much more difficult for each pilot to see what the other is doing (Wiener, 1993a), and crew members may not keep each other advised of what is going on (e.g., typing route changes into the Flight Management System, or FMS). Additionally, preferred styles of interaction with automated aids may vary, and crew members may not be able to follow each other's strategies (e.g., regarding FMS usage or preferred flight mode; Sarter & Woods, 1993).

The anesthetists described at the beginning of this chapter unwittingly fell prey to the same kind of communication pitfalls. Because the blood pressure reading was available to both of the anesthetists independently, each might have thought that there had been a different reading on the device the last time the other person looked at it (Gaba, 1994). Neither one realized that they had been working with the same, invalid reading for 45 minutes. The impact of automated decision aids may be to further isolate operators from each other's situation assessment and decision-making processes and responses, reducing redundancy and human input, and increasing the possibility of errors of omission and commission.

WHO (OR WHAT) IS RESPONSIBLE FOR DECISIONS?

Through tradition or regulation, the human operator has typically maintained ultimate decision-making authority and borne responsibility for the outcomes of decisions. The presence of automated decision aids, however, subtly or blatantly alters the role of the human decision maker, and may obfuscate the issue of responsibility for decisions. When an automated aid is introduced, it changes patterns of cue utilization and the way with which situations are assessed and dealt. Operators learn (and are instructed) that the automated aid is the "best" cue for making a decision. If the system proves initially to be generally reliable, and reflective of the results of the operators' own diagnostic processes, what may (and does) happen over time is that operators check automated output before anything else, and may not look further for traditional cues even when they are available, especially if time is short (e.g., Gigerenzer, Hoffrage, & Kleinbölting, 1991). In some cases, automated cues may replace other information, making a cross-check with traditional cues impossible.

Social psychological research has indicated that as responsibility is shared among more individuals, people are less likely to act quickly, or even to intervene at all, in emergencies (Darley & Latane, 1968). A study of ASRS reports provided evidence consistent with this hypothesis, in that aircraft seemed to be more at risk for collision when under air traffic control (Billings, Grayson, Hecht, & Curry, 1980). Crews seemed to be less likely to watch for traffic when they believed that the responsibility for doing so was shared with ATC. This notion also generalizes to tasks that are shared between humans and automated devices. Early in the evolution of glass cockpits, for example, Foushee (1982) cited the danger that as more and more tasks were shifted from pilot to automated control, crewmembers would feel a loss, or diffusion of responsibility. Vigilance decrements, cited earlier as a potential hazard for and by-product of the use of automation, may also be explained in terms of diffusion of responsibility—individuals may be abdicating responsibility for monitoring tasks to the automation. In fact, automation bias may occur in part because more decision-making agents are involved—operators are less vigilant because they are not solely responsible. Ultimately, as decision-aiding systems become more capable of autonomous functioning and are given more power in the guidance of human activity, human operators may endow them with personas of their own, capable of independent perception and willful action (Sarter & Woods, 1994).

The issue of responsibility in the use of expert systems and decision aids, then, is emerging as complex and confusing. Although responsibility for actions still rests with the user (Mockler, 1989), the legal status of the user of expert system technology is not always clear (Will, 1991), especially in instances in which the user ignores or overrides the advice of the system. Lawsuits may result if system aids are consulted and fail to perform correctly, or give inaccurate or misleading indications, or are incorrectly used. Conversely, operators may be liable for the nonuse of an available system (Zeide & Liebowitz, 1987). For certain sensitive, delicate, or hazardous tasks, for example, it may be unreasonable not to rely on an expert system (Gemignani, 1984).

The question of responsibility is certain to take on more importance and relevance in the future. The presence of automated aids may subtly erode the decision maker's role, and foster an abdication of responsibility to the system. In fact, because one explicit objective of automating systems is to reduce human error, automating a decision-making function may communicate to the operator that the automation should have primary responsibility for decisions. In the future, as automated systems become more prevalent and more autonomous, deskilled experts may no longer have the degree of expertise to judge when the system is performing correctly, or to carry out the necessary degree of supervision (Morgan, 1992). "There is also a problem of over-expectation that obscures responsibility . . . there's a kind of mystique . . . that if the expert system says so, it must be right" (Winograd, 1989, p. 62).

ARE AUTOMATED AIDS MADE FOR HUMAN DECISION MAKERS?

It should be made clear that we are not suggesting that automated decision aids should not be utilized. Rather, we are suggesting that the prevalence of incorrect models of human decision making and of automated decision aids may result in a human–automation system that is, in practice, less effective than the human alone. That is, automated aids are most often designed based on an incomplete model of the factors influencing human decision making. Human decision makers typically possess incomplete or faulty models of the internal functioning, "reasoning" processes, and limitations of automated aids.

Although automated decision aids have been introduced into most domains to enhance human capabilities or to control for human error, they may in fact be creating a decision-making environment that promotes rather than ameliorates certain human tendencies toward biased decision making. To the extent that automated decision aids are endowed with greater expertise or provide the most salient cue, human decision makers are likely to either not attend at all to other information, or to process other cues to remain consistent with automated directives. Competent monitoring of automated systems requires vigilance and expertise, qualities that the systems themselves may be undermining. Additionally, one advantage of automated systems is that they can make more information available to the decision maker, but they may ironically defeat this purpose by masking the information in an opaque interface, or by making verbal communication of information less likely. Finally, the presence of decision aids may encourage an abdication of decision-making responsibility to the automated system. It seems, then, that a variety of factors may be creating conditions that foster the negative rather than the positive capabilities of the human agent. Moreover, we require humans to use decision aids, and design systems that are difficult to ignore, but give decision makers insufficient information on system activities and limitations, and lead them to expect the systems to accomplish decision-making tasks that they are functionally incapable of performing. These factors foster incorrect and inefficient usage of automated decision aids.

It is clear that the impact of automated decision aids on human decision making requires further study, especially with respect to the fostering of automation bias. Among the products of this research should be guidelines for the design of automated decision aids and the training of human users. For a human–computer system to function most efficiently and effectively, the strengths of each should be capitalized on, and care must be taken to offset or compensate for each agent's limitations. Among the strengths of automated systems are their abilities to provide access to large databases of information, to sort through that information much faster and more comprehensively than the human can, and to monitor the consistency of the humans' choices and behavior (Evans, 1987). These capabilities can be utilized by humans to reduce situational ambiguity (e.g., by calling up information that may aid diagnosis) and to guard against human error or bias (e.g., "Do you really want to delete this file?" or " . . . this flight plan?"). Automated systems also have the potential to function as powerful learning tools, if current opaque and untraceable processes are redesigned to be transparent and observable to the learner.

Research guiding system design must also incorporate studies on the appropriate input and timing of information to be presented, as well as on the role of the decision aid. The capabilities of automated decision aids range across many levels. At the most basic level, they can display or highlight raw data that needs to be attended to. They can also provide system monitoring and warning or alert signals, as many current systems do, or display anomalies, trends, or the results of an automated diagnosis with confidence information. Decision aids could also display potential downline consequences or "feedforward" (Wiener, 1993b) for an anomaly or a particular course of action. Another kind of system design would provide wrong decision protection, that is, let the operator know if it senses that a "dangerous" decision is about to be implemented. The system could also supply nothing but an action directive (e.g., "pull lever"; "shut down system").

The choice of intervention level will determine to a great extent how the automated aid will be integrated into the decision-making process, how and when it will be utilized, how much input and analysis will be required by human decision makers, and how much room will be left for human discretion or intervention. The appropriate role for the automated aid may depend on the situation. For example, when collision is imminent, a simple directive may be necessary; in less critical situations, information or guidance might be more appropriate. An additional design issue concerns the integration of automated aiding systems. If systems are operating independently, or are only partially interactive as they are in many automated domains, operators must be given clear information concerning the limitations of the system and potential conflicts between its output and other system requirements. Moreover, the interaction between the level of display and the presence of automation bias needs to be investigated.

IMPROVING THE PSYCHOLOGICAL INTERFACE BETWEEN DECISION AIDS AND DECISION MAKERS

Unless we are willing to allow automated decision aids to become decision makers, there remains an implicit need for people in automated decision making contexts to be good information processors. The psychology of the decision making environment needs to be examined closely to ensure that automated feedback is, in fact, used only as part of a more thorough decision-making process, or to determine if and when automated feedback should supersede vigilant information gathering and processing on the part of the human decision maker.

There is clearly a need for further research investigating the psychological underpinnings of automation bias to discover (a) whether automation bias takes the form of limiting information gathering on the part of the decision maker (i.e., automation is used as a replacement for having to process other relevant data for making an informed decision—the cognitive miser hypothesis); (b) whether people do, in fact, analyze the available relevant information, but the automated feedback is seen as more reliable than decision makers' own diagnosis of the situation (the authority hypothesis); (c) whether automation bias is due to diffusion of responsibility—the idea that as the number of decision makers increases, each human decision maker feels less responsible for actually attending to information and making decisions, as well as less responsible for whatever decisions are ultimately made; (d) the psychological conditions under which automation bias occurs; (e) steps that can be taken to counteract automation bias; and finally (f) the relative trade-offs in speed and accuracy of more vigilant decision-making processes versus exclusive reliance on automated feedback.

Are human decision makers and automated decision aids "made for each other"? We have witnessed the union between the two, but after the initial enthusiasm and "honeymoon" period comes the need to build a realistic working understanding of the newly formed alliance. To build a strong relationship between human decision makers and automated decision aids, attention needs to be given to design issues on the one hand and human psychology on the other. With additional insight into both the consequences of design and the dynamics of introducing artificial cognitive agents into a decision-making context, a stronger alliance can be built between human decision makers and automated decision aids, and we may ultimately be able to say that, in fact, the two are "made for each other."

ACKNOWLEDGMENTS

Preparation of this chapter was supported by NASA grants NCC2-798 and NCC2-837.

REFERENCES

Billings, C. E. (1991). *Human-centered aircraft automation: A concept and guidelines* (Tech. Mem. No. 103885). Moffett Field, CA: NASA Ames Research Center.

Billings, C. E., Grayson, R., Hecht, W., & Curry, R. (1980). A study of near midair collisions in U.S. terminal airspace. *NASA Aviation Safety Reporting System: Quarterly Report No. 11* (NASA TM81225).

Brunswik, E. (1952). The conceptual framework of psychology. *International Encyclopedia of Unified Sciences, 1*(10).

Chambers, N., & Nagel, D. C. (1985). Pilots of the future: Human or computer? *Communications of the ACM, 28,* 1187–1199.

Darley, J. M., & Gross, P. H. (1983). A hypothesis-confirming bias in labeling effects. *Journal of Personality and Social Psychology, 44,* 20–33.

Darley, J. M., & Latane, B. (1968). Bystander intervention in emergencies: Diffusion of responsibility. *Journal of Personality and Social Psychology, 8,* 377–383.

Dreyfus, S. (1989, Spring). In R. Davis, (Ed.), Expert systems: How far can they go? Part One. *AI Magazine,* pp. 61–67.

Dreyfus, H. L., & Dreyfus, S. E. (1986). *Mind over machine.* New York: Free Press.

Echabe, A. E., & Rivera, D. P. (1989). Social representations and memory: The case of AIDS. *European Journal of Social Psychology, 19,* 543–551.

Evans, J. St. B. T. (1987). Human biases and computer decision-making: A discussion of Jacob et al. *Behavior and Information Technology, 6,* 483–487.

Fiske, S., & Taylor, S. (1991). *Social cognition.* Reading, MA: Addison-Wesley.

Foushee, H. C. (1982). The role of communications, socio-psychological, and personality factors in the maintenance of crew coordination. *Aviation, Space, and Environmental Medicine, 53*(11*),* 1062–1066.

Gaba, D. (1994). Automation in anesthesiology. In M. Mouloua & R. Parasuraman (Eds.), *Human performance in automated systems: Current research and trends* (pp. 57–63). Hillsdale, NJ: Lawrence Erlbaum Associates.

Gemignani, M. (1984). Laying down the law to robots. *San Diego Law Review, 21,* 1045.

Gigerenzer, G., Hoffrage, U., & Kleinbölting, H. (1991). Probabilistic mental models: A Brunswikian theory of confidence. *Psychological Review, 98,* 506–528.

Glick, P., Zion, C., & Nelson, C. (1988). What mediates sex discrimination in hiring decisions? *Journal of Personality and Social Psychology, 55,* 178–186.

Hammond, K. R., Steward, T. R., Brehmer, B., & Steinmann, D. O. (1975). Social judgment theory. In M. F. Kaplan & S. Schwartz (Eds.), *Human judgment and decision processes* (pp. 271–312). San Diego, CA: Academic Press.

Hart, A. (1992). *Knowledge acquisition for expert systems.* New York: McGraw-Hill.

Hurst, K., & Hurst, L. (1982). *Pilot error: The human factors.* New York: Aronson.

Kaempf, G. L., & Klein, G. (1994). Aeronautical decision making. In N. Johnston, N. McDonald, & R. Fuller (Eds.), *Aviation psychology in practice.* (pp. 223–254) Aldershot, England: Avebury Technical.

Lord, C. G., Ross, L., & Lepper, M. (1979). Biased assimilation and attitude polarization: The effects of prior theories on subsequently considered evidence. *Journal of Personality and Social Psychology, 37,* 2098–2109.

Martin, D. (1980). *Three Mile Island: Prologue or epilogue?* Cambridge, MA: Ballinger.

Mockler, R. J. (1989). *Knowledge-based systems for management decisions.* Englewood Cliffs, NJ: Prentice-Hall.

Morgan, T. (1992). Competence and responsibility in intelligent systems. *Artificial Intelligence Review, 6,* 217–226.

Mosier, K. L., Palmer, E. A., & Degani, A. (1992). Electronic checklists: Implications for decision making. *Proceedings of the 36th Annual Meeting of the Human Factors Society* (pp. 7–11). Atlanta, GA: Human Factors Society.

Mosier, K. L., Skitka, L. J., & Korte, K. J. (1994). Cognitive and social psychological issues in flight crew/automation interaction. In M. Mouloua & R. Parasuraman (Eds.), *Human performance in automated systems: Current research and trends* (pp. 191–197). Hillsdale, NJ: Lawrence Erlbaum Associates.

NASA ASRS. (1992). *TCAS Incident Reports Analysis* (Quick Response Report No. 235). Mountain View, CA: NASA Aviation Safety Reporting System.

Norman, D. A. (1990). The "problem" with automation: Inappropriate feedback and interaction, not "over-automation." *Philosophical Transactions of the Royal Society of London,* B 327.

Orasanu, J. (1993). Decision making in the cockpit. In E. L. Wiener, B. G. Kanki, & R. L. Helmrich (Eds.), *Cockpit resource management* (pp. 137–172). San Diego, CA: Lawrence Erlbaum Associates.

Orasanu, J., & Connolly, T. (1993). The reinvention of decision making. In G. A. Klein, J. Orasanu, R. Calderwood, & C. E. Zsambok (Eds.), *Decision making in action: Models and methods* (pp. 3–20). Norwood, NJ: Ablex.

O'Sullivan, C. S., & Durso, F. T. (1984). Effect of schema-incongruent information on memory for stereotypical attributes. *Journal of Personality and Social Psychology, 47*, 55–70.

Parasuraman, R. (1986). Vigilance, monitoring, and search. In K. R. Boff, L. Kaufman, & J. P. Thomas (Eds.), *Handbook of perception and human performance, Volume 2* (pp. 43.1–43.39). New York: Wiley.

Parasuraman, R. (1987). Human-computer monitoring. *Human Factors, 29*(6), 695–706.

Parasuraman, R., Molloy, R., & Singh, I. L. (1993). Performance consequences of automation-induced "complacency." *The International Journal of Aviation Psychology, 3*(1), 1–23.

Sagar, H. A., & Schofield, J. W. (1980). Racial and behavioral cues in black and white children's perceptions of ambiguously aggressive acts. *Journal of Personality and Social Psychology, 39*, 590–598.

Sarter, N. B., & Woods, D. D. (1992). Pilot interaction with cockpit automation: Operational experiences with the flight management system. *International Journal of Aviation Psychology, 2*(4), 303–321.

Sarter, N. R., & Woods, D. D. (1993). *Cognitive engineering in aerospace application: Pilot interaction with cockpit automation* (NASA Contractor Report 177617). Columbus, OH: The Ohio State University.

Sarter, N. R., & Woods, D. D. (1994). Decomposing automation: Autonomy, authority, observability and perceived animacy. In M. Mouloua & R. Parasuraman (Eds.), *Human performance in automated systems: Current research and trends* (pp. 22–27). Hillsdale, NJ: Lawrence Erlbaum Associates.

Segal, L. D. (1990). Effects of aircraft cockpit design on crew communication. In E. J. Lovesey (Ed.), *Contemporary ergonomics 1990* (pp. 247–252). London: Taylor & Francis.

Simon, H. A. (1955). A behavioral model of rational choice. *Quarterly Journal of Economics, 69*, 99–118.

Skitka, L. J., & Mosier, K. L. (1994, November). *Automation bias: When, where, why?* Paper presented at the annual meeting of the Society for Judgment and Decision Making, St. Louis, MO.

Swets, J. A. (1986). Indices of discrimination or diagnostic accuracy: Their ROCs and implied models. *Psychological Bulletin, 99*, 100–117.

Swets, J. A., Tanner, W. P., & Birdsall, T. G. (1961). Decision processes in perception. *Psychological Review, 68*, 301–340.

Waterman, D. A. (1985). *A guide to expert systems*. Reading, MA: Addison-Wesley.

Wickens, C. D. (1984). *Engineering psychology and human performance*. Columbus, OH: Merrill.

Wiener, E. L. (1981). Complacency: Is the term useful for air safety? In *Proceedings of the 26th Corporate Aviation Safety Seminar* (pp. 116–125). Denver, CO: Flight Safety Foundation.

Wiener, E. L. (1989). *Human factors of advanced technology ('glass cockpit') transport aircraft* (Technical Report # 117528). Moffett Field, CA: NASA Ames Research Center.

Wiener, E. L. (1993a). Crew coordination and training in the advanced-technology cockpit. In E. L. Wiener, B. G. Kanki, & R. L. Helmreich (Eds.), *Cockpit resource management* (pp. 199–229). San Diego, CA: Academic Press.

Wiener, E. L. (1993b). *Intervention strategies for the management of human error* (NASA Contractor Report 4547). Moffett Field, CA: NASA Ames Research Center.

Wiener, E. L., & Curry, R. E. (1980). Flight deck automation: Promises and problems. *Ergonomics, 23*, 995–1011.

Will, R. P. (1991). True and false dependence on technology: Evaluation with an expert system. *Computers in Human Behavior, 7*, 171–183.

Winograd, T., & Flores, F. (1986). *Understanding computers and cognition: A new foundation for design.* Norwood, NJ: Ablex.

Woods, D. (1994). Automation: Apparent simplicity, real complexity. In M. Mouloua & R. Parasuraman (Eds.), *Human performance in automated systems: Current research and trends* (pp. 1–7). Hillsdale, NJ: Lawrence Erlbaum Associates.

Zeide, J. S., & Liebowitz, J. (1987, Spring). Using expert systems: The legal perspective. *IEEE Expert,* p. 19–22.

11

Supervisory Control and the Design of Intelligent User Interfaces

Bruce G. Coury
Ralph D. Semmel
Johns Hopkins University

INTRODUCTION

The nature of supervisory control is changing. Traditional forms of automation are giving way to expert and knowledge-based systems, and to a new breed of "intelligent agents" that are capable of working independently on behalf of a human operator (Communications of the ACM, 1994). As Maes (1994) pointed out in her discussion of intelligent agents, user interfaces will need to move from the current paradigm of direct manipulation to indirect management, where the user "is engaged in a cooperative process in which human and computer agents both initiate communication, monitor events and perform tasks" (p. 31)

The concept of such a relationship between a human operator and a complex, dynamic system is not entirely new to supervisory control. Sheridan (1992) considered the requirements for human supervision of modern control systems, and explored different forms of the relationship among operator, control system, and controlled process. Central to that discussion is the requirement for computer systems that play specific roles in supervisory control. For instance, he distinguished between the human-interactive computer that communicates directly with the operator and the task-interactive computer that works directly with the controlled process. The modern operator no longer exerts direct control over a system by reading gauges and flipping switches, but instead works with computer-based systems to obtain data and information, plan responses to situations, and execute control actions.

Such an approach has significant implications for the design and development of intelligent supervisory control systems; functional requirements must be decomposed into specific tasks with special-purpose systems built to either take complete responsibility for a task or aid the human operator in the accomplishment of a task. In either case, the user interface must support cooperation and communication among operators and controlling systems. The purpose of this chapter is to explore in more detail the implications of developing intelligent user interfaces for supervisory control, and to describe the kinds of mechanisms that must be put in place to allow an operator to communicate and work cooperatively with a complex system.

Our approach to intelligent user interfaces stems from much of our current work in ship systems automation and the development of intelligent systems for critical operations. A significant part of that effort is being directed toward developing intelligent systems to enhance situation awareness and action planning. In our work, one of the primary goals of the user interface is to direct the user's attention to information that is critical to a situation and the decision-making process. Consequently, much of the discussion of our work concentrates on the development of user interfaces for attention direction, and focuses on how previous research in attention, display formats, and explanation have provided the basis for the design of an intelligent user interface. Specific emphasis is placed on the role of attention in human decision making and problem solving, and the use of advanced query formulation techniques in the retrieval, representation, and conceptualization of situational information.

MOTIVATION FOR NEW DIRECTIONS

The need to develop intelligent user interfaces to perform specific roles in supervisory control is motivated by a number of general trends in automation. First, there are increasing levels of automation, data processing, data fusion, and intelligent control being placed between the operator and the actual process and source of sensory data. As a result, the operator has potentially less direct knowledge of or control over the process or situation, and must rely on intermediate processing and control systems to provide status information or control capabilities.

Second, to achieve high levels of automation, sophisticated artificial intelligence (AI) techniques and engineering analyses are being embedded in monitoring and control systems to detect events, identify situations, and select courses of action. Unfortunately, most of the intended users and operators will not be knowledgeable about or have had experience with expert systems, neural networks, or the types of engineering analysis used by such systems, and would probably find incomprehensible any explanation of the inner workings of those systems. Consequently, some of the current presuppositions about the nature of explanation in knowledge-based systems and the requirements for user/system communication are no longer relevant. In the new forms of explanation, a translator must exist that can decipher the output of various subprocesses and effectively communicate that information to a wide variety of potential users.

Finally, these types of automated systems will fundamentally change supervisory control and operational procedures. In the new types of intelligent supervisory control, operators will rely on intelligent systems and agents to process sensory data, display those data at the user interface in an effective way, recommend and execute actions, and autonomously control system components. In the future, the traditional forms of interaction and supervisory control will give way to a collaborative and cooperative relationship between people and intelligent systems.

As a result, the new directions in intelligent user interfaces for supervisory control will require that the large body of attention and human decision-making research be integrated with explanation and the design of user interfaces. The fact that people employ various forms of reasoning and a wide variety of decision strategies has been recognized by the human performance research community for some time (e.g., Hammond, 1988; Rasmussen, 1986). Unfortunately, there has been insufficient effort to integrate that research into the design of intelligent user interfaces and to develop functionality that can relate the type of reasoning to the cognitive demands of a task. In this chapter we describe the path taken by us to exploit that research and integrate it into the design of an intelligent user interface. We also show how techniques used for information retrieval can provide new and unique ways to reason about data structures and enhance a cognitive approach to attention direction.

Before turning to a detailed description of our approach, we provide a more comprehensive discussion of our rationale and the literature that provides the basis for our intelligent user interface. The discussion begins with our assumptions about the design of intelligent user interfaces and the need to support situation awareness and action planning. We then turn to the relevant cognitive research in attention and decision-making strategies before concluding with a more detailed discussion of how the user interface will work. Throughout the chapter we lay out the framework for an intelligent user interface.

AN APPROACH TO INTELLIGENT USER INTERFACES

Fundamental to our approach is the notion that the user interface in supervisory control is more than a medium for interaction with a system; it is a tool to be used by an operator to understand the state of a problem or situation and accomplish tasks in cooperation with an intelligent system. The user interface is the medium of communication between an operator and a system and must, as a result, provide the kinds of dialogue and knowledge transfer that accompanies any form of effective communication.

The implications of such a notion are clear. No longer is the design of the user interface concerned merely with graphics and text formatting, window management, and dialogue styles. The user interface must become the means for conveying knowledge and information about the state of a system, and the means for interacting with and manipulating a problem. Consequently, to build a user interface for supervisory control, three types of design issues must be taken into consideration (see Table 11.1). *User interaction* refers to the basic concerns of style, format, and dialogue that can be found in most generic user interface guideline documents. Until recently, much of the user interface design research has focused on improving user interaction capabilities (with emphasis placed on improvements to menus and dialogues, window management, and the display of graphical and textual information) and pursuing the goal of consistency in format and style.

Sensor, data, and information management refers to the access, integration, and interpretation of data and information. Although information retrieval has had a long and illustrious research history, only recently have there been efforts to combine the techniques for accessing information and data structures with user interface design for complex systems (Belkin & Croft, 1992; Robertson, Card, & Mackinlay, 1993). An important aspect of user interface design in supervisory control is access to the sensors and sensory data. Sensors provide the perceptual mechanisms of a complex system and the sensory input to the intermediate processing that produces the representation of the situation displayed to the operator. Without access to those sensors and subprocesses and ways to assess the basis for the representation of the situation, the operator is severely handicapped in his or her ability to comprehend the situation, determine the source of data about that situation, or know which data is reliable, accurate, and complete. Consequently, the primary design goal at this level goal is to enhance the retrieval of information and the use of various tools for summarizing, describing, and analyzing data.

Problem management and action planning is concerned with the higher levels of cognitive activity associated with complex decision making, reasoning, and problem solving. To support these activities, the user interface must become an associate in the reasoning process and be designed to aid in the interpretation and evaluation of a problem, direction of attention to critical elements of a situation, and planning of a response to events. Critical to problem management is understanding the way in which

TABLE 11.1

Three Types of Requirements for the Development of Intelligent
User Interfaces

Area	Supports	Basic Concerns	Primary Design Goals
User Interaction	Generic user interface requirements	Window Management Menu & Dialogu Styles Color & Highlighting Graphics & Text Formats	Consistency in format & style
Sensor, Data & Information Management	Use of analysis tools Information retrieval	Analysis Tool Selection & Configuration Tool Management Data Representation & Visualization Information Access & Retrieval	Improved access, integration and interpretation of data & information
Problem Management & Action Planning	Knowledge-based reasoning and problem solving	Interpretation of Problem Generation & Test of Hypotheses Attention Direction Resource Allocation Action Planning User Modeling	Appropriate problem representation Effective scene management Adequate explanation

events are changing over time. Consequently, the user interface will need to represent the history of events and changes in relevant events over time, as well as provide the mechanism for projecting into the future. The primary goals of the user interface for problem management and action planning is to support situation awareness, response selection and scheduling, resource management, and contingency planning.

Notice that each of the design issues also relates to three levels of analysis about human interaction with a system, and they are all similar to the hierarchy of abstraction described by Rasmussen (1986) and the levels of cognitive processing in situation awareness described by Endsley (1988; also chap. 8, this volume). The user interaction level is the lowest level of interaction that can be described by a user interface where key presses, menu choices, and dialogue boxes provide the basic building blocks of communication. Similar to Rasmussen's physical form and physical function levels of abstraction and Endsley's perception of situational elements, the user interaction level provides means to get at system data and invoke system functionality. This level of analysis and design is a necessary ingredient of a user interface, but insufficient for higher-level decision support, information management, and supervisory control.

The sensor, data, and information management level is consistent with Endsley's second level of information integration and Rasmussen's generalized and abstract functional levels. At this level of analysis, data is being fused and integrated to provide information and knowledge about the status and overall performance of a system. The relationship between structure and function becomes critical to a user's understanding of a system and the source of system behavior. Knowing how things are related and the causal structure of a system are critical to accessing and using system information.

At the highest level of problem management and action planning, situation awareness and the ability to project current system behavior into the future become critical to effective decision making and problem solving. Rasmussen refers to this as the *functional purpose* level of his abstraction hierarchy, but realizes in his later work (Rasmussen, 1993) that reasoning at this level must also include intuition, generalization from typical cases, and compensatory strategies. Fundamental to this level of reasoning is the envisioning (de Kleer & Brown, 1983) and mental simulation (Klein, 1989) necessary to determine what the future state of a system might be and the potential behavior of system components.

Although situation awareness is considered a fundamental aspect of problem management and action planning, one must be aware that there are five basic types of awareness that must be considered: spatial, identity, temporal, responsibility, and expectancy. Table 11.2 summarizes the five types of awareness and the critical questions related to each in situation

TABLE 11.2
Components of Situation Awareness

Component	Definition	Important Questions
Spatial Awareness	A person's knowledge of his or her location in space, and the spatial location of other relevant situational entities	Where am I? Where are the other important elements?
Identity Awareness	Knowledge of the type, classification, or importance of relevant situational entities	What is it? What is the basis for its ID? What don't I know about it? How important is it?
Temporal Awareness	Understanding the way in which a situation evolves and changes over time	How fast is it changing? How much time is there to respond? What should I attend to first? What action should I take now?
Responsibility Awareness	Knowledge of the roles and responsibilities of each entity in the situation	Who is responsible for what? Who is in charge? Who is going to make the decision? Who can perform which actions?
Expectancy Awareness	Knowledge of how entities should respond to specific situational factors	Is it behaving the way it should? Did it follow my instructions? Did it respond appropriately to my action? Is this expected behavior

assessment. Given the extensive body of literature concerned with situation awareness in aviation (Endsley, 1988, 1994, also chap. 8, this volume; Sarter & Woods, 1991; Wickens, 1992), it is not at all surprising that much of the work to develop systems to support situation awareness has concentrated on enhancing spatial, identity, and temporal awareness. Knowledge of the spatial location and identity of important situational entities, and how the situation will evolve over time, are critical factors in aviation, especially in combat situations.

As automation increases and the need to manage decision making and problem solving by multiple, distributed intelligent agents becomes an operational requirement, responsibility and expectancy awareness become increasingly important. The distributed nature of such decision making and the constructed nature of the representation of a situation results in an increased reliance on and trust in other cooperative agents. Consequently, knowing who is responsible for which tasks, decisions, and actions becomes critical to effective situation assessment and action planning. In addition, expectations that assigned tasks and decisions will be carried out and that the entities in the situation will behave in specific ways can significantly influence both the effectiveness of the decision making and problem solving, and the cost of unexpected events and inappropriate actions.

The importance of action planning cannot be overlooked. Although there is a tendency to treat situation assessment and action planning as distinctly separate cognitive activities, there is increasing recognition that the assessment of a situation can be significantly influenced by the actions an operator may wish to take (Bodker, 1991; Kirlik, Miller, & Jagacinski, 1993; Suchman, 1987). In these approaches to reasoning and problem solving, "the environment is described as a dynamically varying set of action opportunities competing for the human's limited resources for cognition and action" (Kirlik et al., 1993, p. 930). Actions become the driving force for cognition, determining what is relevant and important in the situation and where processing resources should be directed for assessment and analysis. The implication for the design of intelligent user interfaces is that situation awareness and action planning must be closely coupled.

One can also conclude that a primary goal for an intelligent user interface is to enhance situation awareness by acting as the intermediary between an operator and the processing systems that provide situational information. Such a goal is particularly important when the representation of a situation and recommendations for action are based on a combination of decision and engineering models, quantitative methods, and pattern recognition techniques. Much of the recent development of intelligent user interfaces for knowledge-based systems has recognized the need for such support by concentrating on two basic types of functionality: interface management and explanation. The primary purpose of an interface manager is to generate displays and coordinate interaction with a user. The purpose of the explanation facility is to communicate to the user the basis and rationale for conclusions and recommendations for action. The general requirements for such functionality have been outlined by a number of researchers (Hendler, 1988; Rouse, Geddes, & Curry, 1988; Wexelblat, 1989).

A MODEL OF AN INTELLIGENT USER INTERFACE

The demands of intelligent supervisory control require that traditional user interface and explanation capabilities be enhanced to include the functionality outlined in the model of an intelligent user interface shown in Fig. 11.1. In this model, the Attention Director plays a central role in the user

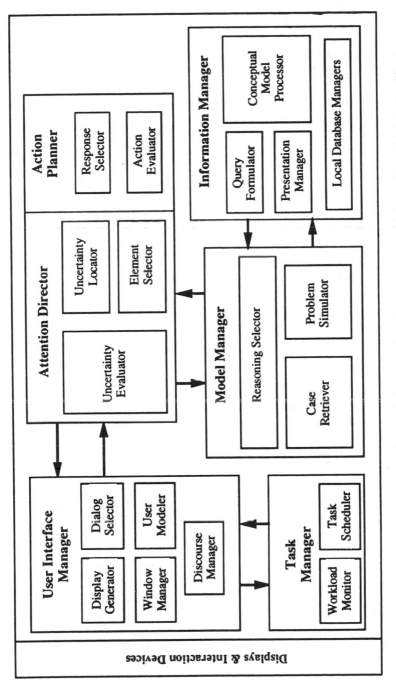

FIG. 11.1. A functional model for an intelligent user interface. The primary components of the model are the Attention Director and Action Planner, Information Manager, Model Manager, Task Manager, and User Interface Manager.

229

interface by evaluating uncertainty and directing the operator's attention to critical information, situation factors, and problem elements. The Attention Director relies on two specialized intelligent agents to handle specific aspects of the problem: a Model Manager and a User Interface Manager. The Attention Director uses the Model Manager to evaluate problem information and to select the type of reasoning process to be used in a specific situation. Although much of the research in explanation has used either a model-based (Dvorak & Kuipers, 1991) or a case-based (Kolodner, 1991) approach, there is a need to develop the Model Manager in light of cognitive research that shows how people employ both types of reasoning (Hammond, 1988; Rasmussen, 1986). In addition, technologies for using meta-level knowledge for accessing databases (Semmel, 1992) suggest ways to manage the reasoning process. The Model Manager also provides the mechanisms for evaluating the history of events and projecting the current system state into the future.

Communication between operators and the intelligent system is handled by the User Interface Manager. This agent acts as the primary medium for interaction between the operator and the system by generating the displays and transforming the reasoning and rationale of subprocesses into a comprehensible form. The User Interface Manager is adaptive and contains knowledge of the operator's level of understanding and knowledge about the problem, thereby allowing it to select the appropriate level of reasoning and abstraction for a particular type of situation and operator. To translate the problem into an adequate form for the operator, a thorough understanding of the reasoning and decision processes employed by a user must be obtained. Consequently, an integral part of the development of the User Interface Manager requires the use of cognitive modeling techniques (e.g., Payne, Bettman, & Johnson, 1988; Pietras & Coury, 1994) to identify and represent the information processing and decision-making requirements of situation assessment and action planning. As a result, the User Interface Manager extends current approaches to explanation (e.g., Feiner & McKeown, 1990) by incorporating user modeling, cognitive modeling, and attention direction capabilities.

The research in explanation for knowledge-based systems has illustrated quite clearly that effective reasoning and decision making is dependent on the availability and use of relevant situational data. In large systems designed to support situation assessment and action planning, data is maintained in databases that must correctly model the world and support efficient retrieval. However, as the complexity of the situation increases, it becomes difficult to automatically direct attention to applicable portions of a database, retrieve relevant cases, or select the appropriate model parameters for envisioning (de Kleer & Brown, 1983) or mental simulation (Klein, 1989). The Information Manager provides the link to the system databases and knowledge structures as well as the mechanism for accessing system data and information.

For an Attention Director and Model Manager to identify and associate appropriate data elements requires knowledge of the database design. Such knowledge, however, exceeds the simple syntactic descriptions found in most data dictionaries and goes beyond the types of indexing discussed in case-based reasoning. In particular, there may be several ways of associating desired data that are distributed and possibly replicated among multiple sources. Moreover, only some of those ways may be semantically reasonable with respect to a specific problem. Thus, the Attention Director may have to select a course of action from several alternatives, and must be able to explain its rationale for making that selection to an operator. To achieve this, meta-level knowledge corresponding to the meaning of the sources is needed to facilitate reasoning.

Our approach to meta-level reasoning is based on a system known as QUICK (for "QUICK is a Universal Interface with Conceptual Knowledge"). QUICK (Semmel, 1992, 1994) enables a client (in this case the Attention Director working through the Information Manager) to view a database as consisting of a single universal relation (Maier, Ullman, & Vardi, 1984; Vardi, 1988). High-level requests submitted to QUICK contain only the attributes of interest and the constraints that must be satisfied. Association information such as attribute location and natural join criteria are not specified. By exploiting conceptual-level design knowledge, QUICK formulates a semantically reasonable query that can be submitted to the database for subsequent execution. QUICK eliminates the need for clients to be aware of the underlying database design through the use of a highly extended entity-relationship (EER) model that is based on association knowledge representation constructs, such as frames and semantic networks (Hull & King, 1987; Markowitz & Shoshani, 1990; Teory, Yang, & Fry, 1986). Employing heuristic analysis techniques, QUICK segments an EER conceptual schema into maximal subgraphs, or contexts, of strongly associated objects (Semmel, 1994). Contexts thus serve as meta-level constructs that represent sets of related conceptual-level objects. From a processing standpoint, the Information Manager accepts high-level requests, identifies the contexts that contain the desired attributes, prunes the contexts of extraneous EER objects, eliminates possible duplication, and maps the resulting subgraphs to semantically reasonable queries.

To demonstrate how the Information Manager can use contexts, consider the simple EER conceptual schema presented in Fig. 11.2. This diagram represents a portion of a corporate database system being developed by the U.S. Army to integrate data that is contained in three information subsystems. The corporate system is being designed to support command, control, and communications at national, theater, tactical, and operational

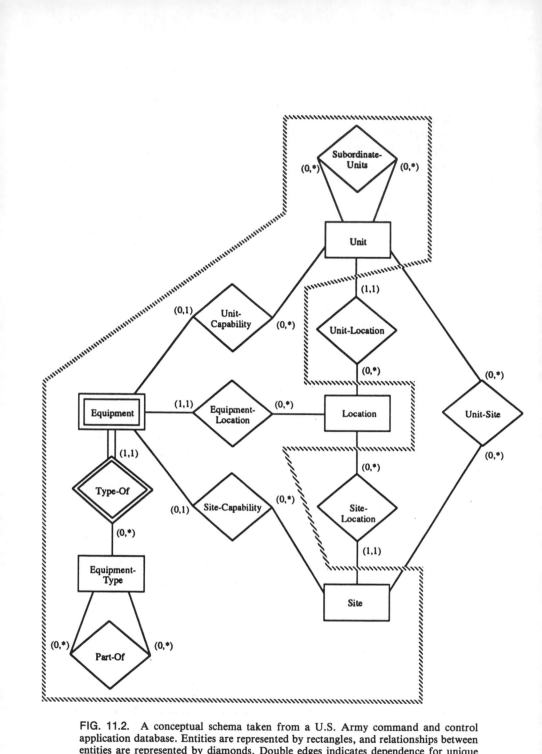

FIG. 11.2. A conceptual schema taken from a U.S. Army command and control application database. Entities are represented by rectangles, and relationships between entities are represented by diamonds. Double edges indicates dependence for unique identification.

levels. In the figure, entity types are represented by rectangles, and relationships between entity types are represented by diamonds; a double edge indicates that an EER object is dependent on another EER object for unique identification. Note that relationships are annotated with structural constraint (MIN, MAX) pairs to indicate cardinality ratio and participation constraints. For instance, the relationship type UNIT-CAPABILITY represents an n:1 cardinality ratio constraint from UNIT to EQUIPMENT, as can be inferred from the MAX components of the structural constraint pairs. Furthermore, UNIT and EQUIPMENT entities have optional participation in the UNIT-CAPABILITY relationship set, as is indicated by the 0 in each of the MIN components of the structural constraint pairs. In other words, each piece of equipment is of a certain type and with a specific capability, and may have a physical location with respect to a site and an organizational unit.

Although an EER diagram is a useful database abstraction mechanism, its primary function is to facilitate design. To perform higher-level reasoning tasks, such as automated query formulation or attention direction, additional knowledge is needed about sets of semantically related conceptual-level objects. Contexts were developed to supply this knowledge. As alluded to previously, a context is a meta-level construct that represents a set of strongly associated EER objects. Contexts can be used by an intelligent interface to determine how distinct EER objects are related. For example, suppose the Information Manager directs QUICK to identify all unit and equipment associations. QUICK first finds the contexts containing the specified objects. In this case, the highlighted context indicated by the dashed line in Fig. 11.2 is identified. Then the found context is pruned of extraneous objects, and the resulting subgraph consisting of the relationship type UNIT-CAPABILITY and its two participating entity types is returned. QUICK then maps the pruned subgraph to the underlying database, formulates a query, and provides the intelligent interface with an augmented subgraph to indicate the reasoning used for information retrieval.

The Task Manager ensures that tasks are being carried out in the appropriate way and in a timely manner. In large-scale distributed systems where multiple operators are working together to supervise a complex system, there must exist a mechanism for coordinating tasks among operators, managing the workload associated with specific requirements, and monitoring local task performance. The Task Manager fulfills that role by transmitting to the User Interface Manager action plans generated by the Action Planner and Model Manager, monitoring progress through those plans, and then assessing the workload associated with the demands of a particular task. If progress is impeded or task execution fails, the Task Manager provides the mechanism for aiding in the adjustment of the task or the management of the workload.

ATTENTION AND UNCERTAINTY

One of the key factors in decision making and problem solving (and of fundamental concern to the Attention Director) is uncertainty. Research has shown that the characteristics of the task and the mapping of task information or data to response categories or plans of action can have a significant impact on information processing strategies and decision performance (Coury, Boulette, & Smith, 1989; Garner, 1974; Hammond, 1988). In addition, that mapping directly affects the uncertainty in a task and determines the need to selectively attend to a specific source of information or to divide attention among multiple sources of information. Consequently, a person's ability to understand the magnitude and locus of uncertainty in a task will be dependent on the way in which in formation and data are displayed.

Uncertainty, as it is used in the context of this research, can arise from two primary sources: ambiguity in situation assessment and ambiguity in response selection. Uncertainty in situation assessment is defined by the mapping of situational data to situational (or state) categories. When the relationships among data produce a direct mapping to a specific state category (i.e., problem attributes allow immediate recognition of a situation), uncertainty is low; when the mapping is not direct, uncertainty is high. The same holds true for response selection: When uncertainty is low, the appropriate response (i.e., plan of action) is immediately evident, but when uncertainty is high, the selection of a response requires analysis and consideration of possible consequences. Such a view of uncertainty is consistent with theories of attention (Garner, 1974), models of operator performance (Moray, 1984; Rasmussen, 1986), inductive reasoning (Holland, Holyoak, Nisbett, & Thagard, 1986), and recognition-primed decision making (Klein, 1989).

In all of the cognitive models of human performance, the need to focus or divide attention is an important element. For instance, Rasmussen's (1986) model of operator performance includes in it two types of search strategies for detecting and correcting failures in dynamic systems: a topographic search strategy and a symptomatic search strategy. Topographic search is analytic and requires focused attention and systematic assessment of the functional status of various system components, a thorough knowledge of the relationship between system function and structure, and an understanding of the effect of component behavior on overall system status. Such a strategy is analogous to model-based reasoning where knowledge of the relationship between structure and function can be used to analyze a unique or unfamiliar problem and determine a response to it. Symptomatic search, on the other hand, is holistic, based on pattern recognition and the direct mapping of situational attributes to failure states. When using a symptomatic search strategy, no direct reference need be made to the status or location of system components or knowledge of the causal relationship between structure and function. Consequently, symptomatic search can be especially effective in routine or highly familiar situations where no in-depth analysis of component behavior is necessary.

Evidence for both types of strategies may be exhibited in a dynamic situation. For instance, an operator may use a symptomatic search strategy to detect and correct routine or familiar deviations from normal state, but resort to a topographic search strategy when confronted with an unfamiliar problem. Because the two strategies have different information requirements, each would be best supported by a different representation of the problem. A symptomatic search strategy requires global indices of system performance, whereas the topographic search strategy requires more precise measures of specific component status and a schematic representation of system structure.

Klein (1989) also proposed a model of decision making for other problem domains that is similar to Rasmussen's symptomatic search. Based on observations of actual decision makers in fire fighting, tank combat, and engineering design, Klein's model has many elements similar to a case-based approach to reasoning and decision making. Familiar situations are equivalent to prototypical instances of a case, triggered by basic situational characteristics that carry with them options for making a decision. Evaluation of a situation and suitable options are based on the initial characterization of the current situation using similar, previously experienced situations. Decision makers rely on their extensive knowledge of and experience with the characteristics of typical and familiar cases, and use those cases to guide information gathering, goal clarification, and solution refinement. The use of mental simulation to visualize the outcome of various problem solutions is a fundamental and unique component of Klein's decision-making process.

One aspect of case-based reasoning that is particularly relevant to the development of intelligent user interfaces is the ability to access cases, then modify them for use in a specific situation (Kolodner, 1991). The ability to retrieve cases is dependent on indexing (i.e., the way in which attribute information about a case is labeled). Kolodner discusses three types of indexing: predictive features, abstractness, and concreteness. Predictive features of a case can be used to predict some or part of the outcome or solution. Abstractness refers to the type of indexing that allows specific cases to be related to broad classes of problems. Concrete indexing retains the specific characteristic of the case. In general, the overriding concern of selecting an index is its utility (i.e., how good the index is in the selection of cases for a particular type of problem). When uncertainty is a predominant characteristic of a situation, the predictive capabilities of the case become extremely important in hypothesis generation, case retrieval, and action recommendations.

Model-based reasoning also provides important insights into the development of our Model Manager. Model-based reasoning relies on a working knowledge of system function and structure, and uses that knowledge to assess or predict system behavior. Well suited to modeling and simulation of physical processes, model-based approaches are commonly found in the areas of diagnosis and control (Dvorak & Kuipers, 1991; Sheridan, 1992). The general idea of a model-based approach is to determine (usually in a hierarchical fashion) which part of the structure of a system is responsible for producing the observed behavior. Such an approach does not rely solely on the mapping of symptoms to faults (and a comprehensive set of cases to handle all contingencies), but instead depends on a simulation or a quantitative model of a dynamic system or problem.

In our research, knowledge of a user's reasoning process guides the development of the Model Manager. Because the purpose of the Model Manager is to provide the ability to select the appropriate reasoning process based on the level and type of uncertainty in a particular situation, we have looked to the research in both case-based reasoning and model-based reasoning in explanation to provide insights into the types of reasoning appropriate to the management of uncertainty. Unfortunately, most of the systems designed to solve problems similar to situation assessment have adopted either a model-based or a case-based approach. There has been no real attempt to relate the human performance literature in reasoning to the selection of a case-based or a model-based approach to explanation. Although the cognitive research indicates that the type of reasoning process will be dependent on the cognitive demands of a task (e.g., Hammond, 1988), there appears to have been no systematic attempt to develop explanation facilities that tie reasoning to task and attention requirements.

HOW IT WILL WORK

Current development efforts have emphasized the Attention Director because of the distinct need to integrate attention and human decision-making research with explanation in the design of an intelligent user interface. A fully functional Attention Director will assess the level and source of uncertainty, communicate that understanding to the user, and work cooperatively with the user to manage uncertainty. The mechanism used to coordinate reasoning approaches will be based on models of human performance that relate decision strategies to levels of uncertainty (e.g., Coury & Boulette, 1992; Coury et al., 1989; Rasmussen, 1986). Indexing the type of reasoning to uncertainty provides the basis for the Attention Director (working in conjunction with the Model Manager) and the User Interface Manager to tie knowledge of uncertainty in the situation to the graphical objects, textual information, and interaction dialogues that appear on the display.

The Model Manager will invoke a case-based reasoning approach when uncertainty is low and there is a unique mapping of data to problem states and response categories. This approach will allow the Attention Director to represent the problem to the user in terms of the relationships among situation attributes, problem states, and appropriate actions. When uncertainty is high and data do not map directly to known situations, an explicit representation of the functional and structural elements of the situation will be required, and a model-based approach will be used. In these situations, the Model Manager will possess a model of the structural and functional associations that describe problem entities and entity associations as well as knowledge of the behavioral relationships among entities. The Attention Director will present the problem to the user in terms of spatial and functional associations with attention drawn to the source of uncertainty.

The development and enhancement of QUICK (Semmel, 1992) for this application will provide the means for the Attention Director and Model Manager to use meta-level knowledge in the identification and retrieval of relevant information and data. From an attention direction standpoint, QUICK will be generalized to work with multiple data and knowledge sources. Specifically, each source could be represented using an extension of QUICK's implementation-independent representation scheme. Then contexts will be formulated and used to focus the Attention Director on relevant portions of appropriate sources. In particular, contexts will be chosen over each knowledge source that covered the set of requested data elements.

In its present configuration, QUICK identifies contexts in a static analysis of the conceptual-level EER schema. Thus, certain paths between source items may be deemed semantically unreasonable. However, there may be cases when a situation dictates that such paths be used. Although experience with QUICK indicates that these cases will be rare, the Attention Director must have the ability to direct QUICK to extend its search scope to include objects external to a particular context and perhaps external to a particular source. There are two mechanisms that facilitate search beyond a single context: a context regeneration mechanism and gateway relationships. A context regeneration mechanism enables a requester to identify EER objects that are not to be considered in a single source. This is accomplished by reevaluating the corresponding EER conceptual schema for the source to produce a new set of contexts that tends to include larger members (i.e., contexts with more EER objects) than those found in the original set. These new contexts then are used to associate requested data elements. For example, the Attention Director, after evaluating the information coming from one source and the corresponding conceptual graph, may determine that other critical information can be found in a larger context for the problem.

Gateway relationship types (Semmel & Winkler, 1994) can be used to indicate explicit associations among different sources. In particular, conceptual-level gateways are established that link meta-level contexts among information sources. Then, when a request is made that requires traversing from one source to another, appropriate contexts in each source are selected via gateways, and multisource conceptual subgraphs are found. For attention direction in a multiple-source system, gateway relationship types are especially important. In particular, there may be situations in which an operator must decide whether to perform an action based on a given hypothesis. However, the data in one source may only support or refute the hypothesis with a low degree of certainty. Thus, the user might need to explore related sources to justify an action based on an increased certainty or to invalidate the action based on a decreased certainty. For example, in assessment of unit capability (in Fig. 11.2) there could be a number of important sources of information that could be linked (i.e., unit data, site data, and maintenance records). A hypothesis based on a unit data source may need, for example, reinforcement through data contained in the maintenance record. Through the use of a gateway relationship type between these two sources, the operator's attention could be directed to the unit data and a final decision on actions to take place could be made. Similarly, if further justification or refutation were necessary, other paths could be explored based on either the original source or new sources.

Another feature supported by QUICK that could be generalized for attention direction is explanation. Currently, QUICK explains its generation rationale by identifying (via simple textual description) the conceptual subgraphs used to formulate a query. In the Attention Director, this capability will be enhanced in the following way: Given a conceptual subgraph, QUICK provides enough information to an Attention Director about the meaning of a query to facilitate disambiguation among multiple conceptual subgraphs that cover the requested attributes. In the simplest case, QUICK and the Attention Director cooperate in an interactive dialogue with an operator. Specifically, the operator employs graphical subgraph depiction and automatically generated textual descriptions to select among valid alternatives. A more ambitious goal entails having the Attention Director reason over QUICK-selected conceptual subgraphs to disambiguate among alternatives.

The User Interface Manager uses the results from the Attention Director and Model Manager to communicate with the operator. In its most basic form, the User Interface Manager will enhance current approaches to intelligent and adaptive user interfaces (e.g., Feiner & McKeown, 1990) by dynamically manipulating graphical schematics of the situation to focus the operator's attention on critical problem elements. For example, when attention is drawn to an entity in the situational picture, the scene will be manipulated by the User Interface Manager (working with the Attention Director) to indicate the source of the uncertainty, and to provide related information and the analysis tools necessary to reduce uncertainty in the situation. The User Interface Manager will possess the basic user models necessary to adapt to the specific needs and goals of an operator.

The development of the Attention Director, Model Manager, and User Interface Manager are dependent on a thorough understanding of the cognitive requirements of situation awareness. Cognitive modeling is the technique being used in this research to capture and represent information processing and decision strategies as well as manage the reasoning process and adapt the user interface to the specific needs of an operator. Such an approach is motivated by the acknowledgment by many researchers (Hammond, 1988; Rasmussen, 1986; Vessey, 1991) of the need to match the problem representation to the cognitive requirements of a task. The cognitive models will provide the mechanism for integrating cognitive research in attention and decision making with the development of explanation capabilities and user interfaces for supervisory control. The model-building effort will also result in a better understanding of the applicability of different types of reasoning (e.g., case-based, model-based, and recognition-primed decision making) to situation assessment and action planning, especially under different types of operational conditions.

The specific approach is based on the cognitive modeling techniques that have been used to assess decision-making performance (Kuipers, Moskowitz, & Kassirer, 1988; Payne et al., 1988), characterize user interaction with computer systems (Kieras & Polson, 1985; Olson & Olson, 1990), and identify decision and planning strategies for information systems (Holtzman, 1989). The work by Pietras and Coury (1994) and Goh and Coury (1994) suggests, however, that cognitive modeling must go beyond low-level models of human–computer interaction and develop a generic methodology for representing cognitive processing in dynamic situations.

FUTURE DIRECTIONS

Our discussion of the development of intelligent user interfaces for supervisory control has emphasized the role of attention in situation assessment and action planning. In our work, we have focused our development effort on the kinds of technological support that will be needed to identify the critical elements of a situation, direct the operator's attention to those elements, and effectively respond to that situation. The development of the technologies relies on our ability to integrate cognitive research in attention into explanation and the design of user interfaces, and to exploit advanced information retrieval technologies for meta-level conceptual reasoning. As a result, this work opens a new line of research and development into ways to evaluate uncertainty, direct attention, and interact with complex systems.

Work cannot end there. Users of intelligent supervisory control systems will not operate in isolation and will need to coordinate their activities across time and space. Consequently, the next step is to extend this work to include multi-agent communication and collaboration. The logical extension of the work is to embed the intelligent user interface within the context of a larger supervisory system that coordinates actions and allocates resources among groups of operators and intelligent systems. This will necessitate the development of intelligent systems for task assignment, resource allocation, and workload management. We envision a system comprised of interacting agents that understand the global requirements of tasks and the roles and responsibilities of operators. Those agents would then use that knowledge to determine task demands and workload requirements for individuals and teams, track task performance, and provide links to system subprocesses and databases. Realization of such a vision and development of intelligent agent and user interface capabilities is highly dependent on successful consideration of many of the issues discussed in other chapters of this book. In particular, the ability to share knowledge between operators and intelligent systems becomes critical to effective supervisory control and fundamental to an operator's trust in a system that fulfills requests, accomplishes tasks, and makes decisions.

In summary, the approach to intelligent user interfaces outlined in this chapter contributes to the development of knowledge-based systems for supervisory control in a number of important ways. The work integrates the cognitive research in attention with the development of explanation capabilities and user interfaces for automated systems. This provides a way to relate the wealth of human performance literature to the design of systems and to the use of models of human performance in the management of the reasoning process, assessment of uncertainty, and direction of attention. From a technology transfer perspective, the enhanced retrieval capabilities and the ability to model a large variety of homogeneous and heterogeneous systems will enable the technology to be used as the basis for sophisticated and intelligent user interfaces.

REFERENCES

Belkin, N. J., & Croft, W. B. (1992). Information filtering and information retrieval: Two sides of the same coin? *Communications of the ACM, 35*(12), 29–38.

Bodker, S. (1991). *Through the interface* . Hillsdale, NJ: Lawrence Erlbaum Associates.

Communications of the ACM (1994). *Special Issue on Intelligent Agents, 37*(7). New York: Association for Computing Machinery.

Coury, B. G., & Boulette, M. D. (1992). Time stress and the processing of visual displays. *Human Factors, 34*(6), 707–725.

Coury, B. G., Boulette, M. D., & Smith R. A. (1989). The impact of uncertainty on classification of multidimensional data with integral and separable displays. *Human Factors, 31*(5), 551–569.

de Kleer, J., & Brown, J. S. (1983). Assumptions and ambiguities in mechanistic mental models. In D. Gentner & A. L. Stevens (Eds.), *Mental models* (pp. 155–190). Hillsdale, NJ: Lawrence Erlbaum Associates.

Dvorak, D., & Kuipers, B. (1991). Process monitoring and diagnosis. *IEEE Expert*, June, 67–74.

Endsley, M. R. (1988). Design and evaluation for situation awareness enhancement. In the *Proceedings of the Human Factors Society 32 Annual Meeting* (pp. 97–101). Santa Monica, CA: Human Factors & Ergonomics Society.

Endsley, M. R. (1994, April). *The role of situation awareness in naturalistic decision making*. Paper presented at the Second Conference on Naturalistic Decision Making, Dayton, OH.

Feiner, S. K., & McKeown (1990). Coordinating text and graphics in explanation generation. In the *Proceedings of the IEEE Conference on AI for Applications* (pp. 290–303). Piscataway, NJ: IEEE Service Center.

Garner, W. R. (1974). *The processing of information and structure*. New York: Wiley.

Goh, S. K., & Coury, B. G. (1994). Incorporating the effect of display formats in cognitive modelling. *Ergonomics, 37*(4), 725–745.

Hammond, K. R. (1988), Judgment and decision making in dynamic tasks. *Information and Decision Technologies, 14*, 3–14.

Hendler, J. A. (1988). *Expert systems: The user interface*. Norwood, NJ: Ablex.

Holland, J. H, Holyoak, K. J., Nisbett, R. E., & Thagard, P. R. (1986). *Induction: processes of inference, learning, and discovery*. Cambridge, MA: MIT Press.

Holtzman, S. (1989). *Intelligent decision systems*. Reading, MA: Addison-Wesley.

Hull, R., & King, R. (1987). Semantic database modeling: survey, applications, and research issues. *ACM Computing Surveys ,19*(3), 201–260.

Kieras, D., & Polson, P. G. (1985). An approach to the formal analysis of user complexity. *International Journal of Man–Machine Studies, 22*, 365–394.

Kirlik, A., Miller, R. A., & Jagacinski, R. J. (1993). Supervisory control in a dynamic and uncertain environment: A process model of skilled human-environment. *IEEE Transactions on Systems, Man, and Cybernetics, 23*(4), 929–952.

Klein, G. A. (1989). Recognition-primed decisions. In W. B. Rouse (Ed.), *Advances in man-machine systems research* (vol. 5, pp. 47–92). Greenwich, CT: JAI Press.

Kolodner, J. L. (1991, Summer). Improving human decision making through case-based decision aiding. *AI Magazine*, pp. 52–68.

Kuipers, B., Moskowitz, A. J., & Kassirer, J. P. (1988). Critical decisions under uncertainty: Representation and structure. *Cognitive Science, 12*, 177–210.

Maes, P. (1994). Agents that reduce work and information overload. *Communications of the ACM, 37*(7), 30–40.

Maier, D., Ullman, J. D., & Vardi, M. Y. (1984). On the foundations of the universal relation model. *ACM Transactions on Database Systems 9*(2), 283–308.

Markowitz, V. M., & Shoshani, A. (1990), Abbreviated query interpretation in extended entity-relationship oriented databases. In F. H. Lochovsky (Ed.), *Entity-relationship approach to database design and querying* (pp. 325–343). Amsterdam: North-Holland.

Moray, N. (1984). Attention to dynamic visual displays in man–machine systems. In R. Parasuraman & D. R. Davies (Eds.), *Varieties of attention* (pp. 485–513). New York:

Academic Press.

Olson, J. R., & Olson, G.M. (1990). The growth of cognitive modeling in human–computer interaction since GOMS. *Human–Computer Interaction, 5,* 221–265.

Payne, J. W., Bettman, J. R., & Johnson, E.J. (1988). Adaptive strategy selection in decision making. *Journal of Experimental Psychology: Learning, Memory and Cognition, 14*(3), 534–552.

Pietras, C. M., & Coury, B. G. (1994). The development of cognitive models of planning for use in the design of project management systems. *International Journal of Human–Computer Studies, 40*(5), 5–30.

Rasmussen, J. (1986). *Information processing and human–machine Interaction.* New York: North-Holland.

Rasmussen, J. (1993). Diagnostic reasoning in action. *IEEE Transactions on Systems, Man, and Cybernetics, 23*(4), 981–992.

Robertson, G. G., Card, S. K., & Mackinlay, J. D. (1993). Information visualization using 3D interactive animation. *Communications of the ACM, 36*(4), 56–71.

Rouse, W. B., Geddes, N. D., & Curry, R. E. (1988). An architecture for intelligent interfaces: Outline of an approach to supporting operators of complex systems. *Human–Computer Interaction, 3,* 87–122.

Sarter, N., & Woods, D. D. (1991). Situation awareness: A critical but ill-defined phenomenon. *International Journal of Aviation Psychology, 1*(1), 45–57.

Semmel, R. D. (1992). QUICK: A system that uses conceptual design knowledge for query formulation. In the *Proceedings of the Fourth International Conference on Tools with Artificial Intelligence* (pp. 214–221). Los Alamitos, CA: IEEE Computer Society Press.

Semmel, R. D. (1994). Discovering context in an entity-relationship conceptual schema. *Journal of Computer and Software Engineering, 2*(1), 47–63.

Semmel, R. D., & Winkler, R. P. (1994). Reverse engineering complex databases to support data fusion. In the *Proceedings of the Fourth Systems Reengineering Technology Workshop* (pp. 192–199). Laurel, MD: JHU/APL Research Center.

Sheridan, T. B. (1992). *Telerobotics, automation, and human supervisory control.* Cambridge, MA: MIT Press.

Suchman, L. A. (1987). *Plans and situated action.* New York: Cambridge University Press.

Teory, T. J., Yang, D., & Fry, J. P. (1986). A logical design methodology for relational databases using the extended entity-relationship model. *ACM Computing Surveys, 18*(2), 197–222.

Vardi, M. Y. (1988, March). The universal-relation data model for logical independence. *IEEE Software,* pp. 80–85.

Vessey, I. (1991). Cognitive fit: A theory-based analysis of the graphs versus tables literature. *Decision Sciences, 22,* 219–240.

Wexelblat, R. L. (1989, Fall). On interface requirements for expert systems. *AI Magazine,* pp. 66–78.

Wickens, C. D. (1992). Workload and situation awareness: An analogy of history and implications. *Insight, 14*(4), 1–3.

12 Team Performance in Automated Systems

Clint A. Bowers
University of Central Florida

Randall L. Oser
Eduardo Salas
Janis A. Cannon-Bowers
U.S. Naval Air Warfare Center Training Systems Division

INTRODUCTION

The tremendous increase in automated systems in the workplace seems to have caught the behavioral sciences unprepared. Despite the almost commonplace use of automated systems in a variety of occupations, there is only a small body of literature that has discussed the effects of these systems on performance. In fact, it is only recently that social scientists have turned their attention to this important aspect of performance (cf. Mouloua & Parasuraman, 1994). Consequently, the need to understand the degree to which automation affects human performance in complex systems is becoming an urgent topic for applied scientists.

It might be argued that even when an adequate database regarding automation and individual performance is obtained, our ability to contribute to enhanced performance in the workplace might still be limited. This is because many of the environments that use this technology are so complex that they require multiple operators. These operators often work interdependently to achieve the desired outcome; thus, they are a team. The nature of the communication and coordination required to foster team performance represents another area of science that is complex, subtle, and not frequently investigated.

It is for this reason that the goal of understanding the effects of automation on team processes and performance seems so daunting. Yet, despite the difficulties, there is a clear need to provide guidance regarding the effects of automation on team performance to system designers and other practitioners in this area. We can contribute most quickly by isolating those variables that seem to have the greatest likelihood of influencing team performance in these systems and using these variables as the foundation for a program of research.

The goal of the current chapter is to provide specific directions for the study of team performance in automated systems. Based on the existing literature and empirical data on automation effects and performance, hypotheses are offered. It is believed that this approach results in a collection of elements for study to arrive at the greatest scientific impact in the least time.

APPROACH

In order to organize our presentation, we will utilize the team effectiveness model (TEM; Tannenbaum, Beard, & Salas, 1992). The TEM, which is illustrated in Fig. 12.1, is an input-process-outcome model used to describe the factors that influence team performance. As is evident from the figure, the model includes organizational and situational characteristics, which refer to elements such as supervisory control, reward structures, and environmental uncertainty. The model also includes four types of input factors: individual characteristics, team characteristics, task characteristics, and work characteristics. Individual characteristics refer to aspects of the individuals that comprise the team, such as attitudes, motivations, knowledge, skills, and abilities. Team characteristics are those variables related to how teams are composed. They include factors such as cohesiveness, power structure, and member homogeneity. Task characteristics are comprised of a variety of factors related to the specific task the team is asked to perform. Variables in this category include task type, task complexity, and workload. The final class of input variables — work characteristics — refers to factors that influence how the team accomplishes their task. Examples are work structure, team norms, and communication structure.

These input factors are thought to converge to determine team processes, which includes the communication, coordination, and cooperation among team members in executing the task. The model also considers several aspects of training, such as the manner in which task analysis is conducted, the actual design of training, and principles of team learning. Finally, several types of performance outcomes must be considered. These include the quality and quantity of outcomes, as well as errors and time to completion.

We next describe the degree to which automation might interact with each of the characteristics described in the TEM in order to identify those areas about which there is sufficient research as well as those areas about which

TEAM EFFECTIVENESS MODEL

(adapted from Tannenbaum, Beard, & Salas, In press)

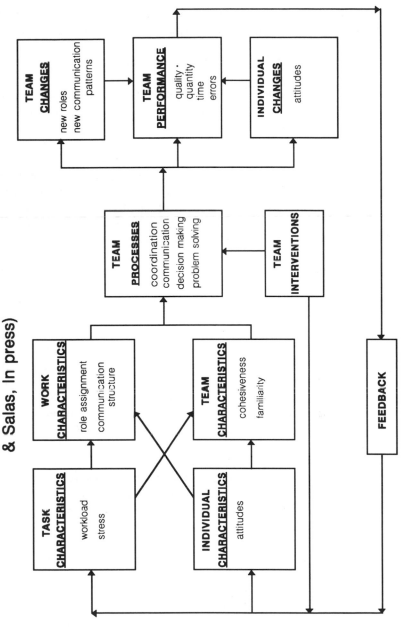

FIG. 12.1. Team effectiveness model (adapted from Tannenbaum et al., 1992).

little is known. In each case, we will summarize the available literature and discuss its implications for team performance in automated systems.

Organizational and Environmental Characteristics

Researchers have discussed the importance of only a few environmental factors. One such factor is the "philosophy of automation" adopted by the workplace (cf. Wiener, 1985). A well-articulated philosophy of automation describes the organization's approach to the acquisition, training, and utilization of high technology. As described by Wiener (1993), the organization's philosophy might be useful in reducing anxieties and misconceptions regarding automated systems. Furthermore, training programs designed to convey the company's philosophy might also serve to reduce informal or inappropriate policies regarding these systems. Although this provides an interesting set of hypotheses, this area has received no empirical investigation.

Closely related to the issue of philosophy of automation is the selection of tasks for automation. Ideally, automated systems are selected to relieve operators from excessive workload. The logical accomplishment of this goal will require a thorough analysis of the task and development of automated systems for critical subtasks. Some scientists have expressed concerns, however, that this reasoning is not frequently used in the design and deployment of automated technology. Rather, it has been suggested that decisions are often made based on the capability of technology rather than the needs of the operators (Graeber, 1989; Wiener, 1989, 1993).

Team tasks are at even greater risk for technology-driven automation development. These tasks are typically much more complicated, and require operators to process several subtasks concurrently, such as performing their individual responsibilities while communicating with other team members. It is likely that a substantial proportion of the workload associated with team tasks arises from a team member coordinating actions with his or her coworkers (Bowers, Braun, & Morgan, in press; Morgan & Bowers, 1995). However, it might not be the case that automating high-workload tasks necessarily results in improved coordination (cf. Thornton, Braun, Bowers, & Morgan, 1992). In fact, many automated systems might have the unintended consequence of interfering with team communication and coordination (Wiener, 1993). Thus, the logic underlying the introduction of automated technology to improve team performance might be faulty. There has not, as yet, been an effort dedicated to automating elements of team process to improve overall performance.

Another variable that is frequently considered under organizational characteristics is supervisory control. In many workplaces, some types of supervisory tasks are being accomplished by automated technologies. The most frequently studied area of automated supervision has been machine-paced work (see Hurrell & Colligan, 1986, for a review). The majority of published studies have indicated that the automation of task pacing is associated with reduced production and/or increased errors. For example, Salvendy and Humphreys compared performance on a marking–stapling task under self- or machine-paced conditions. They found that performance errors in the machine-paced condition were more than 300% more frequent that in the self-paced condition. There were no differences in production between the groups. There is a need to replicate this study in teamwork situations.

Automatic pacing of work might also impact other important team outcomes. For example, machine pacing has been associated with decreased job satisfaction (Frankenhauser & Gardell, 1976; Hurrell & Colligan, 1987), increased stress (Frankenhauser & Gardell, 1976), and increased absenteeism and job turnover associated (Ferguson, 1973; Fried, Weitman, & Davis, 1972).

These results might have important implications for team performance. Contemporary definitions of *team* typically include interdependence as a key element (i.e., Dyer, 1984; Morgan, Glickman, Woodard, Blaiwes, & Salas, 1986; Salas, Dickinson, Converse, & Tannenbaum, 1992). This often takes the form of what Saavedra, Earley, and Van Dyne (1993) referred to as "task interdependence." That is, the output of one member's task is a critical input factor for another member's task. If automatic task pacing does, in fact, contribute to slower individual performance, this delay might be magnified manyfold in large, highly interdependent teams.

Automatic task pacing might also impose a risk of interrupting team process. The negative reaction to task pacing is often due to operator's complaints of boredom (Frankenhauser & Gardell, 1976). It might be that the resulting boredom might place teams at risk to use communication about nontask-related issues as a way of relieving this monotony. Other studies have reported difficulties in communication associated with automatic pacing (cf. Frankenhauser & Gardell, 1976). This might be due to the automatic system being unable to assess and react to teamwork needs in the task. In any event, the available data converge to suggest that automatic pacing is likely to have a negative impact on team process and performance. Unfortunately, there is extremely little data regarding automation and team performance with which to evaluate these hypotheses.

Individual Characteristics

Several individual characteristics might interact with automation to affect team performance. One element that seems important in this regard is trust. It has been suggested that there are likely to be considerable individual differences in the degree to which people trust automatic systems (i.e., Sheridan, 1991). Researchers have been concerned that too much trust in automatic systems might lead to complacency, whereas inordinate skepticism might lead to an unwillingness to use automatic systems to relieve high workload.

Reports of mishaps in highly automated aircraft serve as anecdotal evidence regarding overreliance on automated systems. For example, Poduval (1989) described the case of a DC-10 that overran the runway on landing. The plane landed at a speed that was much higher than advised for the approach. The high speed was apparently due to crew's willingness to allow the automatic throttle control to supervise the approach, even though the system had been reported as malfunctioning. Several other incidents have also been attributed to unwise use of automated systems, contributing to concerns about too much trust in automatic systems.

Despite the concerns about individual trust, there has been relatively little empirical research regarding this important variable. In a survey, Taylor (1989) failed to find evidence of excessive trust of automatic systems, but researchers in this area have not yet investigated of the trust–performance relationship in teams. Based on limited evidence, it is likely that effects of trust in individual performance will also be represented in teams.

Another individual characteristic that seems to be important in understanding automation effects is experience. Experience in a complex system might give operators an advantage in recognizing malfunctions of the automatic system. On the other hand, positive experience with the automatic system might lead to excessive trust and, subsequently, complacency. This is another variable that is likely to be relevant to team composition. Yet, the empirical base simply does not allow conclusions about the role of this variable.

Several other variables, such as various personality factors, have received isolated research attention in automation studies. One such factor is locus of control. This individual characteristic has been a concern of designers of automatic systems because it seems reasonable to believe that subjects with external locus of control are at risk for allowing automatic systems to control the task (Sinclair, 1986). By the same token, individuals with an internal locus of control might feel uncomfortable in a position in which they are accountable for the automatic system. The small body of experimental literature seems to support the notion that locus of control is an important element in performance. For example, subjects with an internal locus of control were more negatively impacted by automatic pacing in an inspection task relative to those with an external locus (Eskew & Riches, 1982).

These individual characteristics have important implications for team performance. Team members are constantly confronted with the need to balance a variety of task demands. A considerable portion of team workload is associated with allocating tasks among the members of the team so as to optimize the team's resources (Bowers, Braun, & Morgan, in press). Automatic systems offer particular promise in this regard. They might serve as convenient "team members" to whom tasks can be assigned, with no negotiation, during periods of high workload. However, the individual characteristics described previously all seem to place teams at risk for overreliance on automatic systems. If members become too trusting of automation, they are apt to "sit and watch" the system rather than taking an active role. In the case of teams, this might be particularly problematic, because team members might lose their situation awareness, or sense of status relative to the goal. In the event that the automatic system fails, the information needed to recreate this situational awareness is likely to be distributed across team members. Considerable communication and time might be required before the team can accurately diagnose their new state and figure out how to react to it (Bowers, Braun, & Kline, 1994; Idaszak, 1989). In high time-pressured, dynamic situations, this delay could lead to severe negative outcomes. Consequently, a goal for training researchers will be to assist teams in allocating workload to the automatic system appropriately, especially teams comprised of members who might be prone to overusing such equipment.

Team Characteristics

One team characteristic that has drawn a considerable amount of interest with the advent of automated technology is team size. When advanced technology systems became available, it seemed reasonable to believe that they might provide a reliable, inexpensive alternative to actual crew members. This issue has been addressed most actively in aviation, where one might hypothesize that crew sizes could drop with no appreciable performance change. After considerable discussion, The President's Task Force on Aircraft Crew Complement (McLucas, Drinkwater, & Leaf, 1981) ruled that modern aircraft could be operated successfully with only two crew members. In fact, Wiener argued that the technology is available to allow commercial aircraft to be flown by a crew of one, although such a change is unlikely (Wiener & Curry, 1980). There has not as yet been an empirical investigation to explore the impact of size changes on eventual team performance.

Automation is also likely to interact with team composition to impact performance. Wiener (1989) argued that the presence of automated systems in aviation has had the effect of reducing the authority gradient between pilots and first officers. In other settings, different types of crews might react differentially to the introduction of automated systems. Wickens and his colleagues have tested the effects of crew composition in a simulated flight task (Wickens et al., 1989). They composed heterogeneous and homogeneous crews based on flight experience. Each crew flew a simulated mission under manual and automatic flight control conditions. The results indicated that, across conditions, hetero- and homogeneous crews demonstrated no difference in flying proficiency. However, the data indicated that heterogeneous crews obtained a greater benefit from automation than did homogeneous crews. Wickens and his colleagues interpreted these results as support for Wiener's hypothesized reduction in the authority gradient resulting from automation. Additional analyses reported by Straus and Cooper (1989) indicate that heterogeneous crews' communication improved in the automated condition. This improved team process might be a manifestation of the reduction in authority differences.

As demonstrated previously, factors related to team characteristics are likely to be important determinants of performance in automated systems. However, these issues have received extremely little research attention, and almost no attention at all outside of aviation psychology. Yet, team composition and staffing will continue to be an important issue in the modern workplace and psychology has little to contribute about how best to build teams to perform effectively in advanced technology systems. This area is in urgent need of empirical research.

Task Characteristics

Automated systems have further altered the type of task performed by operators from one involving direct control of the system to one requiring monitoring and supervisory control and management of the systems (Brown, Boff, & Swierenga, 1991; Lee & Moray, 1992; Rasmussen, 1981; Sheridan, 1987; Wiener & Curry, 1980). Operators in complex systems are often faced with situations that require a transition between evaluation of automated system performance (i.e., monitoring) and appropriate intervention when the automation fails (Lee & Moray, 1992). Harris, Hancock, and Arthur (1993) stated that a challenge for automated systems is to effectively facilitate the operator transition between the passive monitoring and active control. Operators must be able to switch between automation and manual control in order to realize the full benefits of the advanced technology.

Sarter (1991) investigated the nature of operator intervention with automated system during normal and abnormal conditions. During normal conditions operators appeared to perform effectively. However, during abnormal conditions, operators did not always react appropriately to the failure of automated systems. Operators sometimes intervened prematurely, failed to notice system anomalies, or used improper procedures. There is a general agreement among researchers that the nature of these two types of monitoring and supervisory control tasks must be better understood to successfully integrate automated systems into the workplace (Kessel & Wickens, 1982).

Team tasks are likely to compound this problem. The requirement for monitoring of other team members seems to be an important determinant of team performance (McIntyre & Salas, 1995). Automated systems appear to increase these demands for monitoring significantly. Cannon-Bowers and her colleagues suggested that in order for team members to function effectively, they must have an accurate mental model of the equipment, the task, and the team (Cannon-Bowers, Salas, & Converse, 1993). The models of each member must be in order for effective performance to result. The results discussed earlier indicate that there will be an additional requirement for a model of the automatic system. This model, too, must be shared among members to allow effective performance. How best to engender these models remains an important area for research.

Automated systems with decision-aiding capabilities are also expected to dramatically alter the organization and allocation of tasks in complex systems. Thus far, the success of these systems has been limited (Andes & Rouse, 1991; Noble, 1993). For example, Barnett (1990) found no significant difference in either decision optimality or response speed for individuals who received decision aiding at either the situation assessment or response generation stage of decision making process. However, improved judgments with spatial decision aiding were found in one condition characterized by high time stress. Crocoll and Coury (1990) found enhanced performance when individuals were aided with accurate status or recommendation information from the automated system. Research regarding decision aiding in teams will be required before these systems can be useful in many settings. Team decision making imposes additional requirements on decision makers because information is often distributed among several members. Techniques for integrating this information must be derived before automated systems will be useful in this environment.

Although automation may eliminate or substantially reduce some previously performed tasks due to system reliability, a number of new tasks can also be imposed on the crew (e.g., programming flight management systems) as a result of automated systems. Some of the changes in these systems can increase the complexity of the task. A number of investigations have suggested that even experienced operators may be surprised or unaware of automated system capabilities and mode awareness (e.g., Sarter & Woods, 1991; Wiener, 1989). Furthermore, pilots using flight management systems found themselves in situations where things were happening that surprised them (McClumpha, James, Green, & Belyavin, 1991; Sarter & Woods, 1991; Wiener, 1989, 1993) This situation creates ambiguity related to the task, thereby increasing task complexity. This ambuity might be particulalrly troublesome in team situations. Researchers have recently begun to discuss the manner by which teams aquire and maintain situational awareness (Shresta, Prince, Baker, & Salas, 1995). It might well be that the ambiguous distribution of information and system capabilities across team members will result in a decreased ability to form this "mental picture" of the situation.

Similarly, Degani, Chappell, and Hayes (1991) noted that advancements in technology have changed the nature of the information presented to the operators. Although availability of increased information can sometimes facilitate performance, operators must be capable of effectively managing the data in order to avoid potential overload. Braune, Hofer, and Dresel (1991) reported concerns that there may be instances when there is too much or too little information, or that the available information is hard to find. In team performance situations, the need to transfer data and other resources among members is a key subtask associated with effective performance. As the amount of data available to operators increases, so will the demand on team members to parse and distribute information appropriately. This is likely to increase team workload and might have a negative impact on performance.

The effect of automation on workload has received more research attention as compared to other task characteristics. Unfortunately, these studies have yielded mixed results. For example, Curry (1985) and Wiener (1989, 1993) found that pilots were almost evenly divided between those who felt that flightdeck automation reduces workload and those who believed that automation increased workload.

Harris, Hancock, and Arthur (1993) investigated whether the introduction of automation significantly affected workload during situations of low and increasing taskload. They found no significant difference in perceived workload in operators using automated or nonautomated systems during periods of low taskload. As taskload increased, greater workload was reported by operators using automated systems. These results suggest that the workload reduction provided by automatic systems might be quite small and that automation might actually increase workload in some situations.

Laudeman and Palmer (1992) investigated the workload of operators performing in traditional cockpits and automated cockpits. Crews that performed in traditional cockpits perceived lower workload levels. The researchers also compared the workload levels of crews in automated cockpits who either chose or declined to use the automated features. Crews that used the features perceived higher levels of workload than did crews that did not use these systems. These findings are in clear opposition with the goal of automated systems (i.e., to reduce operator workload).

Guide, Wise, Abbott, and Ryan (1993) surveyed pilots to better understand the relationship between automation and workload during various phases of flight. Although many pilots felt automation reduced workload, others thought that it had been dramatically increased, and often at the wrong time (e.g., during periods in which workload was already high). For example, pilots flying automated aircraft viewed workload during approach and landing to be higher than did pilots flying less automated aircraft. These perceptions were supported by a simulator study described by Wiener (1989).

James, McClumpha, Green, Wilson, and Belyavin (1991) and McClumpha et al. (1991) surveyed operational pilots to assess the relationship between automation and workload. The results of the surveys suggested that pilots strongly felt that the overall workload on advanced technology flight decks is much lower than nonautomated flightdecks. One interesting finding of these investigations was that when automated systems were present, the perceived workload was rated significantly lower by captains than by first officers. This result suggests that the effect of automation on workload may be partially position dependent. The workload for captains may be reduced because the automated systems eliminated tasks associated with overall system and crew management. In comparison, the workload for first officers may be increased because of the requirement to operate the automated systems and to monitor the performance of the automated systems.

The findings of the previously discussed investigations indicate that automation might not reduce workload as significantly, or consistently, as hoped. Braune, Infield, Harper, and Alter (1993) reviewed literature related to the effect of automation on workload. The authors summarized the limited research and concluded that the available literature suggested that it was not clear if flightdeck automation helped to significantly reduce workload or if the automation actually simply shifted workload across crew positions. One reason for the mixed results involves the different levels of analysis used by the researchers (i.e., workload issues in general versus workload during specific phases of flight, laboratory versus questionnaire data). Collectively, these results provide intriguing research hyptheses for teams. It seems likely that the effect of automation on team workload is dependent on the nature of the automatic system and the subtask that it impacts (Jentsch et al., 1995). However, until the automation–performance relationship is better understood for individuals, it will be difficult to make reasonable predictions about team effects.

Work Characteristics

An important variable in this category is work assignment. Work assignment refers to the way work is formally and informally delegated and assigned to the various team members. Automation has changed the work assignments found in teams. Wiener (1989) pointed out that the introduction of automation in aviation systems might alter the traditional roles and responsibilities of team members. The nature and degree of assignment changes due to automation is unclear. In one of the few empirical studies, Lozito, McGann, and Corker (1993) found that although the use of the automated system altered the behavior of crews in handling ATC clearance information, it was not associated with assignment of other duties. These results indicate that the behavior changes in operators in response to automated systems might be subtle and relatively specific. Additional studies in this area are required to understand these process changes.

Another interesting characteristic is team norms. Tosi, Rizzo, and Carroll (1990) defined team norms as shared expectations about behavior that apply to all members. Norms are likely to emerge regarding behaviors that are central to the group and are often reflected in rules and procedures. The introduction of automation can affect the development and nature of norms in teams. For example, Degani and Wiener (1994) noted that procedures established for complex systems are often inadequate. It is difficult to mandate a set of stringent procedures for the utilization of automation because interaction with the system occurs in a dynamic, and sometimes unpredictable, environment. Furthermore, although procedures associated with manual control of systems can be specified in considerable detail, procedures related to system management are more elusive. Automation also alters the sequencing of existing procedures. For example, Lozito et al. (1993) found that the use of the automated system altered the order and timing of procedures that crews used when handling ATC clearance information.

In another investigation of procedures and automation, Palmer and Degani (1991) studied the manner in which pilots performed checklists using paper-based versus electronic checklists. The researchers found that pilots using electronic checklists did not use the traditional challenge–response interaction associated with paper-based checklists. The electronic checklist appears to encourage flight crews not to conduct their own checks. Because this represents a substantial departure from the traditional execution of the task, research is required to determine how crews respond to this change.

Some team norms are formally stated in standards and procedures, whereas other norms may be more informal. These informal norms include those established during an initial face-to-face interaction prior to task execution (Farthofer & Kemmler, 1993; Ginnett, 1987), or those based on seniority, status, qualifications, or personality of the individuals. There is a lack of research that examines whether team norms associated with automated systems can also be originated prior to task performance.

Determining the nature of these differences is an important requirement for team training research.

Team Processes

One of the central areas regarding automation and team performance is the degree to which automation negatively effects team process. Scientists in this area have expressed concerns that automation might change the nature of a team's task to such a degree that significant process changes will be required for the team to perform adequately (Brown et al., 1991; Morgan, Herschler, Wiener, & Salas, 1993; Wiener, 1993). For example, Wiener (1989) described several changes associated with advanced technology that are likely to have unintended effects on team communication and coordination: (a) it is physically difficult for crew members to observe the actions of fellow crew members, (b) there is a breakdown of traditional flight roles, and (c) there is a new demand for crew members to help one another interact with certain automatic systems.

Because of these concerns, team process changes have received a considerable amount of attention from researchers, especially in the realm of aviation where recent interest in cockpit resource management has heightened awareness of the importance of effective communication. In an early study, Costley and his colleagues compared the communications in the traditional B737-200 aircraft, the B737-300 with a moderate level of advanced systems, and the highly advanced B757 aircraft (Costley, Johnson, & Lawson, 1989). They concluded that there was a trend toward lower rates of communication with advancing automation of the aircraft. Furthermore, the authors noted a tendency for communication between crew members via the automatic system, rather than person-to-person transmission. However, it is important to realize that there are a number of differences between these aircraft other than automatic systems. It might be that some of these communication differences are attributable to these factors rather than simply automation effects (Wiener, 1993).

Data reported by Clothier (1991) also indicate that automation influences crew performance in some situations. This study investigated the coordination in advanced and standard cockpits using both simulated and actual flights. The results indicated that automated cockpits were associated with higher-quality communication during the simulated flights, yet no difference existed in actual flights. These results might be due to the fact that the simulated flights used a scenario that imposed a great deal of workload. It might be that the advantages of automation are only salient during periods in which the workload savings are particularly pronounced.

Wiener and his colleagues recently reported the initial results of an extensive research effort to investigate the effects of automation on crew coordination and communication (Wiener et al., 1991). They compared the performance of crews flying the traditional DC-9 and the advanced MD-88 aircraft. Interestingly, they found that the more automated aircraft was associated with higher perceived workload. Furthermore, automation did not contribute to improved crew performance. In fact, the data suggested a mild advantage for crews in the traditional cockpit. Subjective evaluations of CRM quality yielded no differences between the two airframes.

A subsequent analysis of these data by Veinott and Irwin (1993) demonstrated that crew members in the advanced technology aircraft communicated at a higher rate. Furthermore, crews in the highly automated aircraft asked more questions, whereas crews in the less automated aircraft showed a greater reliance on command-acknowledgment sequences. This response to automation indicates that rather than reducing workload, automatic systems seem to result in redistributed workload. This altered pattern of communication might also contribute to the perception of redistributed authority in the cockpit.

Bowers and his colleagues recently described a laboratory study in which the effects of automation on communication were assessed during a simulated flight (Bowers, Deaton, Oser, Prince, & Kolb, 1993). Specifically, they were interested in the degree to which automated systems might make resources available that would contribute to more effective decision making. Once again, automation was not related to more effective performance. However, there were some interesting communication differences. In manual aircraft, there was little distinction between the communication of effective and ineffective performers. However, poor performers in the automated condition displayed significantly higher frequencies of unsolicited observations and responses to these observations. Several other communication differences converged to suggest that the nature of effective team processes might be quite different in automatic systems.

As demonstrated previously, the introduction of automatic systems is frequently associated with process changes by teams. Like many other factors in team performance, teams must accommodate to the new system and develop process behaviors that allow optimum performance in the new situation. Clearly, these process behaviors are only now becoming understood. Considerable research will be required before the process–performance relationship in automated systems can be understood. Furthermore, because most of the available research has been conducted in aviation, there is a need to determine the degree to which these results generalize to other tasks.

Another important effect of automation on the communication structures in the team involves the requirement for operators to actively interact with the automated system. Operators must direct the automation to perform tasks by programming the system via data entry and interfaces. As new automation technologies, currently under development and test, are introduced in complex systems, it is likely to further alter the nature of communication structures. One example of such an emerging technology is data link air traffic control. This technology enables crews to receive, review, and enter ATC clearance information in the aircraft system without a requirement for voice transmission.

Lozito et al. (1993) investigated how the introduction of automated air traffic control information via data link affected the communication structure of crews performing simulated scenarios. The results of preliminary investigations involving data link air traffic control demonstrated that crews using traditional verbal communication modalities demonstrated different patterns of interaction with air traffic control as compared to crews using electronic data link systems. Also, crews using traditional verbal communications initiated more exchanges with air traffic control as compared to crews using automated systems. Crews communicating via data links took longer to acknowledge ATC messages than did crews using traditional voice acknowledgment. These results suggest that automation might be an important determinant of the processes teams adopt to complete their tasks.

Although these technologies clearly impact the communication structure between pilots and external agencies, it is likely that the automated data link systems will also change the nature of interaction within the aircraft as well. A somewhat more subtle effect of automation is related to the communication structures involving nonverbal interactions and interactions with the equipment. Segal (1989, 1990) noted that new technology changes the nature of important nonverbal cues demonstrated in complex systems. An important nonverbal source of information is provided when operators interact with displays and controls in a manner that is apparent to other operators. Automated systems, such as multifunctional displays, have the potential of modifying the nature of information provided by this type of nonverbal information. Segal (1989) noted that the information available to other team members may be altered as automation reduces the control actions of operators.

Although it is generally agreed that automation affects the communication structure of teams operating complex systems, a number of important questions remain unanswered. These questions include: How do the communication structures in teams using automated systems differ from the communication structures in teams not using automated systems? What are the best methods to structure communication for teams performing with automated systems? To what extent do communication structures change when crews transition from automated to nonautomated conditions?

AUTOMATION AND TEAM PERFORMANCE: DIRECTIONS FOR THE FUTURE

We began this review with the goal of identifying areas in which there were established trends in the relationship between automation and team performance. In so doing, we also hoped to reveal critical areas for new research. Unfortunately, the review indicated that the current knowledge base regarding automation and team performance is woefully inadequate. Despite the clear implications of individual research, there has been so little research with teams that it is impossible to predict whether similar patterns of results will be obtained. The issue seems to be receiving attention in aviation psychology, which appears to be something of a mixed blessing. Although the progress in team research is certainly gratifying, there is a concern that users will apply these findings to other settings. Yet, there are no data by which to assess the degree to which task type might influence the automation–performance relationship.

The state-of-the-art described previously is especially troubling because teams are being asked to perform critical tasks in automated systems every day. There can be little question that, as far as teams are concerned, system development has been technology-driven. Teams are often placed in these situations to "sink or swim," despite the many published results suggesting that automated systems are not always helpful. This problem results because effects on team process and performance are often only considered after the fact. In other words, by the time researchers can determine the effects of such systems on team performance, the technologies are already firmly entrenched in the larger system. Because changing the automated system is often impossible, the maintenance of team performance becomes a rather formidable problem for training developers.

This reactive approach to the study of automation will ensure that science will trail technology by several years. Thus, there is likely to be little opportunity to prevent the unforeseen, potentially catastrophic consequences of poor performance in these systems. Rather, behavioral scientists will be remanded to the rather unsatisfying duty of attempting to understand why the performance deficit occurred.

Consequently, there is a need for an aggressive program of study dedicated to the effects of automation of team performance. Although there are many research needs in this area, a few preliminary hypotheses can be derived from this review. Some of these are:

1. Team performance is likely to be superior in organizations that have a clearly specified philosophy of automation.
2. Automatic task pacing is likely to have a negative effect on the productivity of manufacturing teams.

3. Excessive trust, external locus of control, and positive experience will all be related to a team's overuse of automatic systems.
4. The effects of automation on team performance will be most salient in high workload situations.
5. Teams in automatic systems will have to communicate differently to maintain effective performance.
6. Teams working with automatic systems might benefit from briefings designed to articulate new team norms.

Perhaps more important, we suggest that scientists in team performance can take a more proactive approach in the area of automation through a paradigm shift. Rather than ask "What does this automatic system do to team performance?" we suggest that the more appropriate question is "How can we develop systems to support team performance?" This distinction is an important one, and is similar to the notion of specialized team architectures discussed by Urban and her colleagues (Urban, Bowers, Cannon-Bowers, & Salas, 1995). This approach has also been recommended for the design of high speed aircraft by Palmer and his colleagues (Palmer, Rogers, Press, Latorella, & Abbott, 1995). There is a sufficient data base in team performance research to allow prediction of a team's needs in certain situations. Based on these results, behavioral scientists can assist in the development of advanced technology that is designed with the expressed purpose of supporting teams. For example, previous research has demonstrated that process behaviors tend to degrade under conditions of high workload. Surprisingly, however, automating other aspects of the task does not necessarily allow team processes to be maintained (Bowers et al., in press). It might be that modeling approaches currently used in research could serve as the basis for automating team processes themselves, allowing teams to perform more effectively.

In conclusion, the requirements for teams to perform in automated systems is almost guaranteed to increase with time. Thus far, the literature indicates that performance in these systems is not guaranteed to be associated with improvements in the quality or quantity of work. In fact, teams might suffer from several unforeseen, unfortunate consequences of these systems. Therefore, there is a need for an aggressive program of research dedicated to the understanding of these systems. We hope that this review serves as a catalyst for such research.

DISCLAIMER

The views expressed herein are those of the authors and do not reflect the official positions of the organizations with which the authors are affiliated.

REFERENCES

Andes, R. C., & Rouse, W. B. (1991). Specification of adaptive aiding systems: Information requirements for designers. *Proceedings of the Sixth International Symposium on Aviation Psychology* (pp. 120–125). Columbus: Ohio State University.

Barnett, B. J. (1990). Aiding type and format compatibility for decision aid interface design. *Proceedings of the Human Factors Society 34th Annual Meeting* (pp. 1552–1556). Santa Monica, CA: Human Factors Society.

Bowers, C. A., Braun, C. C., & Kline, P. B. (1994). Communication and team situational awareness. In R. D. Gilson, D. J. Garland, & J. M. Koonce (Eds.), *Situational awareness in complex systems* (pp. 305–311). Daytona Beach, FL: Embry-Riddle Aeronautical University Press.

Bowers, C. A., Braun, C. C., & Morgan, B. B., Jr. (in press). Team workload: Its meaning and measurement. In M. T. Brannick, E. Salas, & C. Prince (Eds.), *Team performance assessment and measurement: Theory, research, and applications*. Mahwah, NJ: Lawrence Erlbaum Associates.

Bowers, C., Deaton, J., Oser, R., Prince, C., & Kolb, M. (1993). The impact of automation on crew communication and performance. *Proceedings of the Seventh International Aviation Psychology* (pp. 758–761). Columbus: Ohio State University.

Braune, R. J., Hofer, E. F., & Dresel, K. M. (1991). Flight deck information management — a challenge to commercial transport aviation. *Proceedings of the Sixth International Symposium on Aviation Psychology* (pp. 78–84). Columbus: Ohio State University.

Braune, R. J., Infield, S. E., Harper, P. M., & Alter, K. W. (1993). Human-centered requirements for advanced commercial transport flight decks. *Proceedings of the Seventh International Symposium on Aviation Psychology* (pp. 88–92). Columbus: Ohio State University.

Brown, C. E., Boff, K. R., & Swierenga, S. J. (1991). Cockpit resource management: A social psychological perspective. *Proceedings of the Sixth International Symposium on Aviation Psychology* (pp. 398–403). Columbus: Ohio State University.

Cannon-Bowers, J. A., Salas, E., & Converse, S. (1993). Shared mental models in team decision making. In J. C. Castellan (Ed.), *Individual and group decision making* (pp. 221–246). Hillsdale, NJ: Lawrence Erlbaum Associates.

Clothier, C. (1991). Behavioral interactions across various aircraft types: Results of systematic observations of line operations and simulations. *Proceedings of the Sixth International Symposium on Aviation Psychology* (pp. 332–337). Columbus: Ohio State University.

Costley, J., Johnson, D., & Lawson, D. (1989). A comparison of cockpit communication B737-B757. *Proceedings of the Fifth International Symposium on Aviation Psychology* (pp. 413–418). Columbus: Ohio State University.

Crocoll, W. M., & Coury, B. G. (1990). Status or recommendation: Selecting the type of information for decision aiding. *Proceedings of the Human Factors Society 34th Annual Meeting* (pp. 1524–1528). Santa Monica, CA: Human Factors Society.

Curry, R. E. (1985). *The introduction of new cockpit technology: A human factors study* (Tech. Memorandum 86659). Moffett Field, CA: NASA-Ames Research Center.

Degani, A., Chappell, S. L., & Hayes, M. S. (1991). Who or what saved the day? A comparison of traditional and glass cockpits. *Proceedings of the Sixth International Symposium on Aviation Psychology* (pp. 227–234). Columbus: Ohio State University.

Degani, A., & Wiener, E. L. (1994). *On the design of flight-deck procedures* (NASA Contractor Rep. No. 177642). Moffett Field, CA: NASA-Ames Research Center.

Dyer, J. C. (1984). Team research and team training: State-of-the-art-review. In F. A. Muckler (Ed.), *Human factors review* (pp. 285–323). Santa Monica, CA: Human Factors Society.

Eskew, R. T., & Riche, C. V. (1982). Pacing and locus of control in quality control inspection. *Human Factors, 24*(4), 411–415.

Farthofer, A., & Kemmler, R. W. (1993). Leadership behavior in high-tech-cockpits (A 320). *Proceedings of the Seventh International Symposium on Aviation Psychology* (pp. 522–526). Columbus: Ohio State University.

Ferguson, D. (1973). A study of occupational stress and health. *Ergonomics, 16*(5), 649–664.

Frankenhaeuser, M., & Gardell, B. (1976). Underload and overload in working life: Outline of a multidisciplinary approach. *Journal of Human Stress, 2*(3), 35–46.

Fried, J., Weitman, M., & Davis, M. K. (1972). Man-machine interaction and absenteeism. *Journal of Applied Psychology, 56*(5), 428–429.

Ginnett, R. C. (1987). *First encounters of the close kind: The formation process of airline flight crews.* Unpublished doctoral dissertation, Yale University, New Haven, CT.

Graeber, R. C. (1989). Long-range operations in the glass cockpit: Vigilance, boredom and sleepless nights. In A. Coblentz (Ed.), *Vigilance and performance in automatized systems / Vigilance et performance de l'homme dans les systemes automatises. NATO Advanced Science Institutes series. Series D: Behavioural and social sciences, 49,* 67–76.

Guide, P. C., Wise, J. A., Abbott, D. W., & Ryan, L. J. (1993). The opinions of pilots flying automated corporate aircraft with regard to their perceived workload. *Proceedings of the Seventh International Symposium on Aviation Psychology* (pp. 849–853). Columbus: Ohio State University.

Harris, W. C., Hancock, P. A., & Arthur, E. J. (1993). The effect of taskload projection on automation use, performance, and workload. *Proceedings of the Seventh International Symposium on Aviation Psychology* (pp. 890a–e). Columbus: Ohio State University.

Hurrell, J. J., & Colligan, M. J. (1986). Machine pacing and shiftwork: Evidence for job stress. *Journal of Organizational Behavior Management, 8*(2), 159–175.

Idaszak, J. R. (1989). Human operators in automated systems: The impact of active participation and communication. *Proceedings of the Human Factors Society 33rd Annual Meeting* (pp. 778–782). Santa Monica, CA: Human Factors Society.

James, M., McClumpha, A., Green, R., Wilson, P., & Belyavin, A. (1991). Pilot attitudes to flight deck automation. *Proceedings of the Sixth International Symposium on Aviation Psychology* (pp. 192–197). Columbus: Ohio State University.

Jentsch, F., Bowers, C., Redshaw, B., Bergen, H., Henning, J., & Holmes, B. (1995). *Differential effects of automation on team performance, workload, and communications* (Tech. Rep. for the Naval Air Warfare Center Training Systems Division). Orlando, FL: University of Central Florida.

Kessel, C. J., & Wickens, C. D. (1982). The transfer of failure-detection skills between monitoring and controlling dynamic systems. *Human Factors, 24*(1), 49–60.

Laudeman, I. V., & Palmer, E. A. (1992). Measurement of automation effects on aircrew workload. *Third Annual AS/A Program Investigator's Meeting.* Moffett Field, CA: NASA-Ames Research Center.

Lee, J. D., & Moray, N. (1992). Operators' monitoring patterns and fault recovery in the supervisory control of a semi-automatic process. *Proceedings of the Human Factors Society 36th Annual Meeting* (pp. 1143–1147). Santa Monica, CA: Human Factors Society.

Lozito, S., McGann, S., & Corker, K. (1993). Data link air traffic control and flight deck environments: Experiment in flight crew performance. *Proceedings of the Seventh International Symposium on Aviation Psychology* (pp. 1009–1015). Columbus: Ohio State University.

McClumpha, A. J., James, M., Green, R. G., & Belyavin, A. J. (1991). Pilots' attitudes to cockpit automation. *Proceedings of the Human Factors Society 35th Annual Meeting* (pp. 107–111). Santa Monica, CA: Human Factors Society.

McIntyre, R. M., & Salas, E., (1995). Measuring and managing for team performance: Emerging principles from complex environments. In R. Guzzo & E. Salas (Eds.), *Team effectiveness and decision making in organizations* (pp. 149–203). San Francisco, CA: Jossey-Bass.

McLucas, J., Drinkwater, F., & Leaf, H. (1981). *Report of the President's task force on aircraft crew complement.*

Morgan, B. B., Jr. & Bowers, C. A. (1995). Teamwork stressors and their implications for team decision making. In R. A. Guzzo & E. Salas (Eds.), *Team decision making effectiveness in organizations.* New York, NY: Jossey-Bass.

Morgan, B. B., Jr., Glickman, A. S., Woodard, E. A., Blaiwes, A. S., & Salas, E. (1986). *Measurement of team behaviors in a navy environment* (NTSC Tech. Rep. 86–014). Orlando, FL: Naval Training Systems Center.

Morgan, B. B., Jr., Herschler, D. A., Wiener, E. L., & Salas, E. (1993). Implications of automation technology for aircrew coordination performance. In W. B. Rouse (Ed.), *Human/technology interaction in complex systems* (Vol. 6, pp. 105–136). Greenwich, CT: JAI Press.

Mouloua, M., & Parasuraman, R. (1994). *Human performance in automated systems: Current research and trends.* Hillsdale, NJ: Lawrence Erlbaum Associates.

Noble, D. (1993). A model to support development of situation assessment aids. In G. A. Klein, J., Orasanu, R. Calderwood, & C. E. Zsambok (Eds.), *Decision making in action: Models and methods* (pp. 287–305). Norwood, NJ: Ablex.

Palmer, E., & Degani, A. (1991). Electronic checklists: Evaluation of two levels of automation. *Proceedings of the Sixth International Symposium on Aviation Psychology* (pp. 178–183). Columbus: Ohio State University.

Palmer, M. T., Rogers, W. H., Press, H. N., Latorella, K. A., & Abbott, T. S. (1995). *A crew-centered flight deck design philosophy for High Speed Civil Transport (HSCT) aircraft.* (NASA Tech. Memorandum 109171). Hampton, VA: NASA Langley Research Center.

Poduval, A. (1989). Glass cockpits and ab initio pilots. *Proceedings of the Fifth International Symposium on Aviation Psychology* (pp. 97–103). Columbus: Ohio State University.

Rasmussen, J. (1981). Models of mental strategies in process control. In J. Rasmussen & W. B. Rouse (Eds.), *Human detection and diagnosis of system failures* (pp. 241–258). New York: Plenum Press.

Saavedra, R., Earley, P. C., & Van Dyne, L. (1993). Complex interdependence in task-performing groups. *Journal of Applied Psychology, 78,* 61–72.

Salas, E., Dickinson, T. L., Converse, S. A., & Tannenbaum, S. I. (1992). Toward an understanding of team performance and training. In R. Swezey & E. Salas (Eds.), *Teams: Their training and performance* (pp. 3–29). Norwood, NJ: Ablex.

Sarter, N. (1991). The flight management system: Pilots' interaction with cockpit automation. *Proceedings of the International Conference on Systems, Man, and Cybernetics* (pp. 1307–1310). Boston, MA: IEEE.

Sarter, N. B., & Woods, D. D. (1991). The flight management system: "Rumors and facts." *Proceedings of the Sixth International Symposium on Aviation Psychology* (pp. 241–246). Columbus: Ohio State University.

Segal, L. (1989). Differences in cockpit communication. In *Proceedings of the Fifth International Symposium on Aviation Psychology* (pp. 576–581). Columbus: Ohio State University.

Segal, L. (1990). Effects of aircraft cockpit design on crew communications. In E. J. Lovesey, (Ed.), *Contemporary ergonomics 1990* (pp. 247–252). London: Taylor & Francis.

Sheridan, T. (1991). Theoretical approaches to human centered automation. *Challenges in aviation human factors: The national plan* (pp. 15–17). Vienna, VA: American Institute of Aeronautics and Astronautics.

Sheridan, T. B. (1987). Supervisory control. In G. Salvendy (Ed.), *Handbook of human factors/ergonomics* (pp. 1243–1268). New York: Wiley.

Shresta, L. B., Prince, C., Baker, D. P., & Salas, E. (1995). Understanding situation awareness: Concepts, methods, and training. In W. B. Rouse (Ed.), *Human/technology interaction: Volume 7.* Greenwich, CT: JAI Press.

Sinclair, M. A. (1986). Ergonomics aspects of the automated factory. *Ergonomics, 29*(12), 1507–1523.

Straus, S., & Cooper, R. S. (1989). Crew structure, automation, and communications. *Proceedings of the Human Factors Society 33rd Annual Meeting* (pp. 783–787). Santa Monica, CA: Human Factors Society.

Tannenbaum, S. I., Beard, R. L., & Salas, E. (1992). Team building and its influence on team effectiveness: An examination of conceptual and empirical developments. In K. Kelley (Ed.), *Issues, theory, and research in industrial/organizational psychology* (pp. 117–153). Amsterdam: Elsevier.

Taylor, R. M. (1989). Trust and awareness in human electronic crew teamwork. In T. Emerson, J. M. Reising, R. M. Taylor, & M. Reinecke (Eds.), *The human-electronic crew: Can they work together?* (Tech. Rep. WRDC-TR-89-7008). Wright-Patterson Air Force Base, OH: Wright Research and Development Center.

Thornton, C., Braun, C. C., Bowers, C. A., & Morgan, B. B. (1992). Automation effects in the cockpit: A low-fidelity investigation. *Proceedings of the Human Factors Society 36th Annual Meeting* (pp. 30–36). Santa Monica, CA: Human Factors Society.

Tosi, H. L., Rizzo, J. R., & Carroll, S. J. (1990). *Managing organizational behavior.* New York: HarperCollins.

Urban, J. M., Bowers, C. A., Cannon-Bowers, J. A., & Salas, E. (1995). The importance of team architecture in understanding team processes. In M. Beyerlein (Ed.), *Advances in interdisciplinary studies in work teams, Vol. 2.* Greenwich, CT: JAI Press.

Veinott, E. S., & Irwin, C. M. (1993). Analysis of communication in the standard versus automated aircraft. *Proceedings of the Seventh International Symposium on Aviation Psychology* (pp. 584–588). Columbus: Ohio State University.

Wickens, C. D., Marsh, R., Raby, M., Straus, S., Cooper, R., Hulin, C. L., & Switzer, F. (1989). Aircrew performance as a function of automation and crew composition: A simulator study. *Proceedings of the Human Factors Society 33rd Annual Meeting* (pp. 792–796). Santa Monica, CA: Human Factors Society.

Wiener, E. L. (1985). *Human factors of cockpit automation: A field study of flight crew transition* (NASA-Ames Contractor Rep. 177333). Moffett Field, CA: NASA-Ames Research Center.

Wiener, E. L. (1989). *Human factors of advanced technology ("glass cockpit") transport aircraft* (NASA-Ames Contractor Rep. 177528). Moffett Field, CA: NASA-Ames Research Center.

Wiener, E. L. (1993). Crew coordination and training in the advanced-technology cockpit. In E. L. Wiener, B. G. Kanki, & R. L. Helmreich (Eds.), *Cockpit resource management* (pp. 199–229). San Diego: Academic Press.

Wiener, E. L., Chidester, T. R., Kanki, B. G., Palmer, E. A., Curry, R. E., & Gregorich, S. E. (1991). *The impact of cockpit automation on crew coordination and communication: I. Overview, LOFT evaluations, error severity, and questionnaire data* (NASA-Ames Contractor Rep. 177587), Moffett Field, CA: NASA-Ames Research Center.

Wiener, E. L., & Curry, R. E. (1980). Flight deck automation: Promises and problems. *Ergonomics, 23*, 995–1011.

III Applications

13 Cockpit Automation: From Quantity to Quality, From Individual Pilot to Multiple Agents

Nadine B. Sarter
University of Illinois at Urbana-Champaign

INTRODUCTION

Cockpit automation— a mixed blessing! Variations of this statement are ubiquitous in the recent literature on advanced flightdeck technology to indicate that although some of the expected benefits of automation have materialized, its introduction also created unexpected new problems and burdens for the operational community. But is it really cockpit automation per se that is the problem? Is it useful to talk about cockpit automation as though the term denotes a homogenous class of systems when, in fact, it comprises a large variety of systems, some of which differ fundamentally with respect to their objectives, abilities, limitations, and underlying design philosophies? Results of research on pilot–automation interaction seem to suggest that making statements about the merits and risks of cockpit automation in general misses the point that it is the specific design of an automated system and its compatibility and interaction with its human user that determine the nature, safety, and efficiency of overall system operations.

Warnings of potential problems with cockpit automation were voiced as early as the late 1970s (e.g., Edwards, 1977), and concerns have been fueled ever since by incidents and accidents involving automated aircraft (e.g., Lenorovitz, 1990; Sparaco, 1994), by difficulties that pilots experience during training and line operations (e.g., Eldredge, Dodd, & Mangold, 1991; Sarter & Woods, 1992), and by the results of empirical research looking at pilot–automation interaction (e.g., Parasuraman, Molloy, & Singh, 1993; Sarter & Woods, 1994; Wiener, 1989; Wiener et al., 1991).

Since the early days of cockpit automation, some of the major areas of concern have been its potential impact on pilot workload, pilot error, trust calibration, and pilots' manual flying skills (Norman & Orlady, 1989; Wickens, 1994; Wiener, 1988). Another recurrent topic of interest has been the increasing need for and, at the same time, potential loss of situation awareness on glass cockpit aircraft. Although these broad issues continued to be the focus of research from the first studies of cockpit automation in the early 1980s, the specific questions that were asked and the approaches that were taken to study those questions have changed considerably.

In this chapter, important concepts and concerns that evolved in and currently dominate the field of cockpit automation are discussed. In particular, two major trends in the research in this field are laid out: a shift in interest from issues of quantity to those of quality; and a shift in the unit of analysis from individual pilots to the entire crew, from the human element to the automated systems, and finally to the overall system including the human and machine agents in the cockpit and their coordination and cooperation on flight-related tasks.

Early concerns with cockpit automation centered around questions of quantity: Will automation reduce the amount of pilot workload? How much training does it require? How much information can we and do we need to present to pilots? With increasing insight into the nature of problems with pilot–automation interaction, however, it became clear that the more critical questions are related to issues of quality. Instead of asking "How much automation is acceptable?" it is important to define "What kind of automation do we want and need?" Similarly, the solution to problems with training for glass cockpits does not seem to be more training but a change in how pilots are trained for advanced flightdecks. And adding or removing information to or from cockpit CRTs is not as critical as understanding when and how to present information to the pilot.

This shift from issues of quantity to those of quality was paralleled by a shift in the unit of analysis of research on cockpit automation. Early attempts to answer questions about workload or trust in automation focused on the individual pilot in the cockpit. During the last decade, due to an increasing emphasis on the need for efficient crew coordination, this approach was complemented by studies looking at the effects of automation on the interaction and performance of the entire flight crew.

An important milestone in research on cockpit automation was the emergence of a new focal point for research in the early 1980s (see, e.g., Wiener & Curry, 1980)—the design and operation of the machine agent. The goal was to find solutions to observed problems with cockpit technology by developing guidelines for the design of human-centered automation (see Billings, 1991; NASA, 1988) that would assist rather than supplant the pilot.

These approaches and agendas coexist and try to complement one another in the current research environment. Most recently, yet another perspective on cockpit automation has become increasingly popular, one that states that focusing on either the human or the machine element in isolation is not useful in order to understand and eliminate observed problems. Instead, difficulties need to be seen as the result of a mismatch between man and machine and as being related to breakdowns in the cooperation and coordination between all resources within and even outside the cockpit—the two pilots, their automated systems, and the air traffic environment in which they operate. The objective of this school of thought is to find ways to turn these agents into team players that coordinate their activities and keep one another informed about their needs, intentions, actions, and difficulties (see Malin et al., 1991).

CONCERNS WITH PILOT-AUTOMATION INTERACTION: FROM QUANTITY TO QUALITY

In the early days of cockpit automation, both the anticipated benefits of and the predicted difficulties with cockpit automation were conceived of in quantitative terms. For example, it was expected that the introduction of cockpit automation would result in *reduced* workload, *reduced* operational costs, *increased* precision, and *fewer* errors. Anticipated problems included the need for *more* training, *less* pilot proficiency, *too much* reliance on automation, or the presentation of *too much* information (for a more comprehensive list of automation-related questions, see Wiener & Curry, 1980). Actual experiences with advanced cockpit technology confirmed that automation did, in fact, have an effect on most of these domains. However, its impact turned out to be different and far more complex than anticipated. Workload and pilot errors were not simply reduced; less information and more of the same training were no solutions to observed problems. Instead, the introduction of advanced automation seemed to result in changes and new requirements that were qualitative and context dependent rather than quantitative and uniform in nature. In the following sections, some of these effects of cockpit automation on areas such as training, workload, and pilot error are discussed.

Training for Glass Cockpits—The Need for a Different Approach

With the introduction of highly advanced automation technology to the flightdeck, traditional approaches to pilot training were no longer adequate to prepare pilots for their new task of supervisory control of a highly dynamic and complex system. In the 1980s, failure rates in transition training for the new glass cockpit aircraft were at an all-time high (Wiener,

1993), and even today problems are encountered initially when a new automated aircraft is introduced to an airline's fleet. Pilots who successfully complete training report that the transition to these new airplanes is considerably more difficult and demanding than the transition between two conventional aircraft (Sarter & Woods, 1995b).

Are the observed problems indications of a need for more training time to acquire more knowledge about yet another cockpit system? Training-oriented research in other, similarly complex domains (e.g., Feltovich, Spiro, & Coulson, 1991) and a better understanding of the kinds of problems experienced by glass cockpit pilots suggest that it is the nature of training that needs to be changed rather than its duration (see Sarter & Woods, 1995b). Cockpit technology has changed in fundamental ways that require new ways of learning and practice—in other words, training needs to account for the fact that "a 767 is not simply a 727 with some extra boxes" (Wiener, 1993a, p. 200).

Among the most powerful additions to the flightdeck is the so-called flight management system (FMS), which automatically handles a variety of tasks including navigation, flight path control, and aircraft systems monitoring. Major characteristics of this system are its high degrees of complexity, autonomy, and authority. It involves a large number of interacting subsystems that interact in various ways (complexity); it can carry out long sequences of actions without requiring pilot input (autonomy); and the most advanced FMSs currently in operation are capable of taking control of the aircraft and of overriding pilot input (authority). These systems are no longer passive tools that act only in response to pilot input. Instead, they are agents that are capable of changing their status and behavior based on input from various sources—pilot input, sensor input, input by other cockpit systems, or designer instructions.

Due to these fundamental changes in the structure and capabilities of cockpit technology, it is no longer possible to learn about these systems by accumulating compartmentalized knowledge about their individual components and about simple input–output relations. Instead, pilots need to form a mental model of the functional structure of the overall system to understand its contingencies and interactions in order to be able to monitor and coordinate activities with this new cockpit agent.

A mental model helps build expectations of system behavior, and it contributes to the adequate allocation of attention across and within the numerous information-rich cockpit displays to verify anticipated events. It also supports pilots in dealing with novel situations by allowing them to derive possible actions and solutions based on their general understanding of how the system works (Carroll & Olson, 1988). These affordances of a mental model make it a desirable objective for training for advanced cockpit systems that no longer allow for exposure to all aspects of their operation during training—even with more training time.

To support the formation of a mental model, training needs to provide opportunities and encourage pilots to actively explore the available options and dynamics of the automation. Inventing a model based on experimentation has been shown to be preferable to the explicit teaching of a system model (Carroll & Olson, 1988). As Spiro, Coulson, Feltovich, and Anderson (1988) pointed out, "Knowledge that will be used in many ways has to be learned, represented, and tried out in many ways." In contrast, rote memorization is antithetical to the development of applicable knowledge, that is, knowledge that can be activated in context. It rather results in 'inert' knowledge where the user can recite facts but fails to apply this knowledge effectively in actual line operations (see Feltovich et al., 1991).

In addition to changing the approach to and the objectives of initial training, it may be useful to reconsider the purposes of recurrent training. The results of pilot surveys indicate that learning to coordinate and cooperate with cockpit automation is a long process of continuing education (see Sarter & Woods, 1995b; Wiener, 1989). Pilots are surprised by the automation even after more than one year of line experience on some glass cockpit aircraft (e.g., Sarter & Woods, 1992; Wiener, 1989). These surprises occur in situations where the automation either behaves in unexpected ways or fails to do what is expected by the pilot. Although pilots are capable of explaining many of these surprises by discussing them with other pilots or by going back to their manuals, they sometimes develop an inaccurate explanation or cannot find a plausible reason for the observed automation behavior at all. Recurrent training could be an opportunity to support continuing learning in pilots and to prevent the formation of misconceptions in their models by gathering information on the surprises pilots experience and by including frequently reported situations in scenarios for both initial and recurrent training.

In summary, the solution to observed problems with training for automated aircraft is not more training but different training approaches and objectives. Although such modifications and expansions of current training programs can help to better prepare pilots for the operation of glass cockpit aircraft, it is important to avoid the "unfortunate tendency to treat a training department as a dumping ground for inadequate design of hardware or software" (Wiener, 1993b, p. 57).

PILOT WORKLOAD—INCREASED, REDUCED, OR BOTH? THE CONCEPT OF "CLUMSY AUTOMATION"

Early on, one of the major concerns with cockpit automation was the question of whether or not and to what extent the introduction of cockpit automation leads to a net increase or decrease in pilot workload. The problem with this question was not only that methodological problems such as disagreements about the proper measurement of workload made it difficult to answer the question (see, e.g., Roscoe, 1987). More important, it turned out to miss the fact that automation does not have a uniform effect on workload. Its impact depends on factors such as the specific interface design or the phase of flight.

As first shown by Wiener (1989), automation supports pilots most in traditionally low-workload phases of flight, but deserts them or even gets in their way when they need help most in time-critical, highly dynamic circumstances. The basis for this differential effect is the need for pilots to provide automation with information about target parameters, to decide and communicate how the automation should achieve these targets, and to closely monitor the automation to ensure that commands have been received and are carried out as intended. These task requirements do not create a problem during the traditionally low-workload cruise phase of flight for which pilots can preprogram the automation while they are still on the ground. Modifications of their input are rarely required, and there are few transitions in the status and behavior of the automation that need to be monitored. But once the descent and approach phases of flight are initiated, the situation changes drastically. Air traffic control (ATC) is likely to request frequent changes in the flight trajectory. Given that there is not (yet) a direct link between the ATC controller and the automated systems, it is the pilot's task to communicate every new clearance to the machine and to invoke system actions. It is during these traditionally high-workload, highly dynamic phases of flight that pilots report to experience an additional increase in workload.

Wiener (1989), who was the first to describe this problem based on the results of his pioneer work in the field of cockpit automation, coined the term "clumsy automation" to refer to the fact that the effect of automation on pilot workload is a new distribution of workload rather than an overall decrease or increase. This redistribution is the result of an automation design that sometimes fails to support the communication and coordination between pilot(s) and machine. To reduce the problem, attempts are currently underway to develop concepts for system-generated invocation of the automation. The underlying idea is to reduce the load inherent in the need for pilots to manage transitions by having the automation determine when there is a need to become active. The problem, of course, is that automation is not currently capable of accessing all the information necessary to make this decision.

PILOT ERROR—NEW OPPORTUNITIES FOR NEW KINDS OF ERROR WITH ADVANCED TECHNOLOGY

Yet another anticipated benefit of cockpit automation was a reduction in pilot error—but again, operational experience and systematic empirical research proved otherwise. Instead of reducing the overall amount of errors, cockpit automation provided new opportunities for new kinds of error (Woods, Johannesen, Cook,& Sarter, 1994). An example that has recently gained considerable interest is the case of mode errors of omission and commission — a new form of error on the flightdeck that has been shown to be one of the major sources of difficulties with pilot–automation coordination (see Sarter & Woods, 1992, 1994).

Interest in research on the problem of mode error in glass cockpits was triggered by the results of a study by Wiener (1989), who distributed two sets of questionnaires to B757 pilots one year apart to gather information on pilot workload, pilot errors, crew coordination, and pilot training for automation. One of the most interesting results of his study was that about 55% of all pilots responded that they were still being surprised by the automation after more than one year of line experience on the aircraft. In a follow-up study, Sarter and Woods (1992) sampled a different group of pilots at a different airline flying a different glass cockpit aircraft (the B737–300/400) to find out whether Wiener's results could be replicated and, more important, to get detailed pilot reports about the nature of and underlying reasons for automation surprises, which can be seen as symptoms of a loss of mode awareness, that is, awareness of the status and behavior of the automation. In addition, Sarter and Woods (1994) carried out an experimental simulator study to assess pilots' mode awareness in a more systematic and objective way in various situations. Overall, their research confirmed Wiener's findings that automation surprises are experienced even by pilots with a considerable amount of line experience on a highly automated aircraft. Problems with mode awareness were shown to occur most frequently in nonnormal and time-critical situations — even more frequently than suggested by pilots' self-reports, thus indicating a lack of calibration on the part of pilots who underestimate the extent of the problem (see Wagenaar & Keren, 1986).

Mode errors seem to occur due to a combination of gaps and misconceptions in pilots' model of the automated systems and the failure of the automation interface to provide pilots with salient indications of its status and behavior (Sarter & Woods, 1994, 1995a). During normal operations, gaps and misconceptions in pilots' mental model make it difficult or impossible to form accurate expectations of system behavior that provide guidance for pilots' attention allocation. In the case of nonnormal events that can not be anticipated by the crew, the interface needs to attract the pilots' attention to the relevant piece of information, but current systems sometimes fail to do so.

Problems with maintaining mode awareness are aggravated by the fact that there are numerous sources of input to the automation, all of which can trigger changes in system behavior. Input from the copilot, from sensors of the airplane's environment, from other subsystems in the cockpit, or from the system designer can make it difficult for the pilot to keep track of what is going on, particularly if these inputs occur or take effect in highly dynamic or busy phases of flight. This situation has created the opportunity for a new kind of mode error—mode errors of omission. Mode errors of commission were observed earlier in the context of fairly simple devices such as word processing systems (see Norman, 1981), where self-paced actions by an operator are required in order for an error to occur. The user has to take an action that is appropriate for one mode of the device when it is, in fact, in a different mode. With advanced technology, however, mode errors can also take the form of an error of omission where the problem is not an action of the pilot but his or her failure to act when failing to realize that the system has transitioned to a new status or behavior that is not desirable and requires pilot intervention.

This new form of error is disturbing, because of its implications for error detection and recovery. In the case of errors of commission, recovery is more likely as the operator is monitoring the system for the effects of his or her own input and should be alerted by the violation of his or her expectations. Mode errors of omission, however, tend to occur in the absence of an immediately preceding pilot action. Therefore, pilots are less likely to detect an undesired change in system status or behavior, because they have no internal guidance to tell them (where) to look for an event.

It seems that this failure to detect undesired consequences of own or someone else's input to the automation is one of the major difficulties with the coordination between pilots and advanced cockpit systems. The challenge is not to find ways to improve pilots' chances of detecting events such as system failures, which hardly occur anymore with today's highly reliable cockpit systems. If they do occur, they are detected by and sometimes even handled by other automated systems. The problem is not to detect an anomaly; it is to detect that a certain system behavior that may be normal in some circumstances is not desirable in the current situation.

FROM PILOT(S) TO MAN–MACHINE ENSEMBLES

Another important trend in the study of cockpit automation (as in other domains affected by the introduction of advanced technology) occurred with respect to its unit of analysis. Early work focused on the impact of automation on the individual human in the system by trying to assess its effects on *pilot* workload, *pilot* error, or *pilots'* trust in automation.

As a consequence of the growing interest in issues related to crew coordination during the past decade, research on cockpit automation started to extend its scope by looking not only at the individual pilot but at the impact of automation on the performance and cooperation of the entire flight crew. One example for this trend is a study by Wiener et al. (1991), who compared the performance of flight crews who were dealing with the same scenario on either the DC 9 (a conventional aircraft) or the MD 88 (its glass cockpit successor). Another interesting case of crew-oriented research is work by Segal (1990), who showed that the design of advanced automation increases the need for explicit coordination between pilots who represent two independent sources of input to the automation. Automation interfaces tend to be designed in such a way that it is possible for one pilot to give instructions to the automation without allowing the other pilot to infer the nature of the input from observable behavior alone.

For example, during a descent, one pilot may reach for the altitude knob on the flight control unit (FCU) on the glare shield of an aircraft. It is not always possible for the other pilot to tell whether the button is pushed or pulled, which can create a problem, because these two actions have completely different consequences for aircraft behavior. If the button is pulled, the automation will fly the airplane at idle thrust to the lower altitude that has been selected by the pilot on the FCU. If, on the other hand, the button is pushed, the system will use variable thrust to fly the airplane to a system-computed altitude and to honor intermediate altitude constraints that may have been preprogrammed. Given that the selection of the wrong mode can result in an altitude violation, pilots are trained to compensate for this lack of implicit cues by directing the copilot's attention to it and by explicitly mentioning any critical input. This may help detect and recover from a mode error, but it does so at the price of imposing an additional coordination task on pilots.

Another important trend in research on cockpit automation was a shift in focus from the human(s) in the cockpit to the role of the new machine agent(s). A first step in this direction were the proposed automation guidelines by Wiener and Curry (1980), who emphasized the need to support pilots in their tasks by improving the design and operation of cockpit automation. Their work had a strong impact on the development of the National Aviation Safety/Automation Plan and related research agendas (NASA, 1988; see also ATA, 1989) which have the declared goal to develop human-centered automation, that is, automation that is designed to assist the pilot rather than supplant him or her (see Billings, 1991). Most current research on cockpit automation is driven by this objective and shares this focal point.

Recently, a new approach to improving overall system performance has emerged that views problems with advanced technology as the result of an incompatibility between man and machine that can only be resolved by exploring and manipulating the cooperation, coordination, and mutual shaping of all agents on the flightdeck in a coherent way. This approach emphasizes the agentlike character of automated cockpit systems and regards them as a third crew member and another resource that has to be managed and integrated (see, e.g., Javaux, Masson, & DeKeyser, 1993; Sarter & Woods, 1995b).

It may even be justified to think of today's flightdeck systems as more than just one new agent; rather they seem to represent a new group of different agents that are created and activated by the pilot through his or her selection and combination of different modes and levels of automation.

This state of affairs has been ignored for some time, as illustrated by the long-held misperception of pilot–automation interaction that assumed that pilots have to deal with a choice between two options only: to use the automation or to turn the automated systems off and revert to manual operations. This oversimplification is implied, for example, in the recommendation for so-called "turn-off" training, where the basic idea is that pilots should be encouraged to turn off the automation when in doubt about what the system is doing. This recommendation fails to acknowledge that pilots' task is, in fact, much more difficult, because they have to decide on and execute transitions between a large number of modes and levels of automation, each one best suited for a particular set of objectives and circumstances.

The most recently developed approach to pilot–automation interaction is ecological in that it considers problems with coordination and cooperation as the result of a mismatch between the abilities, limitations, and strategies of the human and the machine agent (e.g., Hancock, 1992). This view is strongly influenced by Gibson's concepts of affordances and abilities. Affordances are the contributions of a system to the interaction that is going on between man and machine, whereas abilities are the contributions of the human element. The important fact about the relation between these two factors is that "neither an affordance nor an ability is specifiable in the absence of specifying the other. . . . The concepts are codefining, and neither of them is coherent, absent the other, any more than the physical concept of motion or frame of reference makes sense without both of them" (Greeno, 1994, p. 338).

Acknowledging this mutual interdependence between system and machine properties contrasts with the view that cockpit automation causes problems because it assigns humans the task of supervisory control, for which they are supposedly not wellsuited. From an ecologial perspective, it can be argued that the ability of the human to supervise a system is not independent of the extent to which the system affords this task by means of a design that is compatible with human information processing. As pointed out by Woods (1991), one of the major problems with monitoring current automation technology is that its interfaces are designed for data availability rather than information extraction (see also Bennett & Flach, 1992). In other words, data are present in the cockpit but they are not presented to the operator in a way that is compatible with his or her strategies, requirements, and limitations of information processing.

Support for this ecological perspective also comes from the observation that human and machine agents are not only affected by interventions from outside such as training and design. They also shape one another—the design of a system can have a strong impact on the way a practitioner carries out his or her tasks, and the practitioner exploits the system's flexibility to adapt it to his or her needs and preferences (Woods, 1993). In that sense, both agents converge over time to some extent to become a more cooperative team.

CONCLUSION

This chapter provides an overview of some of the important concepts and concerns that currently dominate the field of pilot–automation interaction. Two major trends that have occurred since the early days of research on cockpit automation are described. The first one relates to a shift in interest from issues of quantity to those of quality. The second trend involves an extension in the unit of analysis of research on cockpit automation. The initial focus on the individual pilot was followed by an increasing interest in the coordination and cooperation of the entire crew. In the early 1980s, the design and operation of the machine agent in the system became the focal point and, most recently, an ecological approach to the study of cockpit automation has evolved, which suggests that it is not useful to look at any cockpit agent in isolation but that it is their cooperation and mutual shaping that determines the nature and safety of operations.

At least two important issues need to be addressed in future research if progress is to be made toward human-centered automation design. First, the current range of research topics needs to be expanded to include the question of the desirable role and behavior of the automation, not just the issue of interface design. This seems critical, given that advanced cockpit systems are agents, not just new devices. Hancock (1992) is one of the few authors (also see Billings, 1991; Sarter & Woods, 1995b) to explicitly address this question. He suggested that the goal should be "mutual dependence where the primary intentionality remains with the human operator" (p. 79). Although it is arguable whether this is the desirable form of human–machine relation, Hancock was certainly right to say that an informed decision needs to be made now in order to avoid that technology development without a sound philosophical basis continues and dictates what problems the research community has to address in its future work.

More research is also needed concerning the processes underlying human attention allocation. For example, the call for "salient" indications of system status and behavior is ubiquitous in the literature on cockpit automation, but little is known about what makes an indication salient. In addition to looking at data-driven processing of information, which is particularly critical in nonnormal situations, it will be important to better understand internally guided task-driven attention allocation which occurs primarily under normal operating conditions. Those conditions require that the pilot is capable of selecting among the presented information by knowing where to look when. The pilot also needs to know what additional information is available, how it can be accessed, and when to call it up. In other words, we need to find out about the human and the machine contributions to the observability of automation.

ACKNOWLEDGMENTS

The preparation of this chapter was supported, in part, under a Cooperative Agreement with NASA–Ames Research Center (NCC 2-592; Principal Investigator: Dr. David Woods, The Ohio State University; Technical Monitor: Dr. Everett Palmer).

REFERENCES

Air Transport Association (ATA) of America (1989). *National plan to enhance aviation safety through human factors improvements*. Washington, DC: Author.

Bennett, K. B., & Flach, J. M. (1992). Graphical displays: Implications for divided attention, focused attention, and problem solving. *Human Factors, 34*(5), 513–533.

Billings, C. E. (1991). *Human-centered aircraft automation: A concept and guidelines* (NASA Tech. Memorandum 103885). Moffett Field, CA: NASA–Ames Research Center,.

Carroll, J. M., & Olson, J. R. (1988). Mental models in human–computer interaction. In M. Helander (Ed.), *Handbook of human–computer interaction* (pp. 45–65). Amsterdam: Elsevier.

Edwards, E. (1977). Automation in civil transport aircraft. *Applied Ergonomics, 8*, 194–198.

Eldredge, D., Dodd, R. S., & Mangold, S. J. (1991). *A review and discussion of flight management system incidents reported to the aviation safety reporting system* (Battelle Report, prepared for the Department of Transportation). Columbus, OH: Volpe National Transportation Systems Center.

Feltovich, P. J., Spiro, R. J., & Coulson, R. L. (1991). *Learning, teaching and testing for complex conceptual understanding* (Southern Illinois University School of Medicine Tech. Rep. 6). Springfield, IL: Southern Illinois University School of Medicine.

Greeno, J. G. (1994). Gibson's affordances. *Psychological Review, 101*(2), 336–342.

Hancock, P. A. (1992). On the future of hybrid human-machine systems. In J. A. Wise, V. D. Hopkin, & P. Stager (Eds.), *Verification and validation of complex systems* (pp. 61–85). (NATO-ASI Series F: Computer and Systems Sciences, V. 10)

Javaux, D., Masson, M., & DeKeyser, V. (1993, July). *Beware of agents when flying aircraft*. Paper presented at the International NATO Workshop on Human Factors Certification of Advanced Aviation Systems, Chateau de Bonas, France.

Lenorovitz, J. M. (1990, June 25). Indian A320 crash probe data show crew improperly configured the aircraft. *Aviation Week and Space Technology*, pp. 84–85.

Malin, J. T., Schreckenghost, D. L., Woods, D. D., Potter, S. S., Johannesen, L., Holloway, M., & Forbus, K. D. (1991). *Making intelligent systems team players: Case studies and design issues— Volume 1: Human-computer interaction design* (NASA Technical Memorandum 104738). Houston, TX: NASA-Johnson Space Center.

NASA (National Aeronautics and Space Administration—Office of Aeronautics and Space Technology) (1988). *Aviation safety/automation—A plan for a research initiative.* Washington, DC: Author.

Norman, D. A. (1981). Categorization of action slips. *Psychological Review, 88*(1), 1–15.

Norman, S. D., & Orlady, H. W. (Eds.). (1989). Flight deck automation: Promises and realities. Proceedings of a NASA/FAA/Industry Workshop, Carmel Valley, CA. (NASA Conference Publication 10036, August 1988)

Parasuraman, R., Molloy, R., & Singh, I. L. (1993). Performance consequences of automation-induced complacency. *International Journal of Aviation Psychology, 3*(1), 1–23.

Roscoe, A. H. (Ed.). (1987). *The practical assessment of pilot workload.* (NATO AGARDograph No. 282). Springfield, VA: National Technical Information Service

Sarter, N. B., & Woods, D. D. (1992). Pilot interaction with cockpit automation: Operational experiences with the flight management system. *The International Journal of Aviation Psychology, 2 ,* 303–321.

Sarter, N. B., & Woods, D. D. (1994). Pilot interaction with cockpit automation II: An experimental study of pilots' model and awareness of the flight management system. *The International Journal of Aviation Psychology, 4*(1), 1–28.

Sarter, N. B., & Woods, D. D. (1995a). How in the world did we ever get into that mode? Mode error and awareness in supervisory control. *Human Factors, 37*(1), 5–19.

Sarter, N. B., & Woods, D. D. (1995b). *Strong, silent, and out-of-the-loop: Properties of advanced (cockpit) automation and their impact on human-machine coordination.* (Cognitive Systems Engineering Laboratory [CSEL] Technical Report, 95-TR-01). Columbus, OH: The Ohio State University, CSEL.

Segal, L. D. (1990). Effects of aircraft cockpit design on crew communication. In E. J. Lovesey (Ed.), *Contemporary ergonomics 1990* (pp. 247–252). London: Taylor and Francis.

Sparaco, P. (1994, January 1). Human factors cited in French A320 crash. *Aviation Week and Space Technology,* pp. 30–31.

Spiro, R. J., Coulson, R. L., Feltovich, P. J., & Anderson, D. K. (1988). Cognitive flexibility theory: Advanced knowledge acquisition in ill-structured domains. In *Proceedings of the 10th Annual Conference of the Cognitive Science Society* (pp. 375–383). Hillsdale, NJ: Lawrence Erlbaum Associates.

Wagenaar, W. A., & Keren, G. B. (1986). Does the expert know? The reliability of predictions and confidence ratings of experts. In E. Hollnagel, G. Mancini, and D. D. Woods (Eds.), *Intelligent decision support in process environments* (pp. 87–103). New York: Springer-Verlag.

Wickens, C. D. (1994, September). *Designing for situation awareness and trust in automation.* Paper presented at the IFAC Conference on Integrated Systems Engineering, Baden-Baden, Germany.

Wiener, E. L. (1988). Cockpit automation. In: E. L. Wiener & D. C. Nagel (Eds), *Human factors in aviation* (pp. 433–461). San Diego, CA: Academic Press.

Wiener, E. L. (1989). *Human factors of advanced technology ("glass cockpit") transport aircraft* (Technical Report 177528). Moffett Field, CA: NASA-Ames Research Center.

Wiener, E. L., & Curry, R. E. (1980). Flight deck automation: Promises and problems. *Ergonomics, 23,* 995–1011.

Wiener, E. L. (1993a). Crew coordination and training in the advanced-technology cockpit. In E. L. Wiener, B. G. Kanki, & R. L. Helmreich (Eds.), *Cockpit resource management* (pp. 199–2230). San Diego, CA: Academic Press.

Wiener, E. L. (1993b). Intervention Strategies for the Management of Human Error (NASA

Contractor Report 4547). Moffett Field, CA: NASA-Ames Research Center.

Wiener, E. L., Chidester, T. R., Kanki, B. G., Palmer, E. A., Curry, R. E., & Gregorich, S. E. (1991). *The impact of cockpit automation on crew coordination and communication: I. Overview, LOFT evaluations, error severity, and questionnaire data* (NASA Contractor Report 177587). Moffett Field, CA: NASA–Ames Research Center.

Woods, D. D. (1991). The cognitive engineering of problem representation. In G. R. S. Weir & J. L. Alty (Eds.), *Human–computer interaction and complex systems* (pp. 169–188). London: Academic Press.

Woods, D. D. (1993). The price of flexibility in intelligent interfaces. *Knowledge-Based Systems, 6*, 1–8.

Woods, D. D., Johannesen, L. J., Cook, R.I., Sarter, N. B. (1994). *Behind human error: Cognitive systems, computers, and hindsight* (State-of-the-art report). Dayton, OH: Crew Systems Ergonomic Information and Analysis Center.

14 Fault Management in Aviation Systems

William H. Rogers
Paul C. Schutte
Kara A. Latorella
NASA Langley Research Center

INTRODUCTION

Fault management gone awry rarely has disastrous consequences in commercial aviation. The Boeing Commercial Airplane Group (1994) reported that 120 hull loss accidents with known causes have occurred over the last 10 years. Of those 120, 13 have involved airframe, aircraft systems, or power plants. Most of these involved older-generation aircraft. Improved fault management may have avoided some of these accidents, but probably not all. Systems on today's aircraft have tremendous reliability and built-in redundancy. Fault management is generally performed very well. But when a fault is not handled properly, people can die, equipment can be lost, and careers can be needlessly ruined. Human error, in this case mismanagement of faults, as pointed out by Rasmussen (1986), should be taken as the starting point rather than the conclusion of investigation. Human error can be caused by fundamental human attentional and cognitive limitations and biases, as well as environmental and contextual factors. Improved fault management must be grounded in an understanding of human performance and the operational environment in which pilots and automation together must perform fault management.

Toward this end, this chapter describes human performance and automation issues related to the management of systems faults on commercial aircraft flightdecks. The basic premise is that fault management can be improved through a better understanding of the operational context within which it occurs, and of the information processes humans use to perform it. In order to focus on these two themes, we provide a framework that describes fault management (FM) as a set of real-time tasks performed across various operational levels, and as information processing stages to which a model of cognitive control can be applied (Rasmussen, 1986). Fault management has typically been defined as a set of distinct tasks: detection, diagnosis, and compensation or recovery (e.g., Johannsen, 1988; Rouse, 1983; Van Eekhout & Rouse, 1981). To this definition we add the task of "prognosis" in order to highlight some important information processing distinctions. Using this framework, we discuss human performance issues relevant to FM, and automation that has been applied to past and current fault management and may improve FM in the future.

We cover only real-time fault management issues, that is, fault management during flight by the flight crew and flightdeck automation; we do not discuss a maintenance or "off-line diagnose and repair" perspective for fault management. The real-time context simultaneously constrains and expands fault management in respect to off-line fault management: Real-time constraints limit options available to the flight crew in terms of fixing or compensating for the problem. The real-time context expands FM in that the initial fault has a cascading effect on aircraft and mission performance that must also be managed. Although most of our discussion is within the commercial aviation context, many of the issues and solutions are applicable to military aircraft and other similar process control domains.

FAULT MANAGEMENT FRAMEWORK

The fault management framework presented here integrates: (a) the operational levels within which real-time fault management occurs, (b) cognitive control levels (Rasmussen, 1983), (c) information processing (IP) stages (Rasmussen, 1976), and (d) the definition of fault management as a set of operational tasks (detection, diagnosis, prognosis, compensation). Prior to presenting the integrated FM framework, each of the four component models is discussed separately.

Operational Levels

Fault management in aviation must be discussed within its operational context in order to understand the richness and complexity of the information processing and physical activities that must be performed. Johannsen (1988) described fault management in such a context as "concerned with solving problems in actual failure situations where undesired deviations from normal operational conditions occur." Woods (1992, p. 251) pointed out that cognition or information processing occurs in the context of "threads of activity." So, too, detection, diagnosis, prognosis, and compensation tasks associated with the management of a particular fault coexist among other systems' management threads and threads related to aircraft management and mission management. We define a *fault* as an abnormal system state or condition that does, or could, affect the operation of the aircraft and achievement of the mission. We define *fault management* as the threads of activity that a pilot performs across the systems, aircraft, and mission to detect, diagnose, prognose, and compensate for the fault. This is similar to Rasmussen's (1983) notion of considering physical systems at different levels of abstraction. Lower levels of abstraction represent the physical systems, and the higher abstraction levels represent functional effects on the aircraft and mission.

The primary mission of a commercial flight is to safely, efficiently, and comfortably move passengers and cargo from airport gate to airport gate. The crew must ensure that the aircraft retains the requisite functionality to accomplish the mission. If the aircraft fails to provide this capability due to a system fault, the crew must engage in fault management. The flight crew is concerned primarily with compensating for the fault so that it minimally affects mission goals. To that end, they must be concerned with control of the aircraft (e.g., maintaining aircraft stability, ensuring that it does not encounter any obstacles, etc.), and the safe completion of the overall mission (e.g., ensuring adherence to the current flight plan, creating and implementing a new plan if the current one can not be safely accomplished, etc.). This differs from fault management for maintenance, where the maintenance crews have as their goal to fix the problem. This goal implies need for more complete or detailed diagnosis at the physical level, whereas the crew is primarily concerned with the functional level.

Given this operational context, we define management of systems faults at three levels (Fig. 14.1). The first level is the systems level. It involves monitoring system parameters for anomalies, diagnosing faulted or abnormal system states, prognosing the effects within the system and possibly on other functionally or physically related systems, and compensating by reconfiguring the system, transferring the function to a back-up system, or simply shutting the system off or operating it in a degraded but functional state. The second level is the aircraft level. It involves maintaining a stable aircraft attitude, monitoring and regulating speed and altitude as necessary to avoid obstacles and other aircraft, and maintaining the desired flight profile. The third level is the mission level. It involves assessing the fault's impact on the mission (e.g., fuel, airport facilities, terrain, and weather). The fault may imply changing the plan for accomplishing the mission. The current flight plan may have to be modified or abandoned.

As an example, we present FM considerations of an engine failure across these operational levels. The engine is a system that affects the stability of the aircraft and therefore the ability to successfully complete the mission as

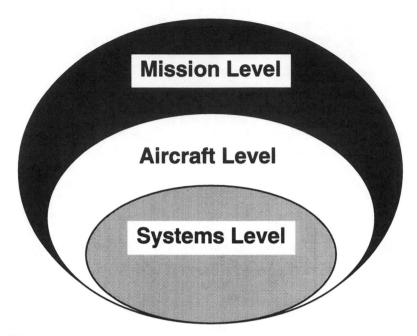

FIG. 14.1.   The three operational levels within which real-time fault management occurs in commercial aviation.

planned. The pilots may need to determine which engine has failed and shut it down (FM at the systems level), but they also need to maintain the stability of the aircraft by adding rudder input and power to another engine (FM at the aircraft level), and they may need to divert to the nearest airport rather than to continue the mission as planned (FM at the mission level).

Consideration of fault management as solely a systems level task simply does not capture the interdependence and complexity of activities across operational levels that result from a systems fault. Our integrated framework therefore describes fault management as it occurs across these three operational levels.

Cognitive Levels of Control

Rasmussen (1983) identified three strata of human performance: skill-based behavior (SBB), rule-based behavior (RBB), and knowledge-based behavior (KBB) (see Fig. 14.2). These are briefly described next.

Skill-based behavior can be characterized as smooth, automated, and tightly integrated patterns of behavior that occur without conscious control. The informational input to SBB is time–space "signals" that, through extensive training or experience, come to indicate unique environmental

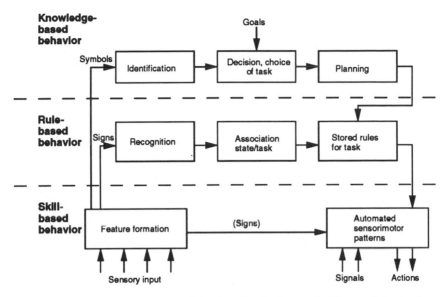

FIG. 14.2. Simplified diagram of the three levels of control of human behavior in problem solving. (Reproduced from Rasmussen, 1983, with permission from IEEE.)

states that trigger well-rehearsed, automated response patterns. In most cases, SBB continues without interruption until completion. Occasionally, SBB utilizes simple feedback control. A feedback signal, a perceived deviation between the actual spaciotemporal state and the desired state, may guide motor control. In general, SBB is quickly and smoothly executed and feedback is used to modulate only slower movements, generally those requiring greater precision.

Rule-based behavior is based on associations of familiar environmental cues with stored rules or procedures for action. These procedures and their associations to environmental triggering cues, or "signs," may be developed over time through experience or may be imparted through formal training. Signs are not individual properties of the environment; rather, they are typically labels for describing patterns of environmental state variables that indicate a course of action. RBB, as opposed to SBB, is goal oriented. Although the goal may not be explicitly expressed, the rule selection and application process is guided by some objective. Feedback plays an important role in RBB because it modifies the strategy (the rule selection and application process) an operator might employ in pursuit of a goal, and selection of goals to pursue.

Knowledge-based behavior requires the operator to function at a higher cognitive level than does RBB or SBB. KBB is required in unfamiliar situations; those in which an operator has no predefined environmental state-behavior (either signal- or sign-driven) mapping. KBB essentially requires the operator to reason to plans for action from first principles, that is, from their knowledge of relationships between environmental states and variables (based on mental models), and their priorities for performance criteria. Environmental information is interpreted and integrated to arrive at abstract, internal representations of functional properties of the environment. These "symbols" serve as inputs to the reasoning process. Although RBB is said to be goal oriented because rules are selected according to some objective, KBB is goal directed — sensory, motor, and cognitive behavior are driven by the explicit formulation of a goal. KBB involves planning, deciding among alternatives, and hypothesis testing.

Rasmussen (1986) noted that Fitts and Posner's (1962) skill-learning phases (the early-cognitive stage, the intermediate-associative stage, and the final-autonomous stage) mirror the KBB, RBB, and SBB hierarchy. Operators in an unfamiliar situation begin by behaving in a goal-directed, knowledge-based manner. After identifying patterns in the environment and associating learned successful behaviors, operators begin to exhibit RBB. Finally, after sufficient practice and refinement, environmental patterns are not identified but directly perceived as signals, and behavior proceeds without attentional control at the skill-based level.

Information Processing

Rasmussen (1986) depicted the relationship between the cognitive levels of control and a normative model of decision phases or information processes as a decision ladder (Fig. 14.3). The boxes in this model describe information processing activities. The circles represent states of knowledge resulting from these information processing stages. Ascending the ladder, from detecting the need for action to resolving the goal to drive behavior is termed *knowledge-based analysis*. Descending the information processing stages on the other side of the ladder, from identification of the target state to execution of the determined action, is termed *knowledge-based planning*. Knowledge-based behavior requires information processing at all these stages to execute a response to environmental information. This is a stylized, discrete human information processing model and it is recognized that processing does not proceed as cleanly and discretely as depicted by the model.

Rasmussen (1986) showed how RBB and SBB can be described as shortcuts in this decision ladder. Rule-based shortcuts result from perception of environmental information in terms of a specific context, that is, in terms of action, in terms of tasks, in terms of procedures, and as deviations from a target state. Skill-based shortcuts stem directly from the first indication that action is necessary to either an immediate preprogrammed response (stimulus-response pairs) or to the immediate identification of the

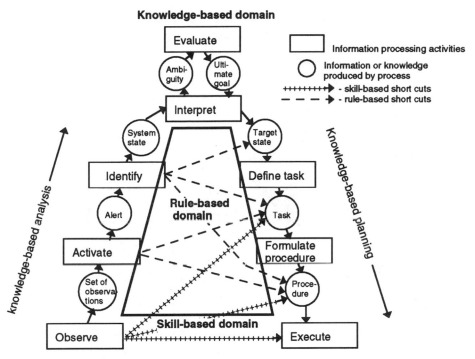

FIG. 14.3.   The relationship between the decision ladder and the three levels of control of human behavior. (Reproduced from Rasmussen, 1986, with permission from Elsevier Science Publishing Co., Inc.)

task, leaving only for the actual response to be formulated. In predictable environments, skill-based and rule-based shortcuts yield the means for temporally and cognitively efficient and reliable responses. However, in unfamiliar environments, this predisposition toward cognitive economy occasionally leads to inappropriate use of these shortcuts. This and other potential frailties of human decision making as related to fault management are discussed in a later section.

Fault Management Tasks

Fault management is defined as four operational tasks: detection, diagnosis, prognosis, and compensation. We have overlaid these tasks on the more elemental information processing stages in Rasmussen's decision ladder (see Fig. 14.4). Detection corresponds to *activate* and *observe*. Diagnosis corresponds to *identify* and *interpret*. Prognosis also includes *interpret*, as well as *evaluate*. Compensation corresponds to *define tasks, formulate procedures*, and *execute*. These categories reflect the general temporal

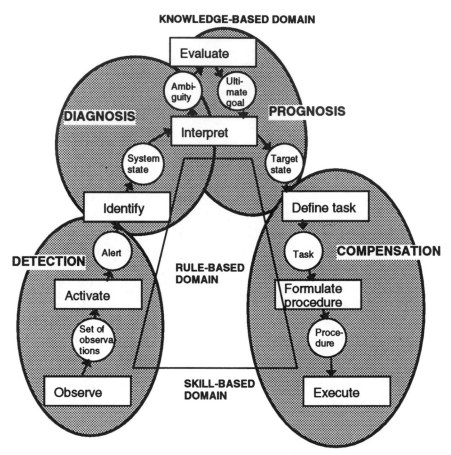

FIG. 14.4.   The four fault management tasks overlaid on Rasmussen's more elemental information processes. (Adapted from Rasmussen, 1986, with permission from Elsevier Science Publishing Co., Inc.)

nature of FM in aircraft, although they oversimplify the true iterative and overlapping nature of much of the processing that occurs. For example, any observation is considered part of the front-end detection task, even though, in fact, additional observation may be required by the other three tasks. Each of the operational tasks is described in more detail next.

Detection.   The detection task involves detecting abnormalities and observing relevant data and information in support of all four FM tasks. In Rasmussen's terms, the detection task indicates that there is something wrong or that there is a need for intervention (*activate*) and collects relevant data in order to have direction for subsequent activities (*observe*).

The detection task determines abnormalities by comparing sensory readings to some model of normality. This model of normality may be static, that is, an absolute value, or may vary with other system/aircraft/mission factors. The detection task determines the states of parameters at the aircraft and mission levels, as well as at the system levels. For example, detection includes observing that the aircraft is yawing due to asymmetric thrust or that the aircraft is behind the scheduled plan. Detection also observes nonsystem information, such as outside air temperature.

Diagnosis.   Although detection serves to describe how the system is abnormal, diagnosis determines the cause of that abnormality. Detection catalogs the symptoms and diagnosis explains them. Although it is generally accepted that diagnosis provides information on the cause of the failure (*identify*), it is equally important to recognize that it determines the consequences of the failure on other systems (*interpret*). For example, an engine flameout may immediately affect other aircraft systems such as the electrical and hydraulic systems. In this case, these systems are not "faulted," but their power supply has been cut off. Diagnosis serves to explain why these systems are not operational so that the crew may attempt to restore their power. In this way, diagnostic information is an input for both the prognosis and the compensation tasks.

Prognosis.   Prognosis is similar to diagnosis in that it describes abnormal states of the systems, aircraft, or mission. Prognosis predicts future states caused by the progression of fault effects over time. Prognosis consists of the *interpret* and *evaluate* processes in Rasmussen's ladder. It interprets the current status of the aircraft and the recent trends of conditions and extrapolates future status based on mental models of systems, aircraft, mission, and their interactions. Prognostications are by definition uncertain. Like diagnosis, prognosis relies on a model of the system but, unlike diagnosis, it cannot be confirmed or refuted by the presence or absence of existing symptoms. For instance, an engine bearing failure could lead to a seizure and subsequent flameout of the engine, or it could lead to a bearing overheat that causes an engine fire. The latter threatens the safety of the aircraft, whereas an engine flameout, especially on a four-engine aircraft, might only require modifying control input and landing procedures and not affect mission planning or safety.

It is apparent from the previous example that prognosis has a significant effect on the priority assigned to managing a particular fault. Even if a failure does not immediately threaten aircraft safety, it may in the future. Preventing it from doing so should be treated with a higher urgency than dealing with a failure that does not pose an immediate or future danger to the aircraft. Prognosis is "thinking ahead of the aircraft" to determine the urgency with which a particular fault must be managed.

Compensation.   The compensation task uses the information derived from the detection, diagnosis, and prognosis tasks to appropriately respond to the fault. It consists of the *define task, formulate procedure*, and *execute* processes in the decision ladder. Typically, the number of response options at the systems and aircraft levels are few. In some cases the response may be to carefully watch a particular system in anticipation that it might become a problem and require compensatory actions. At the mission level, potential response paths can be more numerous, and the activities in replanning or compensating for the fault can be extensive.

The Integrated Framework

Combining these different component models provides an integrated framework in which to describe critical operational issues related to human performance and automation in fault management. This integrated framework is depicted in Fig. 14.5. To summarize, real-time systems fault management in aviation consists of four tasks (detection, diagnosis, prognosis, compensation) that occur across systems, aircraft, and mission levels of operation. The tasks may be described in terms of more elemental information processing stages. The problems that occur in FM performance and automation vary with the particular FM task performed and the operational level involved. Within each task, and across FM as a whole, humans perform at different levels of cognitive control (skill-, rule-, and knowledge-based), which require different amounts of attentional and cognitive resources and are prone to different types of errors. The nature of problems that arise in human performance of fault management and the evolution of automated solutions to these problems are discussed within the context of this integrated framework.

HUMAN PERFORMANCE ISSUES IN FAULT MANAGEMENT

Human performance (HP) in fault management can be described from a variety of perspectives. It can be described as a particular form of decision making or problem solving (for reviews of decision making in general, see Klein, Orasanu, Calderwood, & Zsambok, 1993; Sage, 1981). Human performance in FM can also be ascribed to motivational and individual cognitive style characteristics (Morrison & Duncan, 1988; Rouse & Rouse,

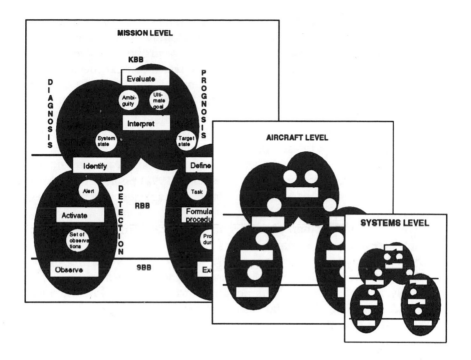

FIG. 14.5. The integrated fault management framework. (Adapted from Rasmussen, 1986, with permission from Elsevier Science Publishing Co., Inc.)

1982). It is beyond the scope of this chapter to address, in depth, these general HP issues in FM from these general perspectives. We discuss general forms of human error that may occur in FM in the context of the levels of cognitive control dimension of our integrated framework. We then review more specific issues in FM in the context of the operational task dimension of the framework.

Although it is rare, pilots occasionally mismanage fault situations. Due to hardware reliability and redundancy, systems faults by themselves are usually not flight critical. When combined with potential mismanagement by the flight crew, they may create a flight-critical situation. Examples from accident reports by the National Transportation Safety Board (NTSB), and incident reports collected by NASA's Aviation Safety Reporting System (ASRS) of cases in which faults were mismanaged are used throughout the following sections to illustrate various human performance issues.

General HP Issues Applied to FM

There are many factors that affect human fault management performance in operational environments. There are contextual factors such as the demands of the particular problem, the constraints and resources imposed by the operational and organizational environment, and the temporal context as defined by how a particular incident or problem evolves (Woods, Johannesen, Cook, & Sarter, 1994). In the aviation context, because faults have implications at the systems, aircraft, and mission level, the context is complex. Faults initiate a temporal process where disturbance chains grow and spread (Woods, 1994). Fault management involves diverse threads of activity (Woods, 1994) among the various operational levels, and pilots must allocate attention to each fault management task and each operational level, and integrate these with performance of other normal flight crew activity threads. As a consequence, as Woods (1992) pointed out, shifting the focus of attention is central to fault management or "disturbance management."

There are known human strengths and limitations in information processing; there are specific characteristics and tendencies that affect humans' ability to perceive, assess, judge, make decisions, solve problems, and respond. On the positive side, humans are good pattern matchers, can accurately assess complex, dynamic situations, and have vast knowledge bases and reasoning skills that allow solution of novel problems. These abilities afford pilots a major role in fault management even though automation has become more and more capable. As Reason (1990) stated, "Human beings owe their inclusion in hazardous systems to their unique, knowledge-based ability to carry out 'on-line' problem solving in novel situations" (p. 182).

Despite these capabilities, humans have many limitations and make errors. Woods et al. (1994) identified three classes of cognitive factors that affect performance: knowledge factors (what knowledge can be used for solving a particular problem in a specific context), attentional dynamics (factors that govern the control of attention and management of workload), and strategic factors (factors that influence the strategy one adopts in dealing with conflicting goals, uncertainty, risk, and limited resources and time). Reason (1990) identified three types of human error: skill-based (SB) slips and lapses, rule-based (RB) mistakes, and knowledge-based (KB) mistakes, based on Rasmussen's levels of cognitive control. Reason characterized these error types by the processing and task context in which they occur, and their relative frequency, observability, and predictability. General forms of human error in FM are described next, using the SB, RB, and KB levels of cognitive control.

Skill-Based Errors. Skill-based errors are errors that result from "failure in the execution and/or storage of an action sequence, regardless of whether or not the plan which guided them was adequate to achieve its objective" (Reason, 1990, p. 9). They are comparable to slips and lapses as defined by Norman (1981). SB errors can occur when attention is allocated to some task or thread of activity other than the one being performed. Inattention to details in a change in the environment may cause a SB error due to the execution of a well-rehearsed but inappropriate schema. Overattention to automated behavior, by imposing attentional checks during schemata execution, may also result in SB errors as they disrupt the automated execution of actions. In general, SB errors may be the result of not adapting familiar automated response patterns to environmental changes, due to the strength of familiar schemata, and the ability for attention to be captured by salient, although not necessarily more important, environmental cues. Because the operator attentional mechanisms that underlie slips and lapses are relatively well understood, SB error forms are largely predictable.

Because systems faults, particularly on modern aircraft, are rare events, pilots typically do not have enough familiarity or experience in managing faults to behave at a skill-based level. There may be cases, however, in which pilots respond to the common aspects of any fault, such as shutting off an alert, in a skill-based level. There are situations too, in which skill-based errors, particularly in compensation, can occur. A highly practiced response routine that is used frequently in normal operations may capture attention and be erroneously substituted for a seldom used but similar fault compensation procedure. In a report from the NASA Aviation Safety Reporting System (ASRS) files, an incident was described in which the advisory message "**R ENG EEC**" occurred, indicating a problem with the right engine electronic engine controller. The procedure called for each of the two EEC (electronic engine controller) switches to be turned off, one at a time. The fuel cutoff switches (which shut down the engines) are located directly in front of the EEC switches, and flight crews are use to shutting off the fuel cutoff switches at the end of each flight. Instead of turning off the EEC switches, the pilot shut off the fuel switches, shutting down both engines in-flight. This error has occurred on at least three occasions. Because of the proximity of the switches and the much more rehearsed activity of shutting off the fuel cutoff switches, it is very likely that this represents a classic skill-based slip. This particular problem was resolved by moving the EEC switches to another location so that these two easily confusable physical actions would be dissimilar.

Rule-Based Errors.   Reason (1990) classified rule-based (RB) errors as mistakes. In contrast to slips and lapses, he stated that "mistakes may be defined as deficiencies or failures in the judgmental and/or inferential processes involved in the selection of an objective or in the specification of the means to achieve it, irrespective of whether or not the actions directed by this decision-scheme run according to plan" (p. 9). RB errors may result from two basic failure modes: misapplication of good rules and application of bad rules. A rule is "good" if it has demonstrated utility within a particular situation. Although a rule itself might be good, its placement in the stack from which rules are selected might be inappropriate for a given situation. Thus, a good rule may be misapplied in the sense that another rule should be applied instead. Additionally, a good rule can be misapplied in the sense that no rule should be applied, that is, there is no rule that provides the correct mapping, and hence knowledge-based processing is required. The other failure mode of RB behavior is to apply bad rules, those with encoding deficiencies or action deficiencies, to a situation. Encoding deficient rules are those in which the rule's conditional phrase is underspecified or inappropriately describes the environment. Action-deficient rules are those in which the consequence of the rule results in inappropriate, inefficient, or otherwise maladaptive responses.

Rule-based processing for deterministic faults has obvious advantages: RB errors are relatively predictable. RB processing is more efficient than is KB processing and, if formalized, allows a good deal of procedural standardization. However, the use of rule-based processing when knowledge-based processing is appropriate (classified previously as the misapplication of a good rule) can have serious consequences for management of novel faults. There is an interesting dilemma in managing novel faults, as pointed out by Reason (1990): "Why do we have operators in complex systems? To cope with emergencies. What will they actually use to deal with these problems? Stored routines based on previous interactions with a specific environment" (p. 183). Not only does there seem to be a natural tendency for humans to migrate to rule-based processing due to cognitive economy, but the explicit strategy of linking fault identification with prescribed procedures in both training and design also induce pilots to operate in a rule-based level. A recent study (Rogers, 1993) evaluated how pilots think about fault management information. Results suggested that pilots primarily want information that will help in detection, initial identification, and response first, and want deeper diagnostic and prognostic information last if at all. This finding was interpreted as indicating a tendency to shortcut the deeper tasks of diagnosis and prognosis, particularly in time-stressed situations. This finding may be the result of the information they receive on current flightdecks, or the frequency that they predict they will need different types of information, but it reflects, nonetheless, that they have a rule-based mind set. NASA Langley Research Center conducted a simulation study to evaluate a fault management aid called "Faultfinder" (Rogers, 1994; Schutte, 1994). The experimental setup provided CRT-displayed checklists; however, there were no checklists for some faults that were introduced. Many of the pilots spent considerable time looking through checklists in an attempt to find one that seemed appropriate for the condition, and, in some cases, pilots used an inappropriate checklist to compensate for a fault.

Knowledge-Based Errors.   Knowledge-based (KB) errors are also classified as mistakes. Similar to RB errors, these occur while attention is focused on the problem. Although both RBB and KBB are classified as problem-solving activities, KBB differs from RBB in that it relies on reasoning (the application of limited cognitive resources to internal representations of the environment) to come to a plan for execution. Although SBB and RBB are controlled largely by feedforward mechanisms embedded in schemata and stored rules, KBB is controlled by feedback, comparing the results of actions and events to hypotheses based on an internal model. Because KBB is feedback driven, it is highly influenced by variations in the task and environmental context. Stress, distractions, fatigue, and other contextual factors are more likely to affect KBB than RBB or SBB. For example, knowledge-based processes are subject to attentional and cognitive tunneling under stress. The crash of the L-1011 in the Everglades in 1972 (NTSB, 1973) illustrates this tunneling: Focused attention by all crew members on a minor fault led to a failure to detect a deviation in flight path that resulted in a classic "controlled flight into terrain" accident. Because KB processing is more susceptible to contextual influences, the forms of KB errors are more difficult to predict than are the forms of SB and RB errors. KB errors are more difficult to detect than are RB or SB errors as well.

Failures in KBB, other than those primarily due to the decremental effects of contextual factors, arise from two basic causes: "bounded rationality" (Simon, 1956) and inaccurate or incomplete internal representations. Human information processing limitations force constrained representations of, and reasoning about, their environments. This results in only a "keyhole" view of the problem space (Reason, 1990). As Woods, Roth, and Pople (1990) pointed out, people do reasonable things given their knowledge, their objectives, their point of view, and their limited resources (time or workload); but errors result from mismatches between problem demands and the available knowledge and processing resources. Bounded rationality results from humans' inability to internally represent and reason with the full complexity inherent in real-world dynamic environments. Poor representations, or "buggy models" (Norman, 1981), of the environment may also be the cause of KB errors. Models may be inaccurate, incomplete, or developed to an uninformative or unusable level of abstraction. Models also erode with time and change with experience. Models of similar devices are confused (Yoon & Hammer, 1988a). Thus, even with unlimited cognitive resources, poor representations may lead to inappropriate actions. KB errors, and to a lesser extent RB errors, are much less likely to be identified and corrected by the initiator than SB errors. Furthermore, when standardized by the proportion of time engaged in the corresponding cognitive level of activity, the number of errors per type increases with the depth of cognitive processing involved (Reason, 1990).

Due to bounded rationality, and the propensity for cognitive economy, KB processing is characterized by cognitive limitations and the use of heuristics. Heuristics are efficient; they save time and effort and avoid overloading limited resources. But heuristic-based biases and other limitations also create opportunities for errors. Wickens and Flach (1988) provided a variety of relevant bias examples: Humans tend to confirm what is already believed to be true (confirmation bias); humans generate only a few of the possible hypotheses that would explain a particular situation, and those are often the most available in memory, not the most plausible; and humans use cue salience rather than cue reliability as a basis for choosing hypotheses. For example, in the Midlands crash of a B-737 (Cooper, 1990) in which the flight crew shut down the wrong engine, there were many cues available indicating which engine failed, but the flight crew apparently did not attend to or use them.

Thus, within the levels of cognitive control model, one might summarize human performance issues in fault management this way: Even though humans are better than automation in novel fault management, a variety of knowledge-based limitations may lead to unpredictable and difficult-to-detect human errors. Rule-based processing is effectively used in the FM domain because most faults are deterministic. RBB allows standardization and proceduralization of the FM tasks, which generally increases efficiency and accuracy. When RB errors do occur, they are more predictable and detectable than are KB errors. But in the natural, and design- and training-induced migration toward rule-based processing, the capabilities that best serve us in novel situations are sacrificed, and humans become susceptible to the misapplication of rules, or to the misuse of rule-based control, when knowledge-based control is appropriate. Finally, although faults are generally too rare for pilots to manage in a skill-based level, there are certain aspects of fault management in which skill-based errors can occur.

More specific human performance issues can be described in relation to each of the fault management tasks. Next we review specific human performance issues in detection, diagnosis, prognosis and compensation. Those discussed are the most relevant to fault management in commercial aviation and are evidenced, where possible, by accidents and incidents. Johannsen (1988) provided a more comprehensive review of types of human errors that occur in FM tasks, mapping of skill-, rule-, and knowledge-based errors onto the tasks of detection, diagnosis, and compensation.

Although several accidents and incidents are used to illustrate various points below, one accident captures a variety of problems that can occur, and the typical "multiple errors" sequence of events that often result in catastrophe. This accident summary, contained in NTSB file 5032, is provided here and is referred to numerous times in the following sections as the "Monte Vista" accident. It involved a DC-9 that departed the end of the runway at Monte Vista, Colorado, in 1989. Fortunately, it resulted in only minor aircraft damage and no injuries or deaths. It should be noted that most of the problems illustrated by this accident are at the aircraft and mission operational levels.

> While in cruise at 35000 ft. the no. 2 generator constant speed drive (CSD) failed and the crew inadvertently disconnected the no. 1 generator CSD (an irreversible procedure). Attempts to start the auxiliary power unit (APU) above the start envelope were made but aborted due to a hot start. A descent was made on emergency electrical power and a landing was made at an airport which was inadequate for the aircraft. The crew chose to make a no flap approach and to lower landing gear by emergency method when both systems were operable by normal means. This decision was influenced by the lack of landing gear and flap position indicators when the aircraft is operated on battery power. The landing speed was fast and the aircraft departed the end of the runway damaging the landing gear and the no. 1 engine. The crew failed to manually de-pressurize the aircraft and the evacuation was delayed until a knowledgeable passenger went to the cockpit and de-pressurized the aircraft. The aircraft operators manual did contain APU starting procedures but not an APU start envelope chart. (NTSB, 1990)

Detection

The flight crew may not detect a fault that requires intervention for a variety of reasons. From a signal detection perspective (e.g., see Sorkin & Woods, 1985), it may be difficult to identify a signal (fault) because it occurs within a noisy environment and it is hard to separate from the noise (i.e., system sensitivity), or the threshold that one sets to make the decision that a signal is present (i.e., the detector's response criterion) may be too high (i.e., it takes a lot of evidence before one is willing to say there is a signal).

In fault management, the nature of human perception and cognition can affect the human's sensitivity, or ability to distinguish the signal from the noise. Humans are generally not well suited to vigilance tasks (e.g., see Craig, 1984), and complacency induced by reliable automation can lead to detection inefficiency (Wiener, 1981). To the extent that systems are highly reliable, and/or humans monitor in a skill-based level, subtle deviations can be easily missed. For example, Molloy and Parasuraman (1992) noted that constant patterns of failures (e.g., highly reliable normal operation with extremely rare failures) lead to inefficient monitoring (see also Parasuraman, Mouloua, Molloy, & Hillburn, chapter 5, this volume). As signal detectors, humans can miss infrequently occurring signals. Furthermore, based on preconceived models of the environment and perceptual biases, humans selectively sample environmental variables, and are often attracted to those that are most salient, not necessarily those that are most reliable or informative (see also Mosier & Skikta, chap. 10, this volume).

The human detection sensitivity problem described previously can be exacerbated by preoccupation with or distraction due to other tasks, as well as by general reaction to stress. There are numerous documented incidents where an aircraft-level cue or symptom wasn't detected because of preoccupation with the fault at the systems level (e.g., the L-1011 Everglades accident described earlier; an ASRS-reported incident in which attention to a flight control sensing of an uncommanded go-around caused the crew to miss a drift in the lateral path and resulted in landing on the wrong runway). Under timestress, humans tend to perceptually and cognitively "tunnel." Embrey (1986) noted that stress causes inadequate search and selective attention becomes narrowed. Information overload (e.g., too many alarms) induces people to restrict sampling to the items that provide the most information (Moray, 1981). This stress-related phenomenon increases the possibility that pilots may not notice information that would potentially indicate a fault.

Environment characteristics may make detection difficult as well. The information that needs to be observed to notice a fault may be unavailable or very subtle. For example, in 1985, a B-747 aircraft developed a slow power loss in its number 4 engine, which normally could be readily detected by a yaw in the aircraft's attitude (NTSB, 1986). Because the fault was slowly evolving, and the autopilot compensated for the yaw without informing the crew that it was doing so, detection was hampered.

According to signal detection theory, detection is characterized by not only factors that relate to the sensitivity of the detection, but also to the more intentional placement of criterion for what is and is not to be considered a signal. In aviation FM, ambiguous, probabilistic, or unreliable mappings between the system state, and displayed sensor data to indicate this state make it difficult for the operator to appropriately set criterion. An engine parameter might be abnormally low, but either the pilot has only a vague model of normality, or he or she doesn't know whether the abnormal value is a reliable indication of a system abnormality. As Woods (1994) pointed out, to perceive a parameter value as abnormal, or as a symptom of an abnormal state, one must not only observe the value, but match it to a model of proper function or expected behavior. Without the precise model of normality, or belief in the observation, the criterion for deciding that an abnormality exists will be set higher.

For example, an ASRS incident was reported that involved a slow, contained compressor failure within the left engine. The failure was accompanied by vibration that was indicated on the instruments as 2.5, versus .7 on the right engine. The crew reported reluctance to rely on the vibration indication because the engine had a history of unreliable vibration indications and the manufacturer had not set any limits that would require a precautionary shutdown. Many incidents have been identified where parameters were not "normal," but the crew was not certain that a system fault that required intervention had occurred. This problem can be thought of as a knowledge-based detection problem; that is, there are no specified rules for determining that an abnormal state exists, and the flight crew often does not have an adequate model of normality to determine the existence of a system fault.

Given the poor sensitivity of the human detector for low-frequency events and the other problems discussed previously, it is not surprising, as discussed in the automation section that follows, that early automated systems provided assistance for detection to advance FM on the flightdeck.

Diagnosis

Diagnosis in complex environments can be extremely difficult. Human diagnosis worsens as the number of components (Wohl, 1983) and connections (Brooke & Duncan, 1981) increase. Flightdeck systems can be highly complex, and the functional relatedness of systems, aircraft, and mission operational levels is extensive. However, in aviation the difficulty of systems-level in-flight diagnosis is minimized because large classes of faults are compensated for in the same manner. For example, regardless of why an engine may fail (e.g., a fan blade failure, a compressor failure, a bearing failure), the crew would normally shut the engine down. There are cases, however, where a more specific diagnosis is required to support differential compensation paths. In the engine example, a more specific diagnosis might be required to decide if it is best to shut the engine down or operate it at reduced power. A more specific diagnosis may also be required to determine if it is appropriate to attempt to restart the engine.

Rasmussen (1986) described two types of cognitive search strategies in diagnosis: topographic search and symptomatic search. Topographic search requires a model of normal operation of the system. Abnormalities are diagnosed in terms of a mismatch between normal operation and actual operation, and the source of the difference is specified in terms of a component or location. Symptomatic search involves observing a pattern of symptoms that represent a questionable system state and the comparison of this pattern to stored patterns of known abnormal states. If one thinks about topographic and symptomatic strategies as model based and rule based respectively (although this is an oversimplification), then generally these strategies are susceptible to the classes of issues described under knowledge-based and rule-based processing. Topographic search is more suitable for novel fault diagnosis, because symptomatic search requires the templates of symptoms for various abnormal conditions to be known. Topographic search, like knowledge-based reasoning, is susceptible to problems induced by "buggy" system models. For example, in the Monte Vista accident, the pilots thought that they had no hydraulic power to lower the flaps or landing gear. The pilots may have had a functional model that erroneously tied lack of electrical power to lack of hydraulic power.

In terms of these diagnostic search strategies, the problem discussed earlier as application of rule-based behavior when no rules applied can be described as forcing the use of symptomatic search when the set of symptoms do not match any sets stored in the library. In the Faultfinder experiment described earlier, the pilots tried unsuccessfully to diagnose a fault by matching the symptoms to a checklist label with a known implicit set of related symptoms. The set of symptoms were similar to those of known engine problems but there were a few distinct and critical symptoms that, if accurate models of the engine and fuel systems had been applied in a topographic search, would likely have led to accurate localization of the problem.

The majority of human diagnostic failures result from the classical knowledge-based causes previously described: bounded rationality and incomplete or inaccurate models. When reasoning must be used to arrive at a diagnosis, a variety of limitations and biases in human cognitive processing come into play (Sanderson & Murtagh, 1990). Fraser, Smith, and Smith (1992) discussed the variety of biases that occur in inductive and deductive reasoning (e.g., confirmation bias), and Woods (1994) discussed problems that arise in use of abductive reasoning in diagnosis (e.g., fixation errors, the failure to generate plausible hypotheses, a preference for single-fault explanations over multiple-fault explanations). Although these problems generally apply to topographic search, they also apply to a form of symptomatic search described by Rasmussen (1986) as "hypothesize and test." Often there is insufficient knowledge or information (an incomplete model) to allow an accurate diagnosis. One manifestation of this problem is the difficulty in separating a sensor failure from a failed system. An ASRS report cited an example of an engine fire warning that indicated the presence of an engine fire. The crew stated that although they were "quite certain this fire warning was false, [they] shut down the engine, completed all appropriate checklists, and diverted." Because they did not have enough information from other cues to build a reliable model of engine health, they opted to act conservatively. In the Eastern Airlines accident (NTSB, 1984) in which an L-1011 lost oil in all three of its engines, just the opposite problem occurred: The flight crew believed they had bad sensor readings when in fact they had a real problem.

In addition, other kinds of contextual factors limit human rationality in emergency situations (e.g., time-stress and cost of failure). Stress can decrease the number of diagnostic tests used (Morris & Rouse, 1985), and increase forgetting and distortion of symptom-state mappings (Embrey, 1986). Moray and Rotenberg (1989) found that operators in FM situations experienced "cognitive lockup," resulting in delayed responses to events on the normally operating systems and serial processing in FM. It must be recognized that in aviation, and other real-time environments where there are time stresses and high costs associated with abnormal situations, humans' performance limitations and biases will be exacerbated. Morris and Rouse (1985) provided an excellent review of task requirements and human performance characteristics related to troubleshooting—diagnosing equipment failures typically without time stress or other ongoing threads of activity. They also extended their discussion to one of training trouble-shooting skills.

Prognosis

Human performance problems in prognosis are very similar to those encountered in diagnosis, which is why they are often discussed together. However, prognosis is more difficult than and different from diagnosis for a variety of reasons. First, it requires predicting events with uncertainty, which involves assessing probabilities. Fraser, Smith, and Smith (1992) described a variety of biases that humans exhibit in probability assessment, including insensitivity to sample size, overestimation of the probability of a conjunctive event, underestimation of the probability of a disjunctive event, ignoring base rates, and incorrect revisions of probabilities. Furthermore, prognosis requires a time-line prediction. Trujillo (1994) provided evidence that pilots are not particularly accurate in predicting the evolution of events over time, particularly those for which the rate of change is not constant. Consistent with this evidence, when prognosing the effect of a fuel leak on the fuel-remaining-at-destination, subjects in the Faultfinder experiment often made large errors. We believe these errors resulted from their inability to combine predictions of the fuel leak and normal fuel burn, especially because both the leak and burn depend on other variables such as throttle setting and altitude.

Another difference between diagnosis and prognosis is that diagnosis, because it involves current states, has residual effects that one can use to confirm hypotheses; in prognosis, the equivalent can only be to wait and see if what evolves is consistent with the prediction. Thus, "hypothesis and test" symptomatic search cannot be applied to prognosis.

Knowledge-based rationality limitations and incomplete or inaccurate models of functional relationships among systems, aircraft control, and mission implications also bound the ability to accurately prognose. For example, in the Monte Vista accident, the pilots probably did not have the knowledge (i.e., a model of the altitude and speed envelope within which the APU will usually start while in-flight) that would have allowed them to predict that an APU start attempt would very likely be unsuccessful, or that it would be more likely to be successful later when they reached a lower altitude. In aviation, prognosis errors are especially critical at the aircraft and missions level, because implications of future effects of faults at these levels can be flight-critical.

Compensation

Rouse (1983) described two types of compensations: compensating for symptoms and compensating for failures. In the context of fault management as a set of tasks that cross three operational levels, it is useful to think of the observable effects of a systems fault at the aircraft and mission levels as symptoms. Thus, the flight crew must compensate for the symptoms and failure at the systems level and the symptoms at the aircraft and mission level. Compensation at the latter two levels often is more difficult than compensation at the systems level. In the Monte Vista accident, for example, the pilots apparently didn't recall the aircraft depressurization procedure. It appears that they also may not have appreciated the inadequacy of the airport at which they chose to land, a critical factor in determining the appropriate compensatory action at the mission level.

Most compensation errors are slips and lapses, such as the incident described earlier in which the pilot turned the fuel cutoff switches instead of the EEC switches. In the Monte Vista accident, the pilot's inadvertent shut down of the wrong generator CSD is also an example of a compensation slip. Compensation errors can also arise from misapplication of a compensatory rule or procedure. At the systems level, checklists provide very rigorous and standard methods for compensating for faults; but observations from the Faultfinder experiment, as well as ASRS reports, suggest that pilots do occasionally select the wrong rule or procedure. These types of errors can occur at the aircraft and missions levels too. For example, in a DC-9 crash in 1985 in Milwaukee, Wisconsin (summarized in NTSB docket 1147), a right engine failure was correctly compensated for initially by use of the left rudder. Shortly thereafter, right rudder deflection occurred and the aircraft stalled, went out of control, and crashed. It is very possible that one pilot was confused about the rule for using rudder input to compensate for asymmetrical thrust, or the inappropriate rudder input could have been a compensatory slip.

Human performance issues in fault management and in the individual tasks of detection, diagnosis, prognosis, and compensation have been briefly reviewed. Although human abilities are critical to management of unusual and novel faults (particularly in stochastic environments such as aviation), human limitations and biases can have catastrophic effects in the management of faults. Aircraft designers have attempted to compensate for these limitations through the use of automation and automated FM aids.

AUTOMATION IN FAULT MANAGEMENT

When discussing automated solutions to fault management in aviation, one must recognize overriding automation issues that must be considered when automating any flightdeck function. Any time new automation is introduced, it changes function allocation. As a result of the change in allocation, and the creation of new pilot tasks in terms of managing and interacting with the new automation, workload can often be increased (Wiener, 1989), and new pilot tasks and interfaces create the opportunity for new types of errors (Wiener & Curry, 1980). Furthermore, many factors affect pilots' trust of automation, which can directly affect cooperative performance (Riley, 1994). Also, introduction of new automation, at least that requiring pilot involvement and cooperative work, requires consideration of interface and training issues. Pilots must be trained on understanding the aid, using the interface, and performing the task with the aid.

Despite concerns about the introduction of new automation and the danger of too much automation on flightdecks, additional automation still offers the potential for improving flight safety and efficiency. Automation in fault management can take on a variety of forms. One form is redundant systems. Because systems faults can have catastrophic effects and humans are imperfect in fault management, many of the flight-critical systems on commercial flightdecks have several layers of redundancy built in, so that no single fault is catastrophic. This has eliminated much of the criticality of fault management in many situations. There are, however, faults that must be dealt with in a timely manner, and some faults occur that, if mismanaged by the flight crew, can still have serious consequences. Consequently, another form of FM automation, which is discussed here, is automation that helps pilots manage systems faults. Yoon and Hammer (1988b), in the context of diagnosis, described three types of aiding automation: that which automatically performs a FM task (e.g., diagnosis); that which performs some processing, such as that provided by expert systems, to facilitate the human-performed FM task; and that which supplies useful information rather than directing the human's processing. Included in this latter type would be advances in graphic displays and smart interfaces that, in addition to providing new types of information, can provide information in a more useful form for the task that must be performed (e.g., see Malin et al., 1991; Rouse, 1978). Another form of fault management automation is that which provides error avoidance and error checking, such as systems that only allow valid data entries, and systems that question the flight crew's compensation for the fault if it is inconsistent with the automation's fault assessment. It should be noted that human performance issues in FM can be addressed through training as well (e.g., see Landerweed, 1979; Mumaw, Schoenfeld, & Persensky, 1992).

Recognizing that humans are not particularly good detectors of rare events, and that knowledge-based processing is more error-prone and KB errors are more difficult to detect and rectify, flightdeck fault management automation has focused primarily on improving performance of the detection task and reducing the requirement for knowledge-based processing. This section reviews past and current commercial flightdeck fault management automation. Although automation that has addressed these aspects of FM generally works well, other aspects of FM still require attention. Remaining problems are described here, and future fault management automated aids that could potentially reduce or eliminate the remaining problems are discussed.

Early Fault Management

Early fault management focused on automatic detection and alerting of faulted conditions within each system. The strategy was to provide a sensor and a unique alert for each fault. The intent, for each system and component, was not only to automatically detect a fault, but also to identify the fault by a distinct alert, minimizing pilots' need to interpret the alert. This theoretically would mitigate the need to allocate human information processing resources to interpretation. Pilots no longer had to be as vigilant about actively monitoring systems health. When a problem occurred, it wasn't as difficult to determine the source of the problem because the alert was unique to a particular fault.

Because there was no attempt to integrate alerts, however, the proliferation of systems and sensors resulted in an overwhelming number of separate fault alerts. Veitengruber, Boucek, and Smith (1977) reported a significant increase in the number of alerting signals on late 1960s and early 1970s aircraft, and very little standardization. They reported that on the DC-10, L-1011, and B-747, there were over 500 visual alerting functions. These alerts were simply too numerous, resulting in too much information to be remembered and utilized in rule-based fault identification. Complex, multiple faults were indicated by many low-level alerts; pilots' attentional and memory capacity was not sufficient to adequately synthesize this information for fault detection and diagnosis. Human fault management performance given this type of aiding, although potentially facilitated, was in some cases degraded due to the pilots becoming inundated with information. The human performance problems attributed by Woods (1992) to attentional control and data overload were evident.

To summarize, the aircraft of the late 1960s and early 1970s improved the detection task by providing automated alerting functions, but in at least some cases the identification problem (i.e., figuring out which system or component was faulted) was exacerbated. Alerts were not abstracted and organized to be perceived as patterns (i.e., signs or signals). This strategy essentially resulted in pilots treating the alerts as symbols, forcing knowledge-based assessment and integration of alerts to identify and diagnose systems faults and aircraft/mission functional ramifications. It became apparent that alerting and fault management on the flightdeck had to be simplified. Pilots needed help in identifying problems and determining responses to faults in a standard, procedure-driven way. Alerts had to convey information while not introducing the potential for further confusion in a high-workload situation.

Current Fault Management

A major research and development effort in the late 1970s and early 1980s (e.g., see Boucek, Berson, & Summers, 1986) aimed to reduce problems associated with the proliferation of fault alerts on early generation aircraft. This effort recognized the importance of simplicity, integration, standardization, and proceduralization in reducing pilot workload and error potential. This led to recommendations to integrate systems alerts so that the number of alerts was reduced and information could be obtained from a single location. It was also suggested that automation not only monitor aircraft system health, but provide improved guidance and status information as well. Fault identification should be easily associated with standardized procedures. Finally, it was suggested that faults and nonnormal states be categorized into three priority levels: high priority, medium priority, and low priority.

As a result of this work, commercial aircraft, beginning with Boeing's B-757 and B-767 and McDonnell Douglas' MD-88, were designed to provide integrated caution and warning systems. These integrated alerting systems, along with standardized nonnormal and emergency procedures, greatly reduced the number of aural and visual alerts, and provided the foundation for standardizing and proceduralizing the whole FM function. Brief description of the fault management systems of two of the latest aircraft, the B747–400 and MD-11, are provided here.

The flightdecks of the Boeing 747–400 and the McDonnell Douglas MD-11 represent two of the most advanced concepts for fault management now in production. Both aircraft have two-crew, glass cockpit, flight management system (FMS) equipped flightdecks. Neither aircraft has had a major accident or incident due to subsystem failure. The FM function on each aircraft is briefly discussed here. It should be noted that these discussions should not be considered to be comprehensive and in some cases are overly simplified.

The MD-11 and 747–400 are revisions of earlier aircraft that had three crew members: a captain, a first officer, and flight engineer (FE). The FE was responsible for systems management, including fault management. The newer aircraft eliminated the FE. Because the systems on the aircraft did not change significantly, the functionality of the FE had to be allocated to either automation or the crew. Some FE duties were flight-critical (e.g., engine operation, fuel dumping) and were allocated to the remaining crew members. However, a number of FM tasks were not flight-critical and that offered designers an opportunity to reallocate these FM tasks. The two flightdecks represent two different allocation designs.

Boeing 747–400. The Boeing approach to the reallocation of FE tasks was to simplify those tasks to the point where they would become manageable by the remaining two crew members. Thus, the crew would still be involved in the FM tasks and may be more situationally aware of various systems' behaviors. FM on the 747–400 is centered around two major components: the Engine Indication and Crew Alerting System (EICAS) and a procedures manual. The EICAS provides automatic monitoring of system health. In general, when an abnormality has been detected, either the crew is alerted to the abnormality with a message such as **ENG 1 OIL PRESS**, or a diagnostic message is automatically generated such as **FIRE ENG 2** or **ENG 3 SHUTDOWN**. These messages appear on a CRT in the center of the flightdeck. If the crew wishes to obtain more information on the problem, they can consult pictorial displays or schematics of systems, and they can, in some cases, find redundant or additional information on the overhead panel. The schematic displays of system status can also greatly assist the pilot in the knowledge-based level of FM when dealing with novel or subtle failures. In moving from the detection and initial diagnosis (identification) to more detailed diagnosis, prognosis, and compensation, the crew generally will consult a procedures manual. The manual, through common labeling of fault conditions, associates a procedure with each EICAS message. Some procedures simply instruct the crew to remain vigilant of the situation. The primary purpose of the procedures is to provide instructions for compensation, although some diagnostic and prognostic guidance is provided as well. The manual provides response guidance not only at the systems level, but also at the aircraft level. For example, the procedure for having two engines inoperative on the 747–400 includes compensatory information at both the systems (e.g., deselecting the air conditioning packs) and aircraft (e.g., flying the final approach with gear down) levels. The procedures manual also assists the flight crew in diagnosis. For example, when the **ENG 1 OIL PRESS** message is displayed, the procedure calls for the crew to check data displayed from a redundant sensor. If that information is normal, then the message is to be considered as a false alarm, effectively diagnosing that the original sensor was faulty. The aircraft also has a powerful computerized flight management system (FMS) that can assist the crew in dealing with mission consequences of system failures. If an engine fails, the FMS can provide information about reduced engine performance that aids the crew in route planning. The FMS can also show fuel constraints on the mission: For example, it will recalculate fuel remaining at destination after a new flight plan has been entered that may include one-engine operation at a lower altitude.

McDonnell Douglas MD-11.   There are many similarities between the MD-11 and the 747–400 integrated alerting systems. Both automatically monitor systems health and support use of rule-based shortcuts by pilots in relating detection and diagnosis (identification) information to prognosis and compensation. They both provide textual alert messages for system abnormalities and have schematics to improve the crew's situation awareness and assist in knowledge-based reasoning (the MD-11 uses an Engine Alerting Display and a Systems Display for these purposes). However, they significantly differ in their approach to the compensation task. On the MD-11, all but flight-critical FM compensations that were normally performed by the flight engineer have been automated using Automated Systems Controllers (ASCs). In the event of a failure, the crew is notified and given information (either textual or pictorial) to show how automation has compensated for the problem. The CRT presents procedural items for compensating at the aircraft and mission levels. Examples of these items (called *consequences*) include **USE MAX FLAPS** and **GPWS OVRD IF FLAPS LESS THAN X**. These consequences also include diagnosis and prognosis information required by the flight crew for performance at the aircraft and mission levels (e.g., **FLIGHT CONTROL EFFECT REDUCED** and **AUTOPILOT 2 INOP**). When the automated FM system is working normally, the crew need not be involved with FM of the systems. They need only be aware of the failure's effects on the aircraft and the mission. There is a chance that an ASC may fail, causing the FM compensation to revert to manual operation. In this case, the ASC alerts the crew that it is returning control to them via a CRT-displayed message. The crew then consults procedures in a manner similar to that used on the 747–400. Thus, although the FM function for the flight crew has been significantly reduced by the ASCs, they comprise another system that requires FM.

These integrated fault management systems simplify the alerting, and present the priority of the condition aurally and/or using text formatting and color coding. Common message phrasing and checklist labels provide a direct, explicit link between the identified condition and the required response. The MD-11's automatic compensation feature is a logical extension of automation in fault management: If there is a unique or many-to one mapping of faulted state to appropriate response, there is no uncertainty in the required response. The main reasons to allow pilots to carry out the response are to provide situational awareness and to avoid the performance degradation due to having been out of the loop and out of practice if and when they must take over.

The design strategy demonstrated in these aircraft for simplification and proceduralization in FM has been very effective. It has allowed pilots to manage most faults in a rule-based mode, which reduces workload and makes pilot errors more predictable and self-correcting. "Routine" faults that have a simple, proceduralized systems-level compensation (e.g., shutting off a fuel pump), and have no effect at the aircraft and mission levels (other than increased monitoring) due to redundancy, present very little difficulty for flight crews of modern aircraft. Thus, research on human performance in FM leading to these modern integrated fault management systems, and the design of the systems themselves, have provided a major safety improvement on modern commercial aircraft.

Future Fault Management

Even with these significant advances in FM, problems still exist. We next outline four classes of problems facing pilots performing FM tasks and, therefore, flightdeck designers in their efforts to facilitate FM through the use of automation. One is that skill-based slips and lapses will always occur, most commonly in compensation. The second problem is that pilots perform FM in a rule-based level when a knowledge-based level is appropriate. The third problem is that subtle, novel, complex, and multiple faults, although extremely rare, will occur, and when they do, human limitations in knowledge-based processing are problematic. The fourth problem is that aiding automation, operational procedures, and training thus far have addressed FM mainly as systems management, and often do not consider FM in the context of aircraft-level and mission-level management. The following sections discuss these problems and potential solutions.

Skill-Based Errors. It is generally accepted that slips and lapses will occur in conducting fault management tasks. Pilots will hit wrong buttons or flip switches into the wrong position, thus shutting down wrong components or putting systems in wrong configurations. Design may reduce these errors by not co-locating controls that may be procedurally confused. Where possible, system actions (e.g., shutting down a generator or pump), should be reversible; if an error is committed, the action should be able to be undone. A major role of automation in terms of these errors is to provide error tolerance and error resistance. This involves detecting errors that occur, and avoiding bad outcomes when errors do occur. Wiener and Curry (1980) recommended that systems be less vulnerable once an error is made and that they provide error-checking mechanisms. Today this is accomplished with multiple crew members and redundant systems, but automation is advancing to the point where there is promise of much greater capability. Automation can be used to detect and alert the crew of their action slips and lapses (Hollnagel, 1991). In its simplest form, automation can check for logical inconsistencies between sensed parameter values and pilot inputs. For example, if right engine parameters are abnormal and the left engine is shut down, the inconsistency could be flagged and brought to the attention of the crew.

Using Rule-Based Instead of Knowledge-Based Processing. For a majority of faults, flight crews perform rule-based FM. Sometimes, rule-based processing appears appropriate but is not (e.g., the fault is novel but similar to known or previously experienced faults). Current flightdeck design, operations, and training leave pilots ill-prepared for such events and, in fact, induce them to look for nonexistent, standard rules to manage the faults. Training on systems knowledge has continually been reduced in recent years. Functional models or representations of systems operations are deemphasized in pilot training (e.g., see Sarter & Woods, 1992). Training involves learning a set of invariant rules and procedures for how to respond to specified messages and conditions. Automation aiding and/or training could help pilots expand the set of rules that may be applied to be more comprehensive and context specific. Aids could also help pilots recognize when processing shortcuts and sensor-response rules may not be appropriate. This could be as simple as making explicit to the pilot that an existing condition does not match a known hard-wired fault state and prescribed checklist. It could also involve sophisticated decision aiding that supplies rule-based and knowledge-based problem solving information (e.g., functional models, what-if scenarios, and confidence ratings).

Knowledge-Based Errors.   As discussed previously, pilot performance varies greatly when faced with novel failures for which there are no rules or procedures. Although these are extremely rare, novel fault management at the systems level could still benefit from aiding. Detection, diagnosis, prognosis, and compensation of novel failures require knowledge-based reasoning. This is a significant challenge because as aircraft become even more reliable and automated FM systems (such as those on modern aircraft) cover more types of failures, pilot experience in using knowledge-based reasoning will tend to decrease. Assisting the flight crew in dealing with these situations can take several forms, ranging from providing topological displays (Kieras, 1992; Rouse, 1978) to affording what-if interaction for hypothesis testing (Yoon & Hammer, 1988a).

Detection, diagnosis, prognosis, and compensation all have knowledge-based components. Knowledge-based processing must be used in each task to deal with unusual, uncertain, subtle, or complex situations, and draws on theoretical models (first principles) to obtain a solution. For detection, this means a refinement of the model of normality that is used in detecting a state. In the Monitaur fault monitoring concept (Schutte, 1989) and the E-MACS fault monitoring display concept (Abbott, 1990), simulation models of the system provide robust values of normalcy for each monitored sensor. Other fault monitoring concepts use models derived by neural networks (Shontz, Records, & Choi, 1993), and abductive networks (Montgomery, 1989). Doyle, Chien, Fayyad, and Wyatt (1993) developed a method for determining sensor/parameter/variable normality based on both system models and past system performance. Although all of these methods provide more sensitive detection systems, careful consideration must be placed on ensuring an acceptable level of false alarms, especially for infrequent events. Studies performed in conjunction with NASA Langley's fault management program (Getty, Swets, & Pickett, 1993; Getty, Swets, Pickett, & Gonthier, 1993) investigated the effects of event rate and response costs and benefits on human performance in false alarm situations.

An abundance of artificial intelligence research is being performed in automated diagnosis (e.g., see Bakker, 1991). Knowledge-based diagnosis generally involves a topographical search through a system model to determine where a failure has occurred. Searches can be hypothesis driven, where a fault hypothesis is introduced to determine if it can produce the current symptoms, or data driven, where symptoms are traced to their origin in the system. The DRAPhyS diagnostic system (Abbott, 1991) is such a hypothesis-driven system. In aiding the human in performing diagnosis, advanced automation can do a great deal of sophisticated diagnosis itself, and it can provide information to improve the human's model of the system, as well as provide more accurate symptoms.

Prognosis has not been widely investigated as a FM task in aviation, so automated aiding concepts at the knowledge-based or rule-based levels are scarce. Prognosis can be viewed as an extension of diagnosis. The aiding approaches described previously for diagnosis can be extended to prognosis. Prognosis has been described here as predicting the consequences of current faults, but the future occurrence of faults and symptoms can also be prognosticated. If pilots knew future states of slowly evolving faults, especially before the fault has reached a critical level, they may be able to preserve equipment and improve aircraft safety by having more time to respond to, or preempt, serious consequences. A recent study (A. C. Trujillo, personal communication) determined which aircraft system parameters would be most useful for predicting slowly degrading system states.

Knowledge-based compensation will likely be required when either there are no procedures appropriate for a particular fault, or procedures give conflicting guidance. Compensation involves developing a plan for a response, which includes assessing the impact of that response on safety. Because of the limited number of responses available to the crew, the greatest area for research lies in providing the flight crew with information that helps assess the consequences of responses and generate alternative plans. For example, it is particularly important to alert the crew if a proposed response may put the aircraft in a more dangerous state.

FM at the Aircraft and Mission Levels. The final area that can be enhanced by improvements in FM, independent of the types of processing that are used, is fault management at the aircraft and mission levels. There are generally a number of options that are available in aircraft control and in mission replanning due to a particular failure. Fault management at the mission level can obviously be critical to safety. For example, the selection of an appropriate airport in fuel-critical or flight-critical situations can mean the difference between a safe landing and a hull loss. In these situations, the flight crew is often time-pressured and cannot consider all of the relevant factors. In the Faultfinder study, the scenarios required pilots to manage mission consequences of the fault, specifically to declare an emergency, contact air traffic control (ATC) and company dispatch, and divert to the closest airport. Although pilots did reasonably well in managing the faults per se, they made many errors in handling all the flight planning, communication, and task management (e.g., scheduling, assigning, and coordinating tasks and task resources) activities necessitated by the faulted condition. It seems clear that this is a ripe area for future research and potential application of automated aiding.

Management of the consequences of faults on these other flight functions has traditionally been left to the flight crew because of the greater uncertainty in terms of the number of alternative responses that may be appropriate for each condition. Furthermore, the appropriate actions depend on dynamic fault and contextual factors such as the stability of the fault, regulations (e.g., extended twin engine operations, or ETOPS, that necessitate landing at the closest suitable airport when certain faults occur), terrain, weather, facilities at airports, and passenger and cargo connection requirements. Assistance in identifying these factors, selecting which are most important to a particular situation, integrating them, and perhaps helping the crew decide among alternatives is a promising area of research. Research in this area has included work on in-flight replanning (Rudolph, Homoli, & Sexton, 1990; Smith, McCoy, & Layton, 1993), and work combining fault management and mission planning aids (e.g., Kaiser et al., 1988). The research on these replanning aids suggests that significant coordination with the ground (company dispatch and ATC) is required in order to effectively evaluate some of the relevant factors such as passenger connections.

Similarly, some research efforts address FM aiding at the aircraft level. That is, they attempt to assist the crew in flying the aircraft in the event of a failure. RECORS (Hudlicka, Corker, Schudy, & Baron, 1990), for example, provides the flight crew with instructions on how to deal with failures that require the use of certain aircraft control effectors in ways that were unintended at design. For instance, it could provide guidance on using asymmetric thrust to control pitch or heading when normal control surfaces have failed. Another approach to automation at the aircraft level (used in some military aircraft) is to have the onboard systems automatically reconfigure the aircraft control laws to make the loss of control surfaces as transparent to the crew as possible (Urnes et al., 1991). On commercial aircraft, allowing the automation this level of authority over flight control is quite controversial (Billings, 1991).

Another potential strategy to aid FM is further integration of alerts and fault information. At the systems level, integration of engine information and alerts with other systems has been recommended (Hornsby, 1992). At the aircraft level, it would be beneficial to integrate alerting systems for aircraft state with those for system faults, because they can share symptoms (e.g., reduced speed could be due to improper control surface position, an engine problem, icing, etc.). A variety of separate alerting systems exist for aircraft state (e.g., windshear, ground proximity, traffic and collision avoidance, stall, and overspeed alerts). The number of separate warnings at the airplane level is beginning to approach the one fault-one alert problem that early generation aircraft experienced at the systems level. Hence, fault management problem solving at the aircraft level, as well as management of other aircraft level problems, would likely benefit from integration of alerts for the same reasons that fault management at the systems level benefited from integrated alerting systems.

SUMMARY

Although FM is typically performed well by humans and appropriately facilitated by advances in commercial flightdeck automation, problems in fault management can cause catastrophic effects. Research and development to improve human fault management and human/automation cooperative FM continues to be important. An understanding of human fault management, and of the context in which it is performed in aviation, are fundamental to continued improvement in FM. FM was discussed as a set of operational tasks (i.e., detection, diagnosis, prognosis, and compensation) that occur at three operational levels (i.e., systems, aircraft, and mission). This chapter presented a fault management framework in which these operational tasks and levels were described in terms of more elemental human information processing stages in which three levels of cognitive control can be applied. Human performance of FM was discussed within this framework. Automation has overcome many of the human limitations in fault management, but has not addressed some, and has actually exacerbated others. Future advances in automation hold much promise for minimizing the number of cases in which poor fault management has catastrophic effects.

ACKNOWLEDGMENTS

The first author was supported for this work under NASA contract NAS1-18799, Task 15. The third author was supported for this work under NASA graduate student researchers program fellowship NGT 50992.

REFERENCES

Abbott, K. H. (1991). *Robust fault diagnosis of physical systems in operation* (NASA-TM-102767). Hampton, VA: NASA Langley Research Center.

Abbott, T. S. (1990). *A simulation evaluation of the engine monitoring and control system display* (NASA-TM-2960). Hampton, VA: NASA Langley Research Center.

Bakker, R. R. (1991). *Knowledge-based diagnosis of technical systems, a three-year progress report* (Memoranda Informatica 91-15). Enschede, The Netherlands: University of Twente.

Battelle. (1990, April). Aviation safety reporting system search request number 1788, Glass cockpit aircraft engine failures. (Available from ASRS Office, 625 Ellis St., Suite 105,

Mountain View, CA 94043)

Billings, C. (1991). *Human-centered aircraft automation: A concept and guidelines.* (NASA-TM-103885). Hampton, VA: NASA Langley Research Center.

Boeing Commercial Airplane Group. (1994). *Statistical summary of commercial jet aircraft accidents, worldwide operations 1959–1993.* Seattle, WA: Author.

Boucek, G. P., Berson, B. L., & Summers, L. G. (1986, October). *Flight status monitor system — operational evaluation.* Paper presented at the 7th IEEE/AIAA Digital Avionics Systems Conference. Fort Worth, TX.

Brooke, J. B., & Duncan, K. D (1981). Effects of a system display format on performance in a fault location task. *Ergonomics, 18*, 53–66.

Cooper, D. A. (1990). *Report on the accident to Boeing 737–400 G-OBME near Kegworth, Leicestershire on 8 January 1989* (Aircraft accident report no. 4/90). London: Air Accidents Investigation Branch, Department of Transport, Royal Aerospace Establishment.

Craig, A. (1984). Human engineering: The control of vigilance. In J. S. Warm (Ed.), *Sustained attention in human performance* (pp. 247–291). New York: Wiley.

Doyle, R. J., Chien, S. A., Fayyad, U. M., & Wyatt, E. J. (1993, May). *Focused real-time monitoring based on multiple anomaly models.* Paper presented at the 7th International Workshop on Qualitative Reasoning. Eastsound, WA.

Embrey, D. E. (1986, July). Approaches to aiding and training operators' diagnoses in abnormal situations. *Chemistry and Industry, 7*, 454–458.

Fitts, P. M., & Posner, M. I. (1962). *Human performance.* Monterey, CA: Brooks/Cole.

Fraser, J. M., Smith, P. J., & Smith, J. W., Jr. (1992). A catalog of errors. *International Journal of Man–Machine Studies, 37*, 265–307.

Getty, D. J., Swets, J. A., & Pickett, R. M. (1993). *The pilot's response to warnings: A laboratory investigation of the effects on human response time of the costs and benefits of responding* (BBN Report No. 7947). Cambridge, MA: Bolt, Beranek, & Newman.

Getty, D. J., Swets, J. A., Pickett, R. M., & Gonthier, D. (1993). *The pilot's response to warnings: A laboratory investigation of the effects of the predictive value of a warning on human response time* (BBN Report No. 7888). Cambridge, MA: Bolt, Beranek, & Newman.

Hollnagel, E. (1991). The phenotype of erroneous actions: Implications for HCI design. In G. R. S. Weir & J. L. Alty (Eds.), *Human–computer interaction and complex systems* (pp. 73–118). New York: Academic Press.

Hornsby, M. (1992). *Engine monitoring display study* (NASA-CR-4463). Hampton, VA: NASA Langley Research Center.

Hudlicka, E., Corker, K., Schudy, R., & Baron, B. (1990). *Flight crew aiding for recovery from subsystem failures* (NASA-CR-181905). Cambridge, MA: Bolt, Beranek & Newman.

Johannsen, G. (1988). Categories of human behaviour in fault management situations. In L. P. Goodstein, H. B. Andersen, & S. E. Olsen (Eds.), *Tasks, errors and mental models* (pp. 251–260). New York: Taylor & Francis.

Kaiser, K. J., Lind, H. O., Meier, C. D., Stenerson, R. O., Enand, S., Scarl, E. A., & Jagannathan, V. (1988). Integration of intelligent aviaonics systems for crew decision aiding. *Proceedings of the Fourth Annual Aerospace Applications of Artificial Intelligence Conference* (pp. 230–241).

Kieras, D. (1992). Diagrammatic displays for engineered systems: Effects on human performance in interacting with malfunctioning systems. *International Journal of Man–Machine Studies, 36*, 861–895.

Klein, G. A., Orasanu, J., Calderwood, R., & Zsambok, C. E. (1993). *Decision making in action: Models and methods.* Norwood, NJ: Ablex.

Landerweerd, J. A. (1979). Internal representation of a process, fault diagnosis and fault correction. *Ergonomics, 22*(12), 1343–1351.

Malin, J. T., Schreckenghost, D. L., Woods, D. D., Potter, S. S., Johannesen, L., Holloway, M., & Forbus, K. D. (1991). *Making intelligent systems team players: Case studies and design issues. Volume 1: Human–computer interaction design* (NASA-TM-104738). Houston, TX: NASA Johnson Space Center.

Molloy, R., & Parasuraman, R. (1992). Monitoring automation failures: Effects of automation reliability and task complexity. *Proceedings of the Human Factors Society 36th Annual Meeting* (pp. 1518–1521). Santa Monica, CA: Human Factors Society.

Montgomery, G. (1989, October). *Abductive diagnostics.* Paper presented at the AIAA Computers in Aerospace VII Conference, Monterey, CA.

Moray, N. (1981). The role of attention in the detection of errors and the diagnosis of failures in man-machine systems. In J. Rasmussen & W. B. Rouse (Eds.), *Human detection and diagnosis of system failures* (pp. 185–198). New York: Plenum Press.

Moray, N., & Rotenberg, I. (1989). Fault management in process control: Eye movements and action. *Ergonomics, 32*(11), 1319–1342.

Morris, N. M., & Rouse, W. B. (1985). Review and evaluation of empirical research in troubleshooting. *Human Factors, 27*(5), 503–530.

Morrison, D. L., & Duncan, K. D. (1988). Strategies and tactics in fault diagnosis. *Ergonomics, 31*(5), 761–784.

Mumaw, R. J., Schoenfeld, I. E., & Persensky, J. J. (1992, June). *Training cognitive skills for severe accident management.* Paper presented at the IEEE 5th Conference on Human Factors and Power Plants, Monterey, CA.

National Transportation Safety Board (NTSB). (1973). *Aircraft accident report: Eastern Air Lines, Inc., L-1011, N310EA, Miami, Florida, December 29, 1972* (NTSB-AAR-73-14). Washington, DC: National Transportation Safety Board.

National Transportation Safety Board (NTSB). (1984). *Aircraft accident report: Eastern Air Lines, Inc., Lockheed L-1011, N334EA, Miami International Airport, Miami, Florida, May 5, 1983* (NTSB-AAR-84-04). Washington, DC: National Transportation Safety Board.

National Transportation Safety Board (NTSB). (1986). *Aircraft accident report: China Airlines B-747-SP, 300 nm northwest of San Francisco, CCA, February, 19,1985* (NTSB-AAR-86-03). Washington, DC: National Transportation Safety Board.

National Transportation Safety Board (NTSB). (1990). [Summary of 1989 DC-9 Monte Vista, CO accident]. Unpublished raw data from NTSB filed Report no. DEN90IA012 (file 5032).

Norman, D. A. (1981). Categorization of action slips. *Psychological Review, 88*, 1–15.

Rasmussen, J. (1976). Outlines of a hybrid model of the process operator. In T. B. Sheridan & G. Johannsen (Eds.), *Monitoring behavior and supervisory control* (pp. 371–383). New York: Plenum.

Rasmussen, J. (1983). Skills, rules, and knowledge; signals, signs, and symbols, and other distinctions in human performance models. *IEEE Transactions on Systems, Man, and Cybernetics, 13*(3), 257–266.

Rasmussen, J. (1986). *Information processing and human–machine interaction: An approach to cognitive engineering.* New York: North-Holland.

Reason, J. (1990). *Human error.* New York: Cambridge University Press.

Riley, V. (1994). *Human use of automation.* Unpublished doctoral dissertation. University of Minnesota, Minneapolis.

Rogers, W. H. (1993). Managing systems faults on the commercial flight deck: Analysis of pilots' organization and prioritization of fault management information. In R. S. Jensen & D. Neumeister (Ed.), *Proceedings of the 7th International Symposium on Aviation Psychology* (pp. 42–48). Columbus, OH: The Ohio State University.

Rogers, W. H. (1994, April). *The effect of an automated fault management decision aid on pilot situation awareness.* Paper presented at the First Automation Technology and Human

Performance Conference, Washington, DC.

Rouse, W. B. (1978). Human problem solving performance in a fault diagnosis task. *IEEE Transactions on Systems, Man, and Cybernetics, 8*(4), 258–271.

Rouse, W. B. (1983). Models of human problem solving: Detection, diagnosis, and compensation for system failures. *Automatica, 19*(6), 613–625.

Rouse, S. H., & Rouse, W. B. (1982). Cognitive style as a correlate of human problem solving performance in fault diagnosis tasks. *IEEE Transactions on Systems, Man, and Cybernetics, 12*(5), 649–652.

Rudolph, F. M., Homoli, D. A., & Sexton, G. A., (1990). *"DIVERTER" decision aiding for in-flight diversions* (NASA-CR-18270). Hampton, VA: NASA Langley Research Center.

Sage, A. P. (1981). Behavioral and organizational considerations in the design of information systems and process for planning and decision support. *IEEE Transactions on Systems, Man, and Cybernetics, 11*, 640–678.

Sanderson, P., & Murtagh, J. M. (1990). Predicting fault diagnosis performance: Why are some bugs hard to find? *IEEE Transactions on Systems, Man, and Cybernetics, 20*(1), 274–283.

Sarter, N. B., & Woods, D. D. (1992). Pilot interaction with cockpit automation I: Operational experiences with the flight management system. *International Journal of Aviation Psychology, 2*, 303–321.

Schutte, P. C., (1989, October). *Real time fault monitoring for aircraft applications using quantitative simulation and expert systems.* Paper presented at the AIAA Computers in Aerospace VII Conference, Monterey, CA.

Schutte, P. C. (1994). [Evaluation of an advanced fault management aiding concept]. Unpublished raw data.

Shontz, W. D., Records, R. M., & Choi, J. J. (1993). *Spurious symptom reduction in fault monitoring* (NASA-CR-191453). Hampton, VA: NASA Langley Research Center.

Simon, H. A. (1956). Rational choice and the structure of the environment. *Psychological Review, 84*, 155–171.

Smith, P. J., McCoy, C. E., & Layton, C. (1993). An empirical evaluation of computerized tools to aid in enroute flight planning. In R. S. Jensen and D. Neumeister (Eds.), *Proceedings of the 7th International Symposium on Aviation Psychology* (pp. 186–191). Columbus, OH: The Ohio State University.

Sorkin, R. D., & Woods, D. D. (1985). Systems with human monitors: A signal detection analysis. *Human-Computer Interaction, 1*, 49–75.

Trujillo, A. C. (1994). *Effects of historical and predictive information on ability of transport pilot to predict an alert* (NASA-TM-4547). Hampton, VA: NASA Langley Research Center.

Urnes, J. M., Hoy, S. E., Wells, E. A., Havern, W. S., Norat, K. F., & Corvin, J. H. (1991). *Self-repairing flight control system. Volume 1: Flight test evaluation on an F-15 aircraft* (WL-TR-91-3025). Dayton, OH: Wright Laboratory, Air Force Systems Command, Wright-Patterson Air Force Base.

Van Eekhout, J. M., & Rouse, W. B. (1981). Human errors in detection, diagnosis, and compensation for failures in the engine control room of a supertanker. *IEEE Transactions of Systems, Man, and Cybernetics, 11*(12), 813–816.

Veitengruber, J. E., Boucek, G. B., & Smith, W. D. (1977). *Aircraft alerting systems criteria study, volume I: Collation and analysis of aircraft alerting systems data* (FAA-RD-80-68). Washington, DC: Federal Aviation Administration.

Wickens, C. D., & Flach, J. M. (1988). Information processing. In E. L. Wiener & D. C. Nagel (Eds.), *Human factors in aviation* (pp. 111–155). New York: Academic Press.

Wiener, E. L. (1981). Complacency: Is the term useful for air safety? *Proceedings of the 26th Corporate Aviation Safety Seminar* (pp. 116–125). Denver, CO: Flight Safety Foundation,

Inc.

Wiener, E. L. (1989). *Human factors of advanced technology ("glass cockpit") transport aircraft* (NASA-CR-177528). Moffet Field, CA: NASA Ames Research Center.

Wiener, E. L., & Curry, R. E. (1980). Flight-deck automation: Promises and problems. *Ergonomics, 23,* 995–1011.

Wohl, J. G. (1983). Connectivity as a measure of problem complexity in failure diagnosis. *Proceedings of the Human Factors Society 27th Annual Meeting* (pp. 681–684). Santa Monica, CA: Human Factors Society.

Woods, D. D. (1992). *The alarm problem and directed attention* (CSEL Report No. 92-TR-05). Columbus, OH: The University of Ohio, Cognitive Systems Engineering Laboratory.

Woods, D. D. (1994). Cognitive demands and activities in dynamic fault management: Abductive reasoning and disturbance management. In N. Stanton (Ed.), *Human factors of alarm design* (pp. 63–92). London: Taylor & Francis.

Woods, D. D., Johannesen, L. J., Cook, R. I., & Sarter, N. B. (1994). *Behind human error: Cognitive systems, computers, and hindsight* (CSERIAC State-of-the-Art Report). Dayton, OH: Crew Systems Ergonomic Information and Analysis Center.

Woods, D. D., Roth, E. M., & Pople, H. E., Jr., (1990). Modeling the operator performance in emergencies. *Proceedings of the Human Factors Society 34th Annual Meeting* (pp. 1132–1137). Santa Monica, CA: Human Factors Society.

Yoon, W. C., & Hammer, J. M. (1988a). Aiding the operator during novel fault diagnosis. *IEEE Transactions on Systems, Man, and Cybernetics, 18*(1), 142–148.

Yoon, W. C., & Hammer, J. M. (1988b). Deep-reasoning fault diagnosis: An aid and a model. *IEEE Transactions on Systems, Man, and Cybernetics, 18*(4), 659–675.

15 Human Factors in Air Traffic System Automation

V. David Hopkin
Human Factors Consultant

John A. Wise
Embry-Riddle Aeronautical University

INTRODUCTION

The type of system exemplified by air traffic control is actually rather common. Examples include telephone switching, military command and control (which is the probably the parent of most current air traffic control systems), railroad systems, automobile traffic systems (e.g., Los Angeles), and electrical distribution systems. Air traffic systems primarily differ from these other systems in terms of the short response time requirements and their four dimensionality. The control of air traffic takes place in three-dimensional space and time; therefore, the reader might usefully consider if the issues discussed also apply (or do not apply) to similar systems.

Air traffic systems are particularly interesting within the context of this book because automation has the *potential* to completely change the current methods of operation. Revisions being discussed range from the computer taking over most of the current controller tasks (e.g., the direct manipulation of aircraft movements) with the controller becoming a monitor, to a completely distributed system where all information is shared by all pilots who perform as their own "controllers" (cf. Wise, Hopkin, & Smith, 1991). Although the mainstream thought is to "only" provide the current air traffic system with enhanced support, the reader should be aware that other modes of air traffic operation are being discussed. Because the majority of the concepts discussed here would be applicable to all the potential forms, the term *air traffic system* is used to describe all forms of current and projected versions of what is currently best known as "air traffic control." The term *air traffic control* is used when the information discussed is relevant to the current mode of operation.

Although there is now quite an extensive literature on the effects of automation on air traffic systems (e.g., Hopkin, 1982, 1989, 1991, 1994a, 1994b; ICAO, 1993; Wise, Hopkin, & Smith, 1991), in fact most air traffic systems remain substantially manual. The main impact of the forms of automation that have already been introduced has been on the information gathered and stored about the air traffic rather than directly on the tasks of the controller. Nothing approaching the scale of automation that already exists in many aircraft cockpits has been installed or is presently contemplated in air traffic systems (Pitts, Kayten, & Zalenchak, 1993), although nearly complete automation of air traffic systems has been envisaged and planned in broad terms (Pozesky, 1989).

Much of the information currently used by the controller is derived now from data that have been gathered, stored, collated, and presented automatically. At the present time, many forms of computer assistance are being planned or have recently been introduced that are intended to support human cognitive functions or to promote the strategic rather than tactical control of air traffic. Full automation of active cognitive tasks, as distinct from tasks that are so routine as to encourage forms of automaticity when humans perform them, is not intended anywhere on a major scale, partly because the legal responsibility for air traffic control safety will remain vested in controllers, who must retain some means to fulfill their legal obligations.

AUTOMATION AND COMPUTER ASSISTANCE

For the purposes of discussing the effects of automation on air traffic systems and particularly on the performance of relevant tasks by the controller, the notion of automation is interpreted as covering both full automation and computer assistance. This is important because the latter generally has more direct effects on task performance. Some of the main differentiating characteristics of automation and of computer assistance in air traffic applications are described later.

Automation is the concept applied to functions that exclude direct human intervention, which are quite numerous in air traffic systems. Because the products of automation are used by the human, who generally cannot intervene in the processes that lead to those products, automation is often applied unselectively if the processes are flawed or inapplicable. The controller may have to use them, and compensate for the flawed products. Where there is some selectivity in the treatment of data in order to accommodate the requirements of different functions, this selectivity is itself automated, and the controller has to intervene actively to modify the automated data treatment. In air traffic control, there has been widespread automation of simple routine functions that are continuous or frequently repeated, but that were originally fulfilled by controllers, albeit in very different forms. An example would be the updating of the altitude of an aircraft on a radar control display label instead of reliance on the spoken messages. This has been accepted to the point of being taken for granted and no longer thought of as automation, but it is. Common examples of automated functions include data gathering, storage, smoothing, compilation, correlation, summarizing, synthesizing, updating, retrieval, and some aspects of data presentation.

Computer assistance as a concept is best reserved for, and applied to, functions where the human remains as the hub, and it is supported by the computer. It is allied to, although not always synonymous with, the concept of human-centered automation (Billings, 1991). The human roles, tasks, and functions are central, and they may be modified but not removed entirely. A criterion of computer assistance is that some form of human participation remains indispensable in that the processes or functions cannot be completed if the human is not present. In air traffic control, the concept of computer assistance usually refers to complex cognitive functions such as problem detection, decision making, problem solving, prediction, planning, scheduling, and resource management. Many forms of computer assistance are not intended for every control but are for specific tasks, functions, and jobs, whereas automation is usually unselective in this respect. In addition, computers usually assist individual controllers rather than teams or supervisors.

Both automation and computer assistance are mainly applied to the air traffic control of commercial, military, and well-equipped general aviation aircraft, especially where traffic can become heavy. There may be little of either where the air traffic control service is intended for other kinds of flying, particularly when the air traffic control facility is not permanently staffed. No matter how sophisticated the automation and computer assistance of some parts of the air traffic system become, nor how equipment is installed in the aircraft or at the large variety of air traffic control facilities on the ground, a unified integrated service to keep all aircraft safely separated has to be provided.

AUTOMATION AS A NECESSITY

Many of the common reasons for automation in other contexts also apply to air traffic systems. They include enhanced safety and efficiency, costeffectiveness, beneficial options resulting from technological advances, improved productivity, and quantifiable gains in task performance. But there is a further reason that, in air traffic systems, is particularly compelling.

In many parts of the world, current air traffic control systems are functioning near, at, or even beyond their planned maximum traffic handling capacity, and all current projections foresee continuing and cumulative increases in air traffic for a long time (McAlindon & Gupta, 1993). To leave current air traffic control systems unchanged is not a viable option, however safe and efficient they may now be. The apparently obvious solution of recruiting more controllers and partitioning even further the region of airspace for which each is responsible is also impractical, because any gains would be more than cancelled by the additional liaison, coordination, and communications workload resulting from the partitioning. Somehow, each controller must maintain and preferably enhance the existing high standards of safety and service while handling more aircraft. What this amounts to is that each controller must spend less time dealing with each aircraft. More and better information is available about the aircraft traffic, but it all has to be seen and understood more quickly. The heavier the air traffic is, the closer the proximity between aircraft, and the fewer the safe maneuvring options for keeping them apart and maintaining a smooth traffic flow. The controller needs help.

In terms of task performance, the main ways of helping the controller to do more in less time are the following:

- Completely remove some existing tasks and functions from controllers. This could be done either by allocating them to another person such as an assistant (which is not usually practical because it adds to the communications load and can conflict with legal responsibilities) or by allocating them wholly to machines and computers (which has already been done extensively for routine data gathering, handling, transfer and updating).
- Improve the data by making them more accurate, precise, consistent, frequently updated, trustworthy, reliable, and generally of better quality. This is happening particularly through better sensors such as satellite-derived data and better communications such as data links.
- Reduce the time needed to perform essential human tasks and functions by simplifying them, shortening them, requiring them to be done in less detail or less often, omitting parts of them, and changing the human–machine interfaces so that information is obtained and understood more quickly and actions are carried out more quickly.
- Change the nature of air traffic control so that it requires fewer human interventions and less critical timing of those interventions, by planning aircraft traffic flow strategically and by reducing the need for direct tactical instructions to single aircraft.

- Share tasks and functions with machines so that some aspects of them become machine functions that are delegated to the machine and done entirely by the machine.
- Use machines to aid and support human activities in ways that have been matched with human capabilities and limitations, and have some flexibility in relation to task demands so that the human–machine relationships are less rigid.

All of these stratagems have been, and will continue to be, adopted in air traffic systems. Associated issues are how to combine them and what the best criteria are for selecting the optimum stratagem or combination of stratagems to resolve different problems.

A TAXONOMY OF HUMAN–MACHINE RELATIONSHIPS

The human–machine relationships with any bearing on air traffic systems can be listed in order to clarify what the actual relationships are, what alternative relationships are feasible, and what problems can be anticipated in matching human and machine. The list is neither exhaustive nor permanent. Further advances in technology and in automation will add other human–machine relationships to the list, such as those that virtual reality makes feasible.

The relationships are listed in the approximate chronological order of their occurrence in the history of human–machine relationships. It is important to emphasise that the addition of new relationships does not remove any existing ones but instead expands the available range, and there may be circumstances in quite highly automated systems where some of the earliest kinds of relationship are still the best.

- The human adapts to the machine.
- The human and the machine compete for functions.
- The human is replaced by the machine.
- The human and machine complement each other.
- The human supports a failed machine.
- The human is adapted to by the machine.
- The human and the machine are symbiotic or hybrid.
- The human and machine duplicate functions in parallel.
- The human and the machine are mutually adaptable.
- The human and the machine are interchangeable.
- The human and the machine have fluid and unfixed relationships to each other.
- The human and the machine together form a virtual world.
- The machine takes over in the event of human incapacitation.

Good human–machine relationships are not serendipitous. None of these relationships will necessarily be effective unless they have been planned to be. Human and machine are not naturally competitive, complementary, symbiotic, or mutually adaptive, but can become so only if the functions, tasks, and workspaces are designed to permit and promote the intended relationships. Most real-life work environments, including air traffic systems, contain examples of more than one of these relationships, sometimes within the same task. This can be acceptable, provided that the distinctions between them have been carefully delimited. It is prudent to presume, until there is evidence to the contrary, that any form of automation or computer assistance will change the nature of the associated human–machine relationships.

RESIDUAL HUMAN ROLES

Whenever the future roles for human controllers in air traffic systems are considered, the need for human intervention in emergencies and unusual circumstances is always raised. Relevant human attributes that could permit effective human intervention are cited, and commonly include intelligence, adaptability, flexibility, a capacity to innovate, and comprehensive professional knowledge and experience of air traffic control. Yet many of these attributes, hitherto treated as exclusively human prerogatives, are also becoming machine attributes. Guidelines applicable to air traffic systems do not yet exist that could guarantee near-optimum matching between human and machine when both possess these attributes. Nor are there adequate principles for reconciling discrepancies between the machine database and the human database of knowledge and experience when they disagree. Suitable criteria to distinguish between optimal and improvable human–machine matching are lacking. Resources may be wasted in trying to improve the unimprovable.

Attempts to intervene in emergencies by exercising human ingenuity flounder if no provision has been made within the human–machine interface to implement the controller's innovative actions, or if there is provision for them but the machine treats them as illegal. An incipient form of human–machine mismatching in future systems is machine prevention of human attempts to exercise the very attributes for which the human has been retained in the system. In system terms, innovative human actions can be wayward and undermine strategic planning retrospectively. It does not assist the human to have an innovative action ruled as illegal, when the human knows very well that it normally is illegal but is nevertheless justified by the emergency.

THE ROLES OF HUMAN ACTIVITY

Casting humans in emergency roles epitomizes the ultimate reliance on the ability of the human to adapt, and to do whatever the machine cannot do (Hopkin, 1982). Although it may be necessary for the human to fulfill these roles, it is no longer sufficient to assign roles to humans for this kind of reason, to group them into tasks and jobs somehow, and to hope that the resultant miscellany is feasible for humans to do and that it is teachable. Successful intervention in an emergency is dependent on the maintenance of knowledge of the current system states and scenario, which the human can achieve consistently only by maintaining direct and active involvement in tasks.

Even to fulfill emergency roles, the human must have work that is sensible in human terms rather than in system ones. For people, continuous inactivity becomes intolerable, and controllers will not stay if they have no work or their tasks seem meaningless. If they have been assigned inactive and unparticipative monitoring roles, they cannot maintain attention indefinitely, yet they must do so if they are to intervene effectively in emergencies. The performance of routine repetitive tasks requiring physical actions helps to entrench memory and understanding, which are reduced whenever such actions are automated because the associated information does not have to be processed by the controller so deeply or in such detail (Narborough-Hall, 1987).

Knowledge that is never used is gradually forgotten (Hopkin, 1988). Skills slowly atrophy without opportunities to practice and maintain them (Bainbridge, 1989). More shallow processing of information can give the controller the impression that the mental picture of the air traffic is more likely to be lost (Mogford, 1994; Whitfield & Jackson, 1982). If this occurs, much of the displayed air traffic scenario becomes meaningless, and the picture of the traffic may have to be painstakingly rebuilt, one aircraft at a time, to restore the self-same conditions that produced the initial loss of picture. This phenomenon is familiar in air traffic control. It is related to, but not the same as, the much more recent concept of situational awareness, which has also been applied to air traffic control (Garland & Hopkin, 1994).

The essential core tasks of air traffic control have to be performed efficiently and safely, with or without computer assistance. Because these tasks are so essential, measures of their performance are often insensitive to other influences, such as changes in the conditions under which they are done, in the computer assistance applicable to them, or even in the professional competence of the individual controller. The adequacy of the performance of these core tasks must always be verified (Wise, Hopkin, & Stager, 1993), because the causes of any impairment in performance would have to be treated very seriously. But if the controller is attempting to maintain professional standards, the introduction of various forms of computer assistance may not affect the measured performance of the core tasks very much unless they make it impossible to achieve professional standards. Otherwise, the quality of the air traffic control service may change very little, although there may be significant changes in the effort required to provide that service and in the detailed manner in which the performance is achieved. Measures of errors, omissions, the time scale of decisions, options considered and discarded, and the performance of tasks that are desirable but not within the essential core may all yield more sensitive indices of the benefits of computer assistance than direct performance measures of core tasks can (Hopkin, 1990).

SOME INFLUENTIAL HUMAN ATTRIBUTES

In matching human and machine to achieve the best performance, there has been a preoccupation with their comparative characteristics. This has often resulted in the allocation of functions to whichever one performed them better, even when both could perform them well or when they were suitable for neither (Fitts, 1951). Comparisons inevitably tend to omit as irrelevant any attributes that cannot be compared because they apply to only one of the alternatives and not both. In practice, this has led to neglect of the influence of human characteristics with no machine equivalent or counterpart (Jordan, 1968). This problem has been aggravated, because comparisons require a common language and terminology, but the only concepts that could be applied to both human and machine were machine concepts (Hopkin, 1982). To describe machines anthropomorphically in human terms for the purposes of comparison seems absurd, but there do not seem to be equivalent machine attributes of, for example, a dislike of enforced idleness or intolerance of boredom (Hopkin, 1980a).

A complication is that human and machine functions can seem identical when described in system or machine concepts but not when expressed in human terms. Suppose that a controller has been accustomed to deciding unaided how a potential conflict between two aircraft should be resolved. The controller selected and gathered relevant evidence, considered options, made a choice, and implemented it. The controller knew which information had already influenced that decision, and viewed any new or revised information in the light of that knowledge as confirming the original decision or warranting its revision. The controller not only knew what evidence had been used, but also knew why it had been used. The outcome in terms of the measurement of performance consists of the actions of the controller to implement the choice. This measured outcome can be identical whether the controller performs it unaided or with computer assistance that formulates alternative options and recommends the best one to the controller. The controller is not forced to accept the recommendation but is expected to accept it most of the time because it really is the optimum in system terms. Expressed in system and activity measures, the unaided task and the task with computer assistance are the same. Subjectively they can seem so different as not to be comparable at all (Wise et al., 1991).

Some of the most powerful influences on computer-assisted task performance are these exclusively human attributes with no machine equivalent. Examples include professional ethos, norms, and standards; the need to gain and keep the trust, respect, and esteem of peers; tacit awareness of colleagues' activities that can affect one's own; locally sanctioned nonstandard procedures within a group or team; the hierarchy of loyalties; and the concordance of attitudes toward the work, the equipment, the job itself, the management, and the conditions of employment (Hopkin, 1980b).

Innovations have to be beneficial, practical, and teachable, but also acceptable. The generally cautious attitudes of controllers toward new equipment have been well publicized, but they are an effective counter to previous overenthusiasm for some forms of computer assistance that promised more than they delivered in terms of improved performance. Different kinds of evidence can sometimes be contradictory, and they have to be reconciled. Some of the first crude and rather garish forms of color coding on electronic displays regularly produced both subjective certainty that task performance had been improved by them and objective measures of task performance that showed no improvement (Narborough-Hall, 1985). The lesson is that performance measurement is only as good as the measures employed. If the color coding in fact aided aspects such as memory and understanding, this could explain the findings, because subjectively this would seem like an improvement; objectively it would not, because the performance measures did not cover these aspects. Technological progress since has presented further, more refined options in the use of color coding on air traffic control displays (Hopkin, 1994a; Reynolds, 1994).

Controllers' attitudes are cautious because they have learned to be cautious from experience. Claimed reductions in task demands and resultant workload often turned out to be actual increases because of the additional data entry tasks on which the computer assistance depended. Apparently, simple air traffic control procedures proved to be far more complex than originally envisaged, and there were difficulties in writing appropriate software that could improve on the controller's unaided performance. Occasionally the performance of a task by the unaided controller was so near to the theoretically attainable maximum performance that no substantial performance benefits could possibly accrue from computer assistance for that task. To counter skepticism and to gain acceptance, the putative benefits of computer assistance now require supporting tangible evidence. There continue to be difficulties in capturing the full functionality of manual tools in computer-assisted form. The electronic flight progress strip provides a recent example of the underestimation of functionality and complexity (Hopkin, 1991; Vortac, Edwards, Jones, Manning, & Rotter, 1993).

HUMAN ERROR

There is now more widespread recognition that the quest to eliminate human error entirely never was realistic, and that it is of more practical importance to identify any errors and contain their effects (Reason & Zapf, 1994). There is also acknowledgement that the kinds of human error that are possible within any air traffic system that sooner or later will occur are largely predetermined by decisions made during the planning, specification, and detailed design of the system, and changes in these system aspects may be necessary to prevent the recurrence of any human errors that could be dangerous (Stager, 1991). In this context, human–machine interfaces that are meaningful only after extensive training may not represent the best approach to the early detection and correction of human errors.

It is usually important not to treat human errors merely by classifying them. Each error deserves individual study. The fact that it could occur at all and was not prevented by the system can be of great significance for safety, and this information can be lost in any taxonomy of errors.

THE ALLOCATION OF FUNCTIONS

The kinds of human–machine relationship that were initially feasible encouraged the allocation of functions to humans or machines, whichever was superior. This principle implies a progressive reallocation of more functions to machines, purely as a corollary of technological advancements. But technological progress has rendered this kind of thinking outdated, because the widening range of human–machine relationships permits the consideration of alternative principles for allocating or sharing functions, to the extent of questioning the necessity for preallocating them at all when most could be fulfilled by either human or machine. This approach offers new solutions to the problem of providing sensible human roles and jobs in systems with extensive computer assistance.

A feature of the human-centered approach is that computer assistance becomes a tool to optimize human roles. One way to use such a tool would be for the human to allocate to the machine whatever the human chooses not to do, and, with appropriate training, controllers could learn to apply such a tool to ensure that excessive human workload would never again be a serious constraint in future air traffic systems, because human workload could be smoothed and adjusted by the flexible use of supporting computer assistance. Training would be needed, partly because it is not intuitively obvious what the best balance between human and machine would be, and partly to counter the human propensity to treat difficult problems as personal challenges not to be delegated for machine resolution. Nevertheless, such flexibility could maintain human skills, avoid periods of idleness or excessive workload, facilitate more flexible adaptation to gross changes in traffic demands, and promote job satisfaction.

UNDERSTANDING OF THE MACHINE BY THE HUMAN

In successful human–machine relationships, the controller must have sufficient understanding of how the computer assistance functions in order to use it correctly (e.g., within its limitations) and not misuse it. This applies particularly when computer-assisted recommendations, intended to be accepted or rejected by the controller, are made to assist problem solving, decision making, or predictions. Acceptance or rejection should not be arbitrary, but nor should it be based solely on the same evidence that has been employed to formulate the recommendation itself. An adequate rationale for the role of the human controller would normally include additional evidence and criteria known only to the controller and known by the controller not to have been included in the computer-assisted recommendation. For this to work, the controller has to have been trained to already know which criteria have been included, or must be able to discover very easily if the controller's evidence for rejecting a recommendation has already been taken account of in its formulation. The controller's knowledge of the system must extend beyond its functionality to cover how far it should be trusted and how any computer failure could be recognized.

Whenever computer assistance proposes solutions to problems for the controller to accept or reject, these proposals have been the subject of extensive development and validation and generally are highly efficient, safe, and the best practical option (Wise et al., 1993). Controllers are trained to accept these solutions under most circumstances and they learn to trust them as they acquire more experience with them. They gradually accept this computer assistance more wholeheartedly and examine the proposed solutions less critically and less rigorously. This creates a problem, for a controller who always accepts the computed solutions may be viewed as highly effective because the computer assistance is being used as it was intended to be used. However, this same invariable acceptance could also signify that the controller does not understand the recommended solutions and fails to question the reasons for them. Some forms of computer assistance have the intrinsic properties of concealing human inadequacy and compensating for human incompetence (Wise et al., 1991). Such incidental consequences of computer assistance for air traffic control task performance illustrate that its effects can range far beyond those directly intended and be far from self-evident.

The most successful forms of human-centered computer assistance must be capable of responding to human intentionality. Ideally, the human controller could tell the system what the human objectives are and enlist its full collaboration in achieving them. Although most of the ensuing collaborative activities would probably be machine actions, the controller would retain both control and responsibility. This kind of computer assistance is not yet imminent. In other contexts, a few forms of assistance are beginning to take this form, but they are not yet common in air traffic systems (Hopkin, 1995).

MEASURES OF PERFORMANCE

Numerous measures of performance are applicable to air traffic systems, and they have to be examined in the context of other measures. A cluster of system measures relate to the safety and efficiency of the service provided. These include traffic handling capacity (often expressed as number of aircraft under control per hour or peak number of aircraft at any one time), delays to traffic for air traffic reasons, maintenance of separation standards, conflict resolutions, utilization of the handling capacity available, the occupancy times of communication channels and the proportion of time that they are in use, and other system measures. There are also measures of controller activity and behavior, such as data inputs, scanning patterns, message contents, the handling and annotation of flight strips, and other controller activities that can be measured as timed events.

These system and activity measures contribute toward but do not comprise measures of controller performance, which must additionally include some form of judgment or assessment. Performance measures thus include errors. They cover not only what the controller did but whether the controller's actions were correct; well timed; properly coordinated; and in full compliance with all rules, instructions, and professional practices. Perhaps most significantly, measures of controller performance should include omissions, and refer to what the controller failed to do as well as to what was done.

Measures of system behavior and activity usually seek to show changes, but measures of human performance are often more judgmental and put more emphasis on improvements and their quantification rather than on changes alone. Performance measures can of course be supplemented by many other kinds of measure to explain and interpret them, including subjective assessments of various kinds, biodata, measures of individual differences, social measures, physiological and biochemical indices, and measures related to management and conditions of employment (Hopkin, 1980b).

TEAM ROLES

Many of the effects of computer assistance on performance are mediated through associated changes in the roles and functions of teams. The complete or partial loss of some team functions through computer assistance does not always become apparent until the changes have been made, because the changes in team functions are not usually intentional. Although much of the work in less automated systems is done by teams, most forms of computer assistance are intended to assist not teams but individuals (Hopkin, 1993b). It would seem prudent to identify all the affected team functions beforehand in order to formulate a consistent policy regarding which should be discarded, which retained, and which replaced with a surrogate.

The most common team functions in air traffic control that computer assistance is likely to affect include the following:

- The criteria for assessing the individual controller's merit, and for planning career development accordingly.
- The factors that influence the development and maintenance of high morale.
- The prime loyalties of the controller to the immediate team, the watch, the facility, management, and the profession.
- The development and communication of professional norms and standards, and the means of conveying them to newcomers.
- The observability of the air traffic control workspace to others, and their ability to understand what is happening and to form judgments about it on the basis of their observations.
- The practicality of mutual assistance and work sharing between controllers.
- Judgments by controllers of each other; for example, professional competence, trustworthiness, and relative strengths and weaknesses as controllers.
- The building of trust between controllers and machines.

- The extent to which others can verify and validate the actions of individual controllers (Wise et al., 1993).
- The practicality and feasible forms of team training (Hopkin, 1994b).

MEANINGFULNESS AND OBSERVABILITY

Computer assistance is being expanded gradually in air traffic control, often associated with new and technically advanced sources of data such as very accurate positional information about aircraft derived from satellites and frequently updated information transponded automatically between air and ground through data links. The fundamental objective is still the safe, orderly, and expeditious flow of air traffic, but the human functions required to achieve this objective can change. Only a very small proportion of the vast amount of information potentially available about each aircraft and about the progress of its flight is needed to exercise control. Changes in air traffic control methods, tasks, and procedures resulting from computer assistance may, if they become gross (such as the change from tactical to strategic control), entail commensurate gross changes in the ways in which aircraft should be represented to the controller. There often seems to be a lag between policies and actuality. A current example is that although more strategic control is planned, nearly all the actual and proposed forms of computer assistance still remain tactical. Current air traffic control workspaces are almost meaningless to those with no knowledge of air traffic control. They contain negligible self-evident information about their purpose, their functions, the tasks, or the procedures. According to current plans, they will not become much more informative for their users or others in future (Hopkin, 1989). Perhaps computer assistance should be employed to render air traffic control functionality more self-evident.

An important characteristic of current air traffic control is its observability. It may be meaningless to the uninformed who can watch it and listen to it, but very meaningful to the controller's peers and supervisor, to the extent that they can read the traffic picture by being at the controller's workstation and seeing and hearing what is done. Many forms of computer assistance in fact alter observability greatly. Some of the consequences of this are mentioned next.

A RESUME OF THE EFFECTS OF COMPUTER ASSISTANCE ON ATC PERFORMANCE

Substantial progress has been achieved in recent years in identifying the origins of human factors problems associated with automation and computer assistance (Hopkin, 1993a). There has also been some progress in tracing their effects on performance, although much remains to be done. It is noteworthy how much wider the range of accepted consequences is now than it would have been only a few years ago. The following are main influences on human factors problems that affect performance:

- The human must do whatever has to be done that machines cannot do.
- The human can do only what the computer allows, and if the human–machine interface makes no provision for an action then it cannot be done, however apposite it may be.
- Many dialogues now must be conducted through the human–machine interface instead of with other people, such as pilots and colleagues.
- The reduced reliance on human speech for air traffic communications is accompanied by a loss of the other kinds of information conveyed through speech, by its pace, pauses, phraseology, vocabulary, accents, and so on, and of the impressions gleaned from speech that influence judgments of competence and confidence.
- Some forms of computer assistance have resulted in more instead of less work for the controller.
- The cognitive effects of reduced direct involvement and reduced need for information processing can include some loss of picture, of situational awareness, and of understanding of the traffic scenario.
- Any form of computer assistance requires some training and some new learning, but it also usually requires the forgetting or unlearning of formerly entrenched and habitual procedures that had reached a state of automaticity in their performance; much less is known about how to train people to forget than about how to train them to remember (Hopkin, 1988).
- With computer assistance, the work of the controller becomes much less directly observable by others, and this has major consequences for the supporting roles that others can fulfill.
- There may be less flexibility in dealing with nonstandard situations and emergencies, particularly if the controller's solutions are interpreted by the machine as illegal actions.
- Many forms of computer assistance have the inherent property of being capable of hiding human inadequacy and covering for human incompetence.
- The issue of the acceptability of the computer assistance to the user is influenced by such factors as responsibility, the retention of skills and of opportunities to use them, and the challenge and interest of the job.

- New sources of human stress can be introduced if the controller has the responsibility for functions but does not fully understand or trust them, and either cannot change them or does not know how to change them.
- Many of the traditional roles and functions of teams in air traffic control are undermined by computer assistance, and some may disappear.
- Changes in the controller's job as a result of computer assistance may require changes in the selection procedures for controllers, as the ability to work well with computer assistance may become a testable requirement.
- Many forms of computer assistance are opaque to the user, in that it is not apparent how they could fail, what they would look like if they did, and how the user could know which functions remained unaffected by the failure and could still be used.
- A significant effort has been devoted to studying the effects of computer assistance on the controller, and too little on its effects on supervisors, assistants, and teams; this imbalance should be redressed.
- Most forms of computer assistance have been developed to suit the controllers in their country of origin, and further human factors complications can be expected if attempts are made to transfer them to other cultures without modification (Kaplan, 1991).

THE UNDERUTILIZATION OF EVIDENCE

In recent years, the importance of human factors considerations in assessing the effects of automation on the performance of air traffic control tasks has become more widely accepted both nationally (FAA, 1991) and internationally (ICAO, 1993). The discipline of human factors can afford to be less defensive, but must deliver what it has promised. It must still proselytize and actively market its wares to counter skepticism and bridge the chasm between the mass of available human factors knowledge and its application. This problem is not unique to air traffic systems, to aviation, or even to complex systems with automation, but seems to apply almost everywhere (Heller, 1991). More strenuous efforts must be directed toward solving it, in parallel with continuing research studies, because much of the cumulative value of the evidence from research is lost if it is never applied.

IS IT ALL FOR NAUGHT?

There will be significant changes in future air traffic systems. As new technologies (e.g., satellite-based navigation systems) provide new abilities (e.g., every pilot may have a display of the location of all other aircraft, or every aircraft may be required to fly a great circle route to its destination), the very nature of air traffic systems may change. Serious consideration is currently being given to the concept of the pilot operating the flight without need for most current prior flight clearances, and the controller's job would be to facilitate the process (RTCA, 1995). This not only poses formidable human factors problems (that would need to be solved as a condition of such a system), but poses a particularly acute dilemma throughout the inevitable transition phases when air traffic control must reconcile flights operating "independently" with those under direct "control." Such issues typify the continuing challenge and interest in applying human factors to air traffic systems, so that human functions remain successfully integrated and reconciled with new air traffic system ideas and technology advances.

REFERENCES

Bainbridge, L. (1989). Development of skill, reduction in workload. In L. Bainbridge, & S. A. R. Quintanilla (Eds.), *Developing skills with information technology* (pp. 84–118). Chichester, England: Wiley.

Billings, C. E. (1991). *Human centered aircraft automation: A concept and guidelines* (NASA TM 103885). Moffett Field, CA: NASA-Ames Research Center.

Federal Aviation Administration (FAA). (1991). *The national plan for aviation human factors*. Washington, DC: Author.

Fitts, P. M. (Ed.). (1951). *Human engineering for an effective air navigation and traffic control system*. Washington, DC: National Research Council.

Garland, D. J., & Hopkin, V. D. (1994). Controlling automation in future air traffic control: The impact on situational awareness. In R. D. Gilson, D. J. Garland, & J. M. Koonce (Eds.), *Situational awareness in complex systems* (pp. 179–197). Daytona Beach, FL: Embry-Riddle Aeronautical University Press.

Heller, F. (1991). The underutilization of applied psychology. *The European Work and Organizational Psychologist, 1*(1), 9–25.

Hopkin, V. D. (1980a). Boredom. *The Controller, 19*, 6–10.

Hopkin, V. D. (1980b). The measurement of the air traffic controller. *Human Factors, 22*, 547–560.

Hopkin, V. D. (1982). *Human factors in air traffic control* (AGARDograph Number 275). Paris, France: N.A.T.O.

Hopkin, V. D. (1988). Training implications of technological advances in air traffic control. In *Proceedings of Symposium on Air Traffic Control Training for Tomorrow's Technology* (pp. 6–26). Oklahoma City, OK: Federal Aviation Administration.

Hopkin, V. D. (1989). Implications of automation on air traffic control. In R. Jensen (Ed.). *Aviation psychology* (pp. 96–108). Aldershot, England: Gower.

Hopkin, V. D. (1990). Operational evaluation. In M. A. Life, C. S. Narborough-Hall, & I. Hamilton (Eds.), *Simulation and the user interface* (pp. 73–83). London: Taylor & Francis.

Hopkin, V. D. (1991). Automated flight strip usage: Lessons from the functions of paper strips. In *Challenges in aviation human factors: The national plan* (pp. 62–64). Washington, DC: A.I.A.A.

Hopkin, V. D. (1993a). Human factors implications of air traffic control automation. In M. J. Smith & G. Salvendy (Eds.), *Human–computer interaction: Applications and case studies* (pp. 145–150). Amsterdam: Elsevier.

Hopkin, V. D. (1993b). Human factors issues in air traffic control and aircraft maintenance. In D. J. Garland & J. A. Wise (Eds), *Human factors and advanced aviation technologies* (pp. 19–28). Daytona Beach, FL: Embry-Riddle Aeronautical University Press.

Hopkin, V. D. (1994a). Color on air traffic control displays. *Information Display, 10*(1), 14–18.

Hopkin, V. D. (1994b). Organizational and team aspects of air traffic control training. In G. E. Bradley & H. W. Hendrick (Eds.), *Human factors in organizational design and management—IV* (pp. 309–314). Amsterdam: North-Holland.

Hopkin, V. D. (1995). *Human factors in air traffic control.* London: Taylor and Francis.

International Civil Aviation Organization (ICAO). (1993). *Human factors in air traffic control.* (Report No. 241-AN/145). Montreal, Canada: International Civil Aviation Organization, Human Factors Digest Number 8.

Jordan, N. (1968). *Themes in speculative psychology.* London: Tavistock.

Kaplan, M. (1991). Issues in cultural ergonomics. In J. Wise, V. D. Hopkin, & M. L. Smith (Eds.), *Automation and systems issues in air traffic control.* (NATO ASI Series F, Vol. 73, pp. 381–393). Berlin: Springer-Verlag.

McAlindon P. J., & Gupta, U. G. (1993). A structured approach to the verification and validation of expert systems. In. J. A. Wise, V. D. Hopkin, & P. Stager (Eds.), *Verification and validation of complex systems: Additional human factors issues* (p.15–23). Daytona Beach, FL: Embry-Riddle Aeronautical University Press.

Mogford, R. (1994). Mental models and situation awareness in air traffic control. In R. D. Gilson, D. J. Garland, & J. M. Koonce (Eds.), *Situational awareness in complex systems* (pp. 199–207). Daytona Beach, FL: Embry-Riddle Aeronautical University Press.

Narborough-Hall, C .S. (1985). Recommendations for applying colour coding to air traffic control displays. *Displays, 6,* 131–137.

Narborough-Hall, C. S. (1987). Automation implication for knowledge retention as a function of operator control responsibilities. In D. Diaper & R. Winder (Eds.), *People and computers, III* (pp. 269–282). Cambridge, England: Cambridge University Press.

Pitts, J., Kayten, P., & Zalenchak, J. (1993). The national plan for aviation human factors. In J. A. Wise, V. D. Hopkin, & P. Stager (Eds.), *Verification and validation of complex systems: Human factors issues,* (NATO ASI Series Vol. F-110, pp. 529–540). Berlin: Springer-Verlag.

Pozesky, M. T. (Ed.). (1989). Special issue on air traffic control. *Proceedings of the IEEE, 77,* 1603–1775

Reason, J. T., & Zapf, D. (Eds.). (1994). Errors, error detection and error recovery. *Applied Psychology: An International Review, 43,* 427–584.

Reynolds, L. (1994). Colour for air traffic control displays. *Displays, 15,* 215–225.

RTCA (1995). *Free flight.* Report of the RTCA Board of Directors' Select Committee on Free Flight.

Stager, P. (1991). Error models for operating irregularities: Implications for automation, in J. A. Wise, V. D. Hopkin, & M. L. Smith (Eds.), *Automation and systems issues in air traffic control* (NATO ASI Series Vol. F-73, pp. 321–338). Berlin: Springer-Verlag.

Vortac, O. U., Edwards, M. B., Jones, J. P., Manning, C. A., & Rotter, A. J. (1993). En route air traffic controllers' use of flight progress strips. Graph-theoretic analysis. *The International Journal of Aviation Psychology, 3,* 327–343.

Whitfield, D., & Jackson, A., (1982). The air traffic controller's 'picture' as an example of a mental model. In G. Johannsen, & J. E. Rijnsdorp (Eds.), *Analysis, design and evaluation of man–machine systems* (pp. 45–52). New York: Pergamon.

Wise, J. A., Hopkin, V. D., & Smith, M. L. (1991). *Automation and systems issues in air traffic control.* Berlin: Springer-Verlag.

Wise, J. A., Hopkin, V. D., & Stager, P. (1993). *Verification and validation of complex systems: Human factors issues.* Berlin: Springer-Verlag.

16 Driver-Centered Issues in Advanced Automation for Motor Vehicles

P. A. Hancock
University of Minnesota

Raja Parasuraman
Evan A. Byrne
Catholic University of America

INTRODUCTION

Traffic accidents involving motor vehicles are the major cause of death for persons up to 38 years of age, and are the leading cause of accidental death up to the age of 78. In 1993, the number of motor vehicle-related deaths totaled 42,000, whereas the number of serious, disabling injuries was over 2 million (National Safety Council, 1994). The estimated cost of these accidents is some $167.3 billion dollars per year, which is a monetary metric that fails to include the manifest human suffering involved. The statistics paint an even grimmer picture when the indirect costs and the economic impact of disabling injuries resulting from road traffic accidents are considered.

Measures introduced in recent years, such as mandatory seat belt usage, lower speed limits, and stricter penalties for drunk driving, have undoubtedly reduced the death toll on the road. Improvements in vehicle and roadway design over the past three decades have also contributed to a 62% decrease in the fatality rate (OCAR, 1994). Nevertheless, the absolute number of accidents continues to grow because of a steady rise in the number of vehicles on the road and the consequent increase in traffic density. Not surprisingly, therefore, many new solutions to the problems of traffic congestion and accidents have been proposed and are being evaluated. Foremost among these are the much-touted Intelligent Vehicle Highway Systems (IVHS), now renamed Intelligent Transportation Systems (ITS).

The overall objective of ITS is to use advanced automation technologies to reduce traffic congestion, accidents, and energy costs, thereby enhancing the safety and efficiency of ground transportation in particular and intermodal transportation in general (IVHS America, 1992). Whether these goals will be achieved in a timely and cost-effective manner is a matter for debate and concern (Lowe, 1993). There is little question that much of ITS is technologically feasible. As discussed elsewhere in this volume (Hancock, chap. 22), technological innovation must be considered in the context of social goals, in this case the public's need for safe and efficient transportation. What is uncertain is the extent to which ITS will be designed with the needs of drivers in mind, and whether ITS will indeed improve road safety (cf., Lund & O'Neill, 1994).

INTELLIGENT TRANSPORTATION SYSTEMS (ITS)

Many of the technologies subsumed under the rubric of ITS involve the incorporation of advanced automated devices into motor vehicles. ITS also includes associated sensor, control, and communication systems both within and external to the vehicle. Driver aids range in scope from near-term systems using currently available technology, such as electronic map displays for in-car navigation, and emerging use of collision-avoidance devices (e.g., VORAD, a device currently being marketed for trucks). Long-term possibilities, such as "platooning" of cars and automatic steering, require the maturation of developing technologies. Wright (1990) estimated that as a proportion of a car's total value, new in-vehicle electronics and automation will climb from the present 6% to 20% by the year 2000. Currently, the driver is isolated from much electronic information because he or she does not need to know most of the moment-by-moment functioning of all of the vehicle and traffic system components. The added computational capability envisaged by Wright and others will be dedicated to information brought specifically to the notice of the driver. Such information will be presented via devices that will provide route guidance, display highway markings and signs inside the vehicle, show traffic patterns, point to available roadside services, give advance warning of possible collision, and so on. What has yet to be fully articulated is how the driver will cope with this information load.

The U.S. Secretary of Transportation has been charged with directing the development of ITS technologies in the next decade as detailed in Section 6054(b) of the Intermodal Surface Transportation Efficiency Act of 1991. This act requires the development of prototype systems by 1997. A key concept provided for in this Congressional mandate is the requirement that development of ITS must include research on human factors to maximize the potential for successful implementation of these systems. We discussed elsewhere in some detail the human factors issues that must be considered (Hancock & Parasuraman, 1992). These considerations have led to the recommendation that user-centered design approaches that keep the driver involved and informed should be followed if ITS is to meet the twin goals of reduced traffic congestion and improved road safety (Hancock, Dewing, & Parasuraman, 1993; Hancock & Parasuraman, 1992; Owens, Helmers, & Sivak, 1993). In this chapter, we examine in greater detail the implications of these developments with respect to the theme of automation and its impact on human performance.

INFORMATIONAL SYSTEMS FOR THE DRIVER

All informational systems for the driver can be viewed as supporting the fundamental purpose of transportation: the safe and efficient passage of individuals and goods. In the present circumstances, there is the tendency to divide systems based on their time horizon: Devices that provide information for immediate guidance are grouped into collision-avoidance and collision-warning systems, although there remain some differences even between these two. Systems that focus on more prolonged time horizons are referred to as navigational or route-guidance systems. We wish to emphasize here that time of action is the basic difference between these forms of system, and thus they actually lie on a continuum. However, for the present purposes, we have divided discussion to accord with contemporary differentiation. Nevertheless, it is our contention that a full systems-based implementation of these technologies must emphasize this temporal commonality.

Prototypes and limited-feature devices for navigation and for collision warning are currently commercially available, although the technology for comprehensive avoidance systems that take evasive control requires further development. We discuss the ergonomic issues concerned with these forms of in-vehicle automation, focusing on route-guidance and collision-warning systems. Such technologies represent an intermediate level of automation on a continuum of complexity, from the relative simplicity of "everyday" automation such as automatic teller machines to the sophistication of automated devices discussed in other chapters in this book—in such environments as the aircraft cockpit (Sarter, chap. 13), manufacturing and quality control (Drury, chap. 19), and medicine (Guerlain & Smith, chap. 18). Much has been learned of human interaction with automation from applications in aviation. It is an open issue whether there are principles of interaction that generalize across domains (Hancock, 1996; Woods, chap. 1, this volume). It may be that certain special issues arise in motor-vehicle automation because of the time-critical nature of many driving tasks, and because of the wide range of abilities that users of motor vehicles possess. Clearly, these are critical issues in ITS implementation.

GUIDANCE AND NAVIGATION SYSTEMS

At present, many of the demonstration ITS projects involve in-vehicle navigation assistance or guidance systems. One prototype that has been extensively tested to date is Travtek (Andre, Hancock, & Smith, 1993; Inman, Fleishman, Dingus, & Lee, 1993; Perez, Fleishman, Golembiewski, & Dennard, 1993; Rillings, 1992). In Travtek and other related systems, database information about an area is accessible to the driver to provide route guidance to areawide destinations and amenities. Travtek and other prototypes within the Advanced Traveler Information System (ATIS) represent a major aspect of ITS development. Under ATIS, guidance and other electronic navigation aids will be provided to the driver in real time for both pretrip and en-route travel planning (Dingus & Hulse, 1993).

ATIS represents a relatively complex set of technologies that will be integrated into the current driver-vehicle system. Much simpler systems that make use of current in-vehicle technology are also possible (Burrus, Williams, & Hancock, 1994; Vaughan, May, Ross, & Fenton, 1994). For example, with the Radio Broadcast Data System (RBDS), a nationwide radio network, it is a simple matter for the car's FM radio to pick up current traffic or emergency information and display it on a digital read-out screen (Younger, 1994).

Systems such as ATIS and RBDS can be considered automated aids to provide enhanced traffic position awareness for the driver. Although the ITS plan does not postulate a safety gain as the primary goal of these devices, they could be considered to be a form of road-safety intervention. Specifically, a driver with this information may be less likely to perform abrupt maneuvers while navigating in unfamiliar areas. The validity of this hypothesis remains to be tested, however, because, with the exception of Travtek, most major empirical evaluations and operational tests of in-vehicle navigation systems are largely still in progress (but see Hofmann, Becker, & Schrievers, 1994). The impact of dynamic traffic and route information on several different aspects of driver behavior requires thorough evaluation. Designers will need to know when, where, why, and through what medium commuters and travelers will use this kind of information. Task analyses (e.g., Wheeler et al., 1994) are beginning to provide answers to these questions, but empirical evaluations of some of the major human factors issues need to be carried out for specific design guidelines.[1]

---

[1]We should note that this is a particularly fast-moving area of technological development, especially in light of much technology transfer from applicable military systems. Therefore, as developments proceed apace, many of the recommendations made here are currently under active scrutiny and evaluation. These efforts are well directed, but our purpose here is to facilitate the transfer of knowledge, principles, and information on human-centered automa-

Protecting the Primary Driving Task

One of the first of these issues asks a seemingly simple question: Where should navigational information be displayed in order not to compromise the primary driving task of vehicle control? Despite several attempts at analysis and some empirical evaluations, a simple answer is difficult to obtain (cf. Antin, 1993; Popp & Farber, 1991; Wierwille, 1993). On the one hand, any navigational display should be presented so that it can be readily and easily fixated by the driver; and given that long periods of fixation away from the primary road environment can be potentially dangerous, the display positioning should allow easy cycling of visual sampling between the outside view and the display. This requirement would appear to mandate placement of the display in front of the driver but not in the direct field of view, i.e., above the instrument panel or on the dashboard (see also Wierwille, 1993). This seems to be the design choice of devices that are currently emerging in the marketplace. Whether such a placement is distracting and diverts attention away from the primary driving task is also a key question. Empirical studies suggest that visual scanning demands are increased with electronic navigation displays (Fairclough, Ashby, & Parkes, 1993), but whether to the detriment of driving performance, and in particular driving response to emergencies, is not known. Drivers may in fact adapt their visual scanning behavior so as to protect the primary driving task. Wierwille (1993) carried out an extensive program of research on the visual sampling behavior of drivers while performing various in-car tasks. In general these studies have found that in-car glances away from the forward scene average about 1 second in duration and rarely exceed 1.6 seconds. For cars fitted with a navigation system, drivers increased their sampling of the roadway as primary driving became more difficult, and at the same time sampling of the navigation display decreased by approximately the same amount.

The self-regulation of in-vehicle display sampling behavior by drivers in response to changing driving task demands needs to be formally considered by designers of ITS systems. The adaptive function of this strategy is clear: Forcing the gaze outside the vehicle will maximize the probability of a critical signal in the roadway environment being detected. However, the question must also be asked whether any potential safety decrement may occur when this strategy is undertaken in an ITS equipped vehicle. Specifically, would the safety benefit provided by in-vehicle navigational display and route guidance information be nullifed when the intended flow or rate of information passage from the display to the driver is interrupted

---

tion to the realm of ground transportation where such approaches to functional automation are relatively new.

or otherwise canceled? If this information is designed to provide the driver with increased positional awareness as he or she drives, what happens to this sense of awareness when the sampling of the display is not frequent enough or cut off entirely? If position awareness is lost as a result, what impact on driving safety may result from the time and strategies required for its reacquisition? Thus, the adaptive strategies currently used by drivers to maintain safe passage in difficult driving environments must be considered by designers of in-vehicle ITS displays to minimize the potential cost to safety. Presenting information in alternative and redundant sensory channels may be required.

Interface Design Issues

Given the heavy visual demands of the primary driving task, the interface to navigation systems must be chosen so as provide the necessary information without excessive fixation away from the outside view. The Guidestar system that is installed in some Oldsmobile models provides turn-by-turn information on a dashboard-mounted visual display, and uses voice prompts to encourage only occasional viewing of the display. Nevertheless, the designers seem to have hedged their bets regarding potential safety problems by only allowing the driver to enter commands to the system when the vehicle is stationary. Zwahlen, Adams, and DeBald (1988) reported a significant deterioration in driving performance when drivers used a touch-panel CRT device to input commands while driving.

The use of voice commands or auditory displays also raises other important issues. Given a specific imperative command (e.g., turn right) it is possible that the driver could ignore traffic control devices, such as a stop sign or stop light, and continue on the preset route in obedience to the in-car message. The overriding of external traffic control devices by in-car commands poses a particular problem in that context-dependent, in-car messaging assumes a much more thorough knowledge of the external environment on behalf of the vehicle than is currently envisaged. However, in recognizing that this form of error has already occurred (Noy, personal communication, 1992), it is clear that the driver can easily attribute more intelligence (cf. Younger, 1994) to the vehicle than it actually has. The critical research issue concerns modality of information presentation and message content that promotes safe interaction with other road users. The addition of cautionary messages like "proceed when safe to do so" provides some clarification, yet the interaction between messaging, sensory modality, and driver decision making remains a problematic issue. If local rules are always given primary consideration, there could be a decrease in ATIS effectiveness, and potential safety benefits may not be realized. Also, it is anticipated that certain vehicles, (e.g., emergency vehicles and mass transit vehicles) will have some control over traffic control devices. Thus a paramedic unit coming up to a stoplight may be able to change it at will in order to facilitate safe passage. How such "special" conditions interact with ITS-equipped and ITS-nonequipped drivers has not been fully articulated. Also, security measures to prevent unauthorized access to such control must also be considered of critical importance.

As aviation display designers have discovered (e.g., Stokes, Wickens, & Kite, 1990), the medium of visual information display has many satisfying solutions. The design process for multiple display systems requires considerable testing to derive a reasonable operational fit between operator and display (e.g., Inuzuka, Osumi, & Shinkai, 1991). A fundamental difference between aviation and automobile displays, however, is that instead of designing for a selected group of individuals who are subject to mandatory training to a high degree of proficiency, a fit must be achieved for the least visually capable (Nicolle, Peters, & Vossen, 1994). In addition, the incorporation of visual displays as part of vehicle and advanced traffic information evolution requires a solution to questions regarding the specific information needs of individual drivers in differing conditions. For example, how often is the speedometer or gas gauge used? This is a question of timing and context, because such displays need not necessarily be omnipresent. Additional questions are: How is it that auditory and visual displays can be presented when the driver needs the information? How can they be suppressed when their presence might conflict with safe vehicle operation?

USER ACCEPTANCE

In addition to interface and format issues, driver acceptance and usage patterns need to be studied. Driver acceptance of navigational devices may not be universal. Kantowitz, Becker, and Barlow (1993) estimated that ATIS technology—like automatic teller machines before them—may require 20 years to be accepted by 40% of the user population. The reliability of the information provided will also impact usability. Hanowski, Kantowitz, and Kantowitz (1994) found that drivers used and benefited from an ATIS during simulated drives if the route information provided was highly reliable, but not if it was unreliable. They also found that usage of and reliance on the ATIS was greater when drivers were navigating in an unfamiliar area.

Individual drivers will differ in the extent to which they attend to, process, and follow navigation information (Spyridakis, Barfield, Conquest, Haselkom, & Isakson, 1991). Some drivers are willing to change their departure and arrival times, as well as routes before and even during their trip, based on congestion information, whereas others will not change their travel plans whatsoever. Those motorists who are willing to adapt to traffic information to varying degrees will be the target market of advanced traffic information systems. For those drivers that currently do not use the existing network of radio and television sources of traffic information, future ITS implementations most likely will have little relative impact. As traffic managers are well aware, the mere provision to the motorist of congestion information does not guarantee its use. Hence, knowing the proportions of large metropolitan populations that will be willing to receive and adapt to advanced traffic information networks will be critical to the effective use of in-vehicle guidance systems. For example, Burrus and Hancock (1995) found that compliance with RBDS messages grew as the communicated time delay on the route increased. Furthermore, they found that the content of the message and the context of driving conditions (either free-flowing or congested) interacted to affect drivers' decisions to follow or ignore in-vehicle sources of information.

COLLISION-WARNING AND AVOIDANCE SYSTEMS

Many accidents occur when drivers fail to detect vehicles or other road users, either completely or in sufficient time to avoid collision. Collision-warning and avoidance systems have been proposed as a potential remedy to this type of accident, and this is perhaps the major selling point of ITS from the viewpoint of safety, because roughly half of all traffic deaths occur as a result of multivehicle collisions. Collision-warning systems will represent one of the most visible of all ITS technologies to reach the commercial marketplace in the coming years. They promise to provide the driver with timely information about the potential for imminent collision. Put in this way, the problems seem potentially solvable. Unfortunately, the problem of collision avoidance is not at all straightforward and whether it has a practical solution has still to be demonstrated. From a systems perspective collision avoidance must be considered together with other common facets of ITS development, notably the guidance and navigation systems discussed previously. These systems can provide online recommendations about preferred route guidance and the appropriate ways in which to avoid stoppage and slow down. In a fundamental way as we have noted, collision-avoidance warning systems perform exactly the same function, except that the time constraint provides a quantitatively different challenge to design.

In many ITS arenas (e.g., advanced traffic management systems) it may be possible to import solutions that have been developed in other operational domains, such as aviation. However, this technology transfer might not work in collision avoidance. For example, the comparable warning system in aviation, the Traffic Alert and Collision Avoidance System (TCAS), suffered in its early development from problems of excessive false alarm rates and mistrust or lack of usage by pilots (Wiener, 1988). The general lessons learned from that experience can inform the design of in-vehicle warning systems. However, many design issues are specific to road transportation because, unlike commercial aviation, the time window for perception and action in the vehicle falls in the millisecond range. Another important difference is that, unlike the aviation environment — where individual elements (aircraft) operate under positive control, orchestrated by the overt influence of the air traffic control system — the overall control structure of the roadway environment operates covertly from the perspective of the driver, with controlling features imbedded in signs, signals, and emerging control technologies such as changeable message signs.

In a detailed assessment of research needs for in-vehicle collision avoidance devices, Hanowski et al. (1994) identified a dozen major issues, including effectiveness of visual, auditory, tactile, proprioceptive, and multimodal warnings; coupling between warning function and modality; spatial location of warning information; temporal sequence of warning; labeling and segmentation; effects of false alarms; and driving changes as a result of warning availability. The scope of the human factors challenge should be clear, because as yet only limited information is available for these and related questions. In this section we examine some of the more general issues. The discussion is necessarily restricted to a few collision types, principally two-vehicle collisions occurring for vehicles traveling in the same or opposite directions. Of course, many other collision types are possible, and a comprehensive assessment of collision-avoidance systems will need to take into account collision typology (Massie, Campbell, & Blower, 1993).

Detection Algorithms for Collision Warnings

Rear-end collisions constitute about 23% of all crashes that are reported to the police and account for about 5% of all associated fatalities (Knipling, Wang, & Yim, 1993). A recent analysis of a controlled sample of such crashes found that the principal causal factor was "driver inattention"; the next most important factor was "following too closely" (Knipling et al., 1993). This clearly suggests that a well-designed collision-warning system could potentially reduce crash rate if it alerted an inattentive driver and induced the driver not to follow so closely. But there are several critical design issues that must be considered before either of these outcomes could be achieved reliably and consistently. The first issue concerns the detection algorithms to be used for a collision-warning system.

The simplest possibility, applicable to rear-end collisions, is a time-to-headway criterion, that is, the time it would take the driver's (following) vehicle to reach the lead vehicle or object, based only on the following driver's speed. This is applicable to situations in which the lead vehicle is stationary (e.g., stopped at a traffic light, or stalled) and in low- or zero-visibility conditions; and, indeed, rear-end crashes into a stationary vehicle are more common than crashes in which the lead vehicle is moving (Knipling et al., 1993). Nevertheless, a general detection algorithm based only on a time-to-headway criterion would be overly sensitive and prone to false alarms. More complex algorithms would combine time-to-collision (TTC) measures based on headway and relative closing speed with other factors such as visibility, traffic density, and so on (see Hogema & van der Horst, 1994).

Farber and Paley (1993) carried out an interesting simulation analysis of rear-end collisions on the freeway and the potential benefits that might accrue from a warning system using a detection criterion based on the relative closing rate of the lead and following vehicles. They applied a quasi Monte Carlo procedure to speed, closing rate, and headway data obtained from loop detectors installed on I-40 in Albuquerque, New Mexico. Data for a total of 35,689 vehicle pairs were analyzed. The detection criterion used was

$$D = (FCS\text{-}LCS)^2/2a + (FCS\text{-}LCS)*RT$$

where D = warning distance, FCS = following vehicle speed, LCS = lead vehicle speed, a = following vehicle deceleration, and RT = following driver response time (to the onset of the lead car braking and to the collision warning signal).

In this model, the warning signal is sounded if the actual distance between vehicles is less than D. Without a collision-warning system, Farber and Paley (1993) estimated a crash rate of 173 for every million lead vehicle stops. With the warning system, they estimated that the crash rate could be reduced by at least 50%. Knipling et al. (1993) carried out a similar Monte Carlo analysis using slightly different detection algorithms and estimated that between 37% and 74% of rear-end crashes could be theoretically prevented with a headway detection system.

Farber and Paley (1993) noted that their model raises but cannot answer several significant questions, such as the driver's response to rare events like collision warnings, the impact of false alarms on driver behavior, and driver acceptance of collision-warning systems. Human factors analyses and tests are needed to address these issues (see also Knipling et al., 1993). Nevertheless, the analyses carried out by Farber and Paley and by Knipling and colleagues are important first steps in establishing figures of merit against which the effectiveness of different collision-warning systems can be compared. To date, only a few empirical evaluations of different detection criteria for collision-warning systems have been carried out. Janssen and Thomas (1994) compared different collision-avoidance detection criteria on vehicle-following performance under simulated normal and low-visibility conditions. They found beneficial effects for a system that used a 4-second TTC criterion; after being triggered, the system increased the usual counterforce on the accelerator pedal, thus providing a rapid kinesthetic signal to the driver. Drivers with this system had fewer short following headways (< 1 second) and lower overall driving speed. Schumann and colleagues (Schumann, Godthelp, Farber, & Wontorra, 1993) evaluated the potential to modify the vehicle control dynamics by modifying the driver's control input. To accomplish this, different tactile feedback formats were contrasted with auditory warnings as methods to signal to drivers whether or not remain in their lane when attempting passing maneuvers. They found that drivers were more responsive to proprioceptive cues (such as steering wheel vibration or force feedback, e.g., resisting the driver's control input to change lanes) as compared to auditory warnings.

False Alarms

The selection of a detection criterion must balance the need for early detection with the avoidance of false alarms. The issues of nuisance alarms, false alarms, and reliability of any in-vehicle warning system for collision avoidance will be crucial for future ITS deployment, and much more research into these areas is required (Horowitz & Dingus, 1992; Knipling et al., 1993).

The initial consideration should be the cost of a miss with respect to that of a false alarm. Missed signals (collisions) have a phenomenally high cost, yet their potential frequency is undoubtedly very low. Indeed some drivers may drive for decades without taking the sort of evasive action mandated by such a system. However, if a system is designed to minimize misses at all costs, then the problem of frequent false alarms is immediately encountered. It is well known from operation in other vehicles (e.g., single seat high-performance aircraft) that false alarms in warnings are highly distractive and disturbing to the pilot. There is at present no simple solution to the trade-off that such a ratio implies other than the simplistic affirmation of deriving a perfect detection algorithm.

Of course, the presence and occurrence of warnings may themselves change driver behavior. That is, a conservative warning system seeking to avoid all collisions may be triggered frequently, but the driver may react to these warnings and drive more conservatively so that false alarms are minimized. In essence this is the antithesis of risk homeostasic conception (Wilde, 1982). What is likely is that collision warning systems will influence driver behavior in general, not just on those occasions when the warning is appropriate.

A low false alarm rate, and arguably a zero false alarm rate, would appear to be critical for acceptance of collision-warning systems by drivers. Yet the failure to supply sufficiently advance warning would perhaps be even more insidious (but see Sorkin, Kantowitz, & Kantowitz, 1988). Farber and Paley (1993) suggested that too low a false alarm rate may also be undesirable, given that police-reportable rear-end collisions are very rare events (perhaps occurring once or twice in the lifetime of a driver). If the system never emits a false alarm, then the first time the warning sounds would be just before a crash, and it is possible that the driver may not respond with vigilance to such an infrequent event. Farber and Paley (1993) speculated that an ideal detection algorithm might be one that gives an alarm in collision-possible conditions, even though the driver would likely avoid a crash. Although technically a false alarm, this type of information might be construed as an aid in allowing improved response to an alarm in a collision-likely situation. This idea is similar to that of graded warnings, and to the concept of "likelihood" displays as espoused by Sorkin et al. (1988). Either method of signaling possible collisions might be particularly useful in training drivers to use warning systems. For the experienced driver, however, the only acceptable system would be one with a true false alarm rate that is indistinguishable from zero.

Signal Detection Theory and the Posterior Probability of True Alarms

Signal detection theory (SDT) could be used to evaluate the effectiveness of warning systems that use different detection algorithms and have different detection and false alarm rates. SDT provides quantitative indices of the sensitivity (bias-free accuracy) and bias of a detection or diagnostic system (Swets & Pickett, 1982). Decisions about the presence or absence of a critical event or signal are presumed to be based on an observation statistic X that is distributed normally, with two overlapping distributions postulated for signal (S) and noise (N) events. For a given observation, the detection system compares X against a criterion C. If C is exceeded, a positive response (R) is emitted; if not, a decision is made that noise alone was presented. $P(R/S)$ and $P(R/N)$ represent the probabilities of correct detections (hits) and false detections (false alarms), respectively. The accuracy of the detection system is given by the separation of the signal and noise distributions, d'; response bias by C or the likelihood ratio measure β. Given a pair of hit and false alarm probabilities $\{P(R/S), P(R/N)\}$, d' and

β can be computed from the corresponding normalized values (Swets & Pickett, 1982). SDT could be used to evaluate the effectiveness of different detection algorithms, for example, those proposed by Farber and Paley (1993). Detection of a potential collision in a vehicle equipped with a warning device also represents an instance of a joint human–computer monitoring system—both the human and the machine detector directly observe the environmental stimuli indicating possible collision, and the human makes a joint decision combining both this input with the output of the warning system. SDT has also been applied to the issue of determining the joint accuracy of such detection systems (Parasuraman, 1987; Sorkin & Woods, 1985).

For a collision-warning system d' will be determined by such factors as sensor noise and the effectiveness of the detection algorithm, whereas β can be freely set by the system designer to achieve different hit and false alarm rates for a system with a given value of d'. The technology exists for system engineers to design sensitive warning systems with high values of d'; and these systems will generally be set with a decision threshold that minimizes the chance of a missed warning while keeping the false alarm rate below some low value. The importance of keeping the false alarm rate as close to zero as possible has already been mentioned previously. However, despite the best intentions of designers, the availability of the best sensor technology, and the development of very sensitive detection algorithms, one fact may conspire to limit the effectiveness of alarms: the low a priori probability or base rate of potential collision scenarios. For the human driver, that a given warning system may have a very high hit rate and a very low false alarm rate is less important than the inverse of these conditional probabilities, that is, $P(S/R)$ and $P(N/R)$. In particular, the posterior probability $P(S/R)$ is of vital importance to the human monitor of any automated alarm system. This represents the probability that a given alarm response represents a true condition (e.g., potential collision). Bayes' Theorem can be used to derive $P(S/R)$:

$$P(S/R) = P(R/S) + [P(R/S) + P(R/N)(1 - p)/p]$$

where p is the a priori probability of the signal S.

If the base rate p is low, as it often is for many real events, then the posterior odds of a true alarm can be quite low even for very sensitive warning systems with very high hit rates and low false alarm rates. This can best be seen with a example. Assume a very sensitive collision-warning system with a d' of 4.65, and assume that the decision threshold β is set to 1. This gives a near-perfect hit rate for the system of .99 and a false alarm rate of only .01. Despite these very impressive detection statistics, the human operator may well find that the posterior odds of a true alarm with such a system can be quite low. Table 16.1 shows the values of the posterior

TABLE 16.1.
Values of the Posterior Probability of a Signal Event (S) Given an Alarm
Detection Response (R), and the Associated Odds of a True Alarm for
Several Values of the a Priori Probability p

	A Priori Probability p	Posterior Probability P(S/R)	Odds of a True Alarm
A.			
$d' = 4.65$	.0001	.0098	1 in 102
β set to 1	.001	.09	1 in 11
P(R/S) =.99	.01	.5	1 in 2
P(R/N) = .01	.1	.92	10 in 11
B.			
$d' = 4.65$	.0001	.068	1 in 15
β set to 24	.001	.42	3 in 7
P(R/S) =.95	.01	.88	8 in 9
P(R/N) = .0013	.1	.988	100 in 101

probability *P(S/R)* and the associated odds of a true alarm for several values of *p* ranging from .0001 to .1. As part A of Table 16.1 shows, even with such a sensitive detection system, when the a priori probability is low, say .001, only 1 in 11 alarms that the system emits represents a true alarm. The situation is a little better if the decision threshold is set a little stricter, sacrificing hit rate for a lower false alarm rate (part B of Table 16.1). Now three out every seven alarms represent a true alarm for a *p* of .001. Little wonder, then, that many human operators tend to ignore or to turn off alarms—they have cried wolf once too often (Sorkin, 1988). As Table 16.1 indicates, consistently true alarm response occurs only when the a priori probability of a signal is relatively high. There is no guarantee that this will be the case in many real systems. More generally, these results suggest that designers of collision-warning systems must take into account the decision threshold at which these systems are set. Swets (1992) discussed procedures for effective choosing of appropriate values of *β* for diagnostic systems. Decision thresholds for collision-warning systems must be set not just for high hit and low false alarm rates, but for relatively high values of posterior true alarm probabilities as well.

There are several considerations in addition to the question of quantitative assessment of warning system algorithms. Should developers focus specifically on collision-detection systems such as rear-end collision protection (Knipling et al., 1993), or should they employ some form of general protective envelope approach (Hancock, 1993)? Can alarm systems be individualized such that they respond to likely accident conditions for the pertinent driver age group? The structure and function of a collision-warning system implies first the development of some complex, multi-array detection system that itself is of considerable engineering challenge. Yet having derived a veridical warning signal, its customization for consumption by drivers of widely different abilities is uniquely a human factors question. It is this arena that promises enhanced safety, yet also represents the most complex portion of ITS development (see Lund & O'Neill, 1994, for a particular critique).

Information Presentation Formats

Given that the related problems of the appropriate detection algorithm, the false alarm rate, and the posterior odds of true alarms of warning systems can be solved, the next issue concerns the form in which collision information is provided to the driver. Several researchers have suggested that the interface to collision-warning systems, as for navigation systems, should be multimodal, using both visual and auditory sensory channels (e.g., Hirst & Graham, 1994). The systems that are emerging in the marketplace appear to follow this recommendation. For example, the VORAD headway detection system, which uses front and side radars, issues a yellow warning light when a target is detected, a red light when the closing speed or distance is unsafe, and an audible tone when collision danger exists. Hirst and Graham (1994) tested several warning presentation formats on driver braking performance in a simulator. All systems used a 4-second TTC criterion. They found that subjects preferred a system that combined a traffic light visual display (as in the VORAD system) with a nonspeech auditory display (500 Hz tone). Subjects also braked earlier to potential collision targets with this system than they did when using other systems.

Driver-in-the-Loop Collision-Avoidance Systems

The devices considered thus far are informational systems that preserve the control authority of the driver to take or not to take evasive action in response to warning information. However, future systems may automate to some degree the control function, thereby qualifying as partial or full avoidance systems. These technologies will raise the fundamental question that has been considered in aviation and other domains and discussed elsewhere in this book, that of driver versus automated control. Should the system only inform the driver, should it usurp control, or should there be shared (or assisted) control? Will an informational system have time to detect potential conflicts and represent them with the preferred avoidance strategy in the time horizon available? Given an actual imminent collision situation, how is the automated vehicle expected to respond? For the alert driver, there are several options, such as running off the road onto a soft shoulder, and this may be considered a preferred action given the expected cost of collision. However, given the multiplicity of environmental contingencies, how is an automated decision system expected to perform the tasks of detection, evaluation, and response initiation within the available time frame? The computational power presently required to accomplish such actions, were they actually possible, would certainly cost some orders of magnitude more than the value of the car. However, these capabilities may become achievable within the coming years at a reasonable cost. In the absence of such availability, a useful design recommendation is to keep the driver in the loop given that drivers currently are able to avoid most obstacles set in their path.

The accident statistics quoted at the beginning of the chapter, which point to increasing crash rates over the years, may seem to contradict the view that most drivers can avoid collisions quite well. Yet collisions and accidents are actually rare events in the lifetime of any driver. That is, compared to the opportunities for collision on the road, the number of actual collisions is extremely low. It is difficult to speculate on the total potential for collisions for any given set of drivers or driving conditions, because this is at best an estimated baseline. However, with this perspective, it is clear that human drivers are able to control and navigate current road systems and avoid the overwhelming majority of such adverse events. Clearly any surrogate automated controller must exceed human response capability under all operational driving conditions for acceptable total replacement, because it has been shown that human response is particularly averse to systems that purport to provide surrogate control but in reality prove to be unreliable (Wiener, 1988). It is important not to remove this established human capability prior to the unequivocal establishment of a proven superior system. The idea of retaining driver-in-the-loop architecture, with the expectation of the final arbitration of avoidance action remaining with the human component, is predicated on a human-centered approach to ITS system design. This human-centered approach presents numerous human factors-related problems, including the question of whether human response time can accomplish avoidance under even the most favorable of interface design conditions. As always, the alternative systems architectures, founded principally on traffic engineering concerns, hope to solve the problem via automation techniques with relatively little direct concern for the proximal user of the system—the driver. Although such control strategies have always had their appeal to the design engineer, it has been shown with increasing frequency that they are prone to unacceptable failure.

Adaptive Systems

The design philosophy of keeping the driver in the loop might suggest to some that everything that can be automated should be automated, leaving the human as the "backup system of last resort" (Kantowitz & Sorkin, 1987). The lessons from aviation, medical systems, process control and similar domains have taught the fallacies of such a strategy (see Woods, chap. 1, this volume). This is particularly true of driving where the driver is liable to retain active control of steering, at least for the immediate and foreseeable future. Thus, if task allocation is not to become a default approach (Chapanis, 1965) or based simply on descriptive, machine-oriented comparisons (Meister, 1971), new approaches that will stand up to the rigors of the fast, real-time application characteristics of collision-avoidance systems will need to be developed.

One potential solution lies in the application of adaptive systems (Hancock & Chignell, 1987, 1989; Parasuraman, Bahri, Deaton, Morrison, & Barnes, 1992; Rouse, 1988; see also Scerbo, chap. 3, this volume). Adaptive human–machine systems represent a particular subset of hybrid systems that emphasize the use of mutually adaptive capability on behalf of both human and machine to promote flexible and rapid response to uncertain conditions and unusual task demands. As such, they may be particularly relevant in transient, unstable emergency conditions as typified by the imminent collision scenario. In the field of aviation research, adaptive systems have been found to promote improved operator response to unexpected system failures (Parasuraman, 1993). In principle, such benefits could also accrue in road transportation, although the shorter time frames involved for decision and action may pose additional design challenges. The use of adaptive systems has been suggested for application to numerous human factors problems raised by ITS development, in particular workload regulation (Verwey, 1990; Verwey & Kaptein, 1994). Collision avoidance also appears to be a major area where such development could be immediately evaluated. However, as in other application areas for adaptive automation (see Scerbo, chap. 3, this volume), the promise of this approach to function allocation remains to be realized (see Duley, Scallen, & Hancock, 1994).

Ecology of Driver–Environment Interactions

Although there are many practical questions as to the construction of collision-avoidance warning systems, behind such questions lies a more fundamental concern with how we can understand human interaction with complex technical systems in general. The time constraints involved in collision avoidance bring this discussion into particular prominence. Traditional approaches to human–machine interaction are predicated upon the information processing paradigm that has predominated for some years and has provided many valuable insights into human behavior. However, when considering human responses in unusual driving conditions (e.g., impending collision), a more detailed analysis is required of the ecological or environmental circumstances to which the individual must react. This is especially true for the dynamic and uncertain environments that connote a collision-avoidance situation.

Over 50 years ago, Gibson and Crooks (1938) pointed to the commonalities between driving and forms of locomotion such as walking and running. Each of these respective tasks requires the individual to navigate toward a goal while avoiding obstacles and areas that might also slow or inhibit passage. One of the valuable constructs they advanced was the notion of a field of safe travel. This was represented as the field of possible paths that a vehicle can take unimpeded. It is bounded by the terrain, principally the roadway, but also by moving objects; in the present case, other vehicles. Steering is defined as keeping the vehicle headed into the center of this field of safe travel. A companion construct is the minimum stopping zone. Acceleration is the function that subsumes goal achievement. Deceleration also contributes to the same function but principally through obstacle avoidance. During most normal driving, the stopping zone is inside the field of safe travel, and Gibson and Crooks indicated that drivers often act to keep the ratio of the field and zone close to constant. The ratio between the field and the zone is an index of cautiousness.

These observations imply a view that is advocated here, namely that accident-likely scenarios cannot be considered as simple extensions of normal driving. Accidents and their prevention must be considered as an allied but independent investigation from that of the typical driver's task. Consider the following example. Curbs, shoulders, and shallow ditches are frequently excluded from the field of safe travel, but in an emergency it is precisely these paths that are sought by the driver. This can be seen in everyday avoidance on the freeway in potential rear-end collision situations. Thus, the field of safe travel is, in reality, the field of "safest" travel given the current conditions. In consequence, driving is not simply locomotion in a free environment. First, the vehicles that act as locomotion prosthetics have velocity capabilities that exceed unaided human capability. In addition, most other moving obstacles are not insensate, but are also purposeful, goal-oriented agents. Therefore, normal driving requires mutual cooperation and, paradoxically, some forms of collision require also require mutual "interaction." If driving requires the maintenance of mutual exclusion, then collision-avoidance warning systems must also inform the driver as to the status of these fields, and the progressive incursion into these fields by additional objects and users of the roadway.

The work of Gibson and Crooks (1938) has relevance for the design of contemporary collision-avoidance warnings and in-vehicle automation in general. However, a number of important problems must first be solved.

First, although the definition of the objective qualities of the identified minimum safe stopping zone is well articulated, the connection among the field of safe travel, the concept of "affordance," and the linkage to action is less clear. This is critical, because the design of an effective warning is predicated on the development of interfaces that "directly" display integrations of such fields and zones (e.g., see Hansen, 1994). A further problem that has yet to be even addressed in an adequate manner is how we start to structure a sensor array that can convey the information concerning the safest field of travel, given the constraints of a collision imminent environment. Because we are uncertain how the human operator achieves this evaluation, it is more than problematic as to how to construct a definitive algorithm to achieve the same objective. In fact, it is with respect to this problem that we must begin to identify the cost-effective nature of fully automated versus adaptive human–vehicle collision avoidance control systems. Even given this information, how best to display it to the driver for maximum effectivity of response is still, as we have noted, unclear. What is obvious is that the intrinsic time constraints defeat simple-minded alphanumeric displays and suggest use of heads-up and multimodal display configurations. One manner with which to specify preferred zones of progress would be through color representations of fields directly projected onto the windshield of the vehicle. How this aim is accomplished with the dual requirement of a 100% collision identification rate, together with a requirement for a zero false alarm rate (in other words, flawless observations), is also as yet undetermined.

The theoretical takeover of all functions by automated control appears to offer the best of all possible worlds, a vision of an automated utopia (or "autopia"). The fact is, however, that we live in a society with rampant litigation; hence even marginal failure of such a system, or worse subsequent collision resulting from initial automated avoidance, appears to be mired in the questions of responsibility. Also, it is very clear that operators of any system are unwilling to have control literally lifted from their hands at any stage of operation. User accceptance may well depend on a better definition of the implicit and explicit "contract" between the driver and the automated system. Public transportation is widely accepted even though this form of transit involves an explicit loss of control on the part of the user. Because the division of control between the system and the user is explicitly defined at the outset, and the user consents to this division, the loss of control is tolerated. However, if consent is not given, users will tend not to tolerate loss of control. In this sense, some proposed ITS components such as vehicle platooning may be more easily accepted than collision avoidance technology, because the former is analogous to the public transportation case, and consent can be sought and granted. With collision avoidance technology, consent is not easily obtainable given the time constraints, and consequently drivers may not easily accept any system that usurps control without their consent.

ADVANCED OPTIONS

Many of these proposed ITS components and other advanced collision avoidance systems being envisaged are the antithesis of the driver-in-the-loop philosophy advocated here that fosters negotiation of the field of safe travel and consideration of the driver's need for control and consent. Collision avoidance systems that take full control of the vehicle from the operator are the final expression of automation in ITS technology for reducing crash risk. The Advanced Vehicle Control Systems (AVCS) are considered a mechanism to "organize" the vehicles on the roadway in order to promote safety and increase traffic flow (IVHS America, 1992). One system proposed is the concept of automated chauffeuring, where vehicles are taken automatically from on-ramp to off-ramp. Other systems are based on headway detection systems that provide automated cruise control and braking. The technological foundations for AVCS systems are considered to be antilock braking systems, traction control, active suspension, and four-wheel steering currently found on some late-model vehicles. For collision avoidance, the levels of automation in ITS systems range from warnings to full automated takeover of the driving vehicle. A general design principle expressed in proposed schemes for the development of ITS systems is for the intensity of automated action to increase as time for reaction (time before collision) decreases (IVHS America, 1992; OCAR, 1994).

LOW-TECHNOLOGY OPTIONS

The hoopla and hype associated with ITS both in the popular press and in the general scientific media, which generally play up the high-tech nature of these systems, may blind one to the availability of some low-tech solutions to the problems of road safety. The use of radio for guidance and navigation information, as in RBDS, is one example. Other examples include systems that increase visibility and conspicuity. The ability of the driver to observe and detect vehicles, pedestrians, and other objects, and the environmental features that promote the driver to negotiate safe passage in the road, are critical for the prevention of automobile crashes. There are some areas where the development of automated devices may promote this effort.

Windshields and mirrors comprise the fundamental transducer for the transmission of environmental information, whereas the driver's visual system is receiver of this information (Flanagan & Sivak, 1993; Kebler, 1993). These vehicle components are generally considered to be static, fixed and not amenable to automation. There are, however, several applications in which automated technology may provide safety advances in this area. For example, automated defrosters, windshield wipers, and outside lighting in response to variations in environmental conditions are methods available to ensure optimal visibility for the driver. Somewhat more advanced technology proposals also exist with the goal of increasing driver visibility (see IVHS, 1992; OCAR, 1994).

More advanced but nevertheless economical solutions that still fall under the low-tech category involve using automated devices to restrict an unfit driver's access to the vehicle and thereby prevent their entry to the roadway. These fitness for duty (FFD) devices, known also as ignition-interlock devices, require the driver to perform a perceptual/motor task, the outcome of which determines whether the automobile is allowed to start. These devices have been applied as a means to prevent driving while under the influence of alcohol. Two primary forms exist for the detection of impairment in an alcohol condition, those based on breath-alcohol correlations and those based on perceptual-motor performance (see Kinghorn & Bittner, 1992). With increasing research and development to determine reliable screening tasks associated with driving behavior, these devices have the potential for more widespread use beyond alcohol impairment to the assessment of performance variations due to fatigue, medication usage, drowsiness, or other factors likely to affect driver state.

Another approach that does not require extensive investment in technology is one related to training and driver capability. As we have noted, accident-related scenarios occur very infrequently in the life of a driver. As a result, drivers have little experience on which to base their avoidance actions. Consequently, some collisions may arise from unfamiliarity rather than from the failure to avoid per se. This argues that training drivers, most probably in simulation, may be a good method to promote collision reduction. This form of training could occur during licensing, thus expanding the concept of ITS beyond the bounds of the vehicle. Irrespective of the merits of this strategy, drivers will need to have training on any in-vehicle system that permits total or partial driver control. These are only a limited subset of low-technology options that require some exploration alongside their more seductive high-tech companions.

DESIGNING FOR ALL USERS

The goal of an efficient, safe transportation system will be met only if the design of technology fits the needs of the "average" driver and is complemented by sufficient attention to individual and group differences. Age represents one of the more important driver variables. ITS technologies will be used by drivers with a wide range of abilities and skills, including groups such as young adults (16–25 years) and mature drivers (over 65 years), who tend to be overrepresented in vehicle crashes. As a result, many of the important human factors considerations involve design of systems that are safe and effective for older drivers (Hanowski et al., 1994).

The fastest-growing segment of the United States population are older adults. Drivers aged 75 years and older, especially women drivers, represent the fastest-growing segment of road users (McKelvey, Maleck, Stamatiadis, & Hardy, 1988). In 1990, 13% of licensed drivers were over 65, and these drivers accounted for approximately 10% of total drivers involved in crashes. When accident data are evaluated on the basis of total vehicle miles driven, older adults show a disproportionate crash risk (Cerrelli, 1989). How will ITS technology cope with the necessity to cater to such individuals while serving a broader traveling public? In a more positive way, how can ITS open driving to an increased percentage of handicapped and disabled drivers and serve the important goal of enhanced mobility for these individuals?

If designed appropriately, guidance systems could prove beneficial to older adults showing evidence for difficulty remembering spatial relationships for critical route events (Lipman, 1991). ITS technology may also assist the older driver through automated devices that compensate for perceptual deficits or aid decision making. Examples include headway information, side object detection, route guidance in unfamiliar areas, and so on. However, there is a safety consideration for all drivers concerning the method and degree of interaction with these devices and how this impacts on the primary task of driving. For example, several studies have shown that interacting with in-vehicle navigation aids and even cellular phones can affect driving performance (McKnight & McKnight, 1992, 1993). In addition, even when ITS information is presented on an centrally located in-vehicle display, older drivers may take longer to extract information than younger drivers (e.g., Hayes, Kurokawa, & Wierwille, 1989; Imbeau, Wierwille, & Beauchamp, 1993). To mitigate the possibility of unacceptably long in-vehicle glance durations, display elements should be integrated to promote feature extraction by older adults.

It is important to emphasize the confluence of diverse factors that make the testing and evaluation of older drivers the critical arena for ITS development. In addition to the relative disproportional growth of this segment of the driving public, there is good reason to believe that an ITS system that works for older drivers will work for the rest of the driving public, and, conversely, an ITS system that fails to serve the older driver will leave a large and increasing segment of the driving population at risk. A more thorough analysis of the accident record, however, shows that despite their acknowledged self-regulation of risk, older drivers are still overrepresented in accidents involving a number of common driving maneuvers, for example, merging, overtaking, and left turns (Hancock, Caird, Shekhar, & Vercruyssen, 1991). Also, older drivers are overrepresented in accidents involving maneuvers such as backing, perhaps partially due to physical difficulties associated with twisting and turning to see what is behind the car. This is surely a case in which low-tech solutions such as better mirrors, medium-technology solutions such as existing backing warnings, and some low-level object detection system will help older drivers avoid tragedies such as backing over children that "they never saw." It is also this group of drivers that will benefit most from well-designed in-vehicle collision avoidance systems.

One particularly useful way to identify significant problems and resultant analytical methods, hypotheses, and models directed to their solutions is to examine the practical problems of building an ITS interface for the older driver. Some problems have already been identified that pertain to the timing of warnings, whether too soon or too late. These questions can be addressed within the context of older driver testing in which customization of warning interval can be adjusted with respect to the actions of each individual driver. For younger drivers, warnings that were too early would quickly come to be ignored. However, the fact that older drivers need a longer foreperiod of warning will again test the boundaries of the envelope of the technology, making the system that works for the older driver work for everyone. In general, older individuals are more reluctant to use technical innovations; hence, acceptance by older drivers is critical. There is evidence that older adults tend to use advanced technology less than younger adults do (Brickfield, 1984), although whether this is indeed an age effect or a cohort effect is unclear. Such acceptance mandates reliability and an obvious safety benefit, as trust that is lost by an ITS system, or indeed any system, is not regained easily (Lee & Moray, 1992; Riley, 1994).

CONCLUSIONS

Advanced automation technology is being introduced in motor vehicles and the road transportation system at a more rapid pace than automation has ever been implemented before in any large-scale system. The long-term goal of this effort is to reduce traffic congestion, accidents, and energy costs, thereby enhancing the safety and efficiency of automobile transportation (IVHS America, 1992). Currently available in-vehicle devices include collision-warning systems and navigational or route-guidance systems. Although these systems are conventionally divided into two temporal categories, with collision-warning being considered short term and navigation assistance more long term, a systems-based implementation of these technologies must emphasize their temporal commonality. Several critical human factors issues — such as protection of the primary driving task, information-presentation format, false alarms, and user acceptance — are critical to the design and effective use of these devices. Advanced options for in-vehicle automation should be based on a driver-in-the-loop architecture. Finally, automated devices for motor vehicles will need to be designed to match the wide range of abilities and skills of drivers.

As the federal government, industry, and the academic community all struggle with the questions associated with a rational architecture for ITS, it is encouraging to see that the basic predicate of the proposed system is user services. Although this is promising news, a more careful examination of the competing architectures shows that overwhelming emphasis is given to the technical capabilities to support such services. With little exception, the concerns of the driver, the traffic manager, the pedestrian, and other system users are given lesser consideration. Thus, the focus is on of the provision of services and the technology, not on whether or how they may or may not be used. As this process continues, the hard lessons learned from other large-scale system operations and their successes and failures need to be championed and integrated into ITS development. Ultimately, if the lofty goals of ITS are to be realized, a user-centered approach will be critical.

REFERENCES

Andre, A., Hancock, P. A., & Smith, K. (1993). Getting from there to here with TRAVTEK. *Ergonomics in Design, 2,* 16–17.

Antin, J. F. (1993). Informational aspects of car design: Navigation. In B. Peacock & W. Karwowski (Eds.) *Automotive ergonomics* (pp. 321–337). Washington, DC: Taylor & Francis.

Brickfield, C. F. (1984). Attitudes and perceptions of older people toward technology. In P. K. Robison & J. E. Birren (Eds.), *Aging and technological advances* (pp. 31–38). New York: Plenum.

Burrus, M., & Hancock, P. A. (1995, August). *Evaluation of driver response to an in-vehicle ITS technology.* (Draft Report). Minneapolis, MN: Minnesota Department of Transportation.

Burrus, M., Williams, G., & Hancock, P. A. (1994). *The safety of the RDS: A comparison study of three delivery systems.* Paper presented at the 12th Triennial Congress of the International Ergonomics Association, Toronto, Canada.

Cerelli, E. (1989). *Older drivers. The age factor in traffic safety.* (NTIS No. DOT-HS-807-402). Washington, DC: National Highway and Traffic Safety Administration.

Chapanis, A. (1965). On the allocation of functions between man and machines. *Occupational Psychology, 39,* 1–10.

Dingus, T. A., & Hulse, M.C. (1993). Human factors research recommendations for the development of design guidelines for advanced traveler information systems. *Proceedings of the Human Factors and Ergonomics Society, 37,* 1067–1071.

Duley, J. A., Scallen, S. F., & Hancock, P. A. (1994). The response of experienced pilots to interface configuration changes for adaptive allocation. In M. Mouloua & R. Parasuraman (Eds.), *Human performance in automated systems: Current research and trends* (pp. 134–141). Hillsdale, NJ: Lawrence Erlbaum Associates.

Fairclough, S. H., Ashby, M. C., & Parkes, A. M. (1993). In-vehicle displays, visual workload, and usability evaluation. In A. G. Gale (Ed.), *Vision in vehicles IV* (pp. 245–254). Amsterdam: North-Holland.

Farber, E., & Paley, M. (1993, April). *Using freeway traffic data to estimate the effectiveness of rear end collision countermeasures.* Paper presented at the Third Annual IVHS America Meeting, Washington, DC.

Flanagan, M., & Sivak, M. (1993). Indirect vision systems. In B. Peacock & W. Karwowski (Eds.), *Automotive ergonomics* (pp. 205–217). Washington, DC: Taylor & Francis.

Gibson, J. J., & Crooks, L. E. (1938). A theoretical field analysis of automobile driving. *American Journal of Psychology, 51,* 453–471.

Hancock, P. A. (1993). Evaluating in-vehicle collision avoidance warning systems for IVHS. In E. J. Haug (Ed.), *Concurrent engineering: Tools and technologies for mechanical system design* (pp. 947–958). Berlin: Springer-Verlag.

Hancock, P. A. (1996). On convergent technological evolution. *Ergonomics in Design, 4,* 22–29.

Hancock, P. A., Caird, J. K., Shekhar, S., & Vercruyssen, M. (1991). Factors influencing drivers' left-turn decisions. *Proceedings of the Human Factors Society, 35,* 1447–1451.

Hancock, P. A., & Chignell, M. H. (1987). Adaptive control in human–machine systems. In P.A. Hancock (Ed.), *Human factors psychology* (pp. 305–345). Amsterdam: Elsevier.

Hancock, P. A., & Chignell, M. H. (Eds.). (1989). *Intelligent interfaces: Theory, research, and design.* Amsterdam: Elsevier.

Hancock, P. A., Dewing, W., & Parasuraman, R. (1993, April). Human factors in intelligent travel systems. *Ergonomics in Design, 1,* 12–39.

Hancock, P. A., & Parasuraman, R. (1992). Human factors and safety issues in Intelligent Vehicle Highway Systems (IVHS). *Journal of Safety Research, 23,* 181–198.

Hanowski, R. J., Bittner, A. C., Rowley, M. S., Wilson, J. C., Raby, M., Byrne, E. A., & Parasuraman, R. (1994). *An information clearinghouse conceptual design: A system for meeting crash avoidance research needs relevant to older drivers.* (Final Report). Seattle, WA: Battelle Seattle Research Center.

Hanowski, R. J., Kantowitz, S. C., & Kantowitz, B. H. (1994). User acceptance of unreliable route guidance information. *Proceedings of the Human Factors Society, 38,* 1062–1066.

Hansen, J. P. (1994). Representation of system invariants by optical invariants in configural displays. In J. M. Flach, P. A. Hancock, J. K. Caird, & K. Vicente. (Eds.), *Local applications of the ecological approaches to human–machine systems* (pp. 208–233). Hillsdale, NJ: Lawrence Erlbaum Associates.

Hayes, B. C., Kurokawa, K., & Wierwille, W. W. (1989). Age-related decrements in automobile instrument panel task performance. *Proceedings of the Human Factors Society 33rd Annual Meeting* (pp. 159–163). Santa Monica, CA: Human Factors Society.

Hirst, S., & Graham, R. (1994). The presentation and timing of collision avoidance warnings. *Proceedings of the International Ergonomics Association, 4,* Toronto, Canada.

Hoffmann, O., Becker, S., & Schrievers, G. (1994). Comparative assessment of a navigation and route guidance system in simulator experiments and field trials. *Proceedings of the First World Congress on Applications of Transport Telematics and Intelligent Vehicle-Highway Systems* (pp. 1677–1684). Paris: Intelligent Transportation Systems.

Hogema, J., & van der Horst, R. (1994). Driver behavior under adverse visibility conditions. *Proceedings of the First World Congress on Applications of Transport Telematics and Intelligent Vehicle-Highway Systems* (pp. 1623–1636), Paris.

Horowitz, A. D., & Dingus, T. A. (1992). Warning signal design: A key human factors issue in an in-vehicle front-to-rear-end collision warning system. *Proceedings of the Human Factors Society, 36,* 1011–1013.

Imbeau, D., Wierwille, W. W., & Beauchamp, Y. (1993). Age, display design and driving performance. In B. Peacock & W. Karwowski (Eds.), *Automotive ergonomics* (pp. 339–357). Washington, DC: Taylor & Francis.

Inman, V. W., Fleishman, R. N., Dingus, T. A., & Lee, C. H. (1993). Contribution of controlled field experiments to the evaluation of TRAVTEK. *Proceedings of the IVHS America, 1993 Annual Meeting* (pp. 659–666). Washington, DC: IVHS America.

Inuzuka, Y., Osumi, Y., & Shinkai, H. (1991). Visibility of head up display (HUD) for automobiles. *Proceedings of the Human Factors Society, 35,* 1574–1578.

IVHS America (1992). *Strategic plan for IVHS in the United States.* Washington, DC: Author.

Janssen, W., & Thomas, H. (1994). In-vehicle collision avoidance support under adverse visibility conditions. *Proceedings of the International Ergonomics Association, Vol. 4., Ergonomics and Design* (pp. 179–181). Toronto, Canada: IEA.

Kantowitz, B. H., Becker, C. A., & Barlow, S. T. (1993). Assessing driver acceptance of IVHS components. *Proceedings of the Human Factors Society, 35,* 1062–1066.

Kantowitz, B. H., & Sorkin, R. D. (1987). Allocation of functions. In G. Salvendy (Ed.), *Handbook of human factors* (pp. 355–369). New York: Wiley.

Kebler, F. R. (1993). Light diffusion characteristics and visibility interferences in automobile windshields. In A. G. Gale (Ed.), *Vision in vehicles IV* (pp. 69–78). Amsterdam: North-Holland.

Kinghorn, R. A., & Bittner, A. C., Jr. (1992). *Commercially available ignition interlock devices: Background, status, and potential.* Seattle, WA: Battelle Human Affairs Research Center.

Knipling, R. R., Mironer, M., Hendricks, D. L., Tijerina, L., Everson, J., Allen, J. C., & Wilson, C. (1993). *Assessment of IVHS countermeasures for collision avoidance: Rear-end crashes* (NTIS No. DOT-HS-807-995). Washington, DC: National Highway and Traffic Safety Administration.

Knipling, R. R., Wang, J. S., & Yin, H.M. (1993). *Rear-end crashes: Problem size assessment and statistical description.* Washington, DC: National Highway and Traffic Safety Administration.

Lee, J., & Moray, N. (1992). Trust, control strategies, and allocation of function in human–machine systems. *Ergonomics, 35,* 1243–1270.

Lipman, P. D. (1991). Age and exposure differences in acquisition of route information. *Psychology and Aging, 6,* 128–133.

Lowe, M. (1993, December 12). Smart car 54, where are you? *The Washington Post,* p. C5.

Lund, A. K., & O'Neill, B. (1994). *IVHS—can it deliver on safety?* (Technical Report). Arlington, VA: Insurance Institute for Highway Safety.

Massie, D. L., Campbell, K. L., & Blower, D. F. (1993). Development of a collision typology for evaluation of collision avoidance strategies. *Accident Analysis and Prevention, 25,* 241–257.

McKelvey, F. X., Maleck, T. L., Stamatiadis, N., & Hardy, D. K. (1988). Highway accidents and the older driver. *Transportation Research Record 1172* (pp. 47–57). Washington, DC: National Research Council.

McKnight, A. J., & McKnight, A. S. (1992). *The effect of in-vehicle navigation information systems upon driver attention.* Landover, MD: National Public Services Research Institute.

McKnight, A. J., & McKnight, A. S. (1993). The effect of cellular phone use upon driver attention. *Accident Analysis and Prevention, 25,* 259–265.

Meister, D. (1971). *Methods of performing human factors analyses.* New York: Wiley.

National Safety Council. (1994). *Accident facts.* Itasca, IL: Author.

Nicolle, C., Peters, B., & Vossen, P.H. (1994). Towards the development of ATT guidelines for drivers with special needs. *Proceedings of the First World Congress on Applications of Transport Telematics and Intelligent Vehicle-Highway Systems* (pp. 1655–1650), Paris: Intelligent Transport Systems.

OCAR (1994). *SMART (systematic methodology for assuring a reduction in vehicular traffic crashes) Program Plan. Final Draft.* Washington, DC: National Highway Traffic Safety Administration, Office of Crash Avoidance Research.

Owens, D. A., Helmers, G., & Sivak, M. (1993). Intelligent vehicle highway systems: A call for user-centered design. *Ergonomics, 36,* 363–369.

Parasuraman, R. (1987). Human–computer monitoring. *Human Factors, 29,* 695–706.

Parasuraman, R. (1993). Effects of adaptive function allocation on human performance. In D. J. Garland & J. A. Wise (Eds.), *Human factors and advanced aviation technologies* (pp. 147–157). Daytona Beach, FL: Embry-Riddle Aeronautical University Press.

Parasuraman, R., Bahri, T., Deaton, J. E., Morrison, J. G., & Barnes, M. (1992). *Theory and design of adaptive automation in aviation systems.* Warminster, PA: Naval Air Development Center.

Perez, W. A., Fleischman, R. A., Golenbiewski, G., & Dennard, D. (1993). TRAVTEK field study results to date. *Proceedings of the IVHS America 1993 Annual Meeting* (pp. 667–673). Washington, DC: IVHS America.

Popp, M. M., & Farber, B. (1991). Advanced display technologies, route guidance systems, and the position of displays in cars. In A. G. Gale (Ed.), *Vision and vehicles III* (pp. 219–225). New York: Elsevier Science.

Riley, V. (1994). A theory of operator reliance on automation. In M. Mouloua & R. Parasuraman (Eds.), *Human performance in automated systems: Current research and trend* (pp. 8–14). Hillsdale, NJ: Lawrence Erlbaum Associates.

Rillings, J.H. (1992). TRAVTEK. *Proceedings of the IVHS America 1992 Annual Meeting* (pp. 569–578). Newport Beach, CA: IVHS America.

Rouse, W. B. (1988). Adaptive aiding for human/computer control. *Human Factors, 30,* 431–443.

Schumann, J., Godthelp, H., Farber, B., & Wontorra, H. (1993). Breaking up open-loop steering control actions the steering wheel as an active control device. In A.G. Gale (Ed.), *Vision in vehicles IV* (pp 321–332). Amsterdam: North-Holland.

Sorkin, R. D. (1988). Why are people turning off our alarms? *Journal of the Acoustical Society of America, 84,* 1107–1108.

Sorkin, R. D., Kantowitz, B. H., & Kantowitz, S. C. (1988). Likelihood alarm displays. *Human Factors, 30,* 445–459.

Sorkin, R. D., & Woods, D. D. (1985). Systems with human monitors: A signal detection analysis. *Human-Computer Interaction, 1,* 49–75.

Spyridakis, J., Barfield, W., Conquest, L., Haselkom, M., & Isakson, C. (1991). Surveying commuter behavior: Designing motorist information systems. *Transportation Research-A, 25A (V3),* 17–30,

Stokes, A., Wickens, C., & Kite, K. (1990). *Display technology: Human factors concepts.* Warrendale, PA: Society of Automotive Engineers, Inc.

Swets, J. A. (1992). The science of choosing the right decision threshold in high-stakes diagnostics. *American Psychologist, 47,* 522–532.

Swets, J. A., & Pickett, R. M. (1982). *Evaluation of diagnostic systems.* New York: Academic Press.

Vaughan, G., May, A., Ross, T., & Fenton, P. (1994). A human factors investigation of an RDS-TMC system. *Proceedings of the First World Congress on Applications of Transport Telematics and Intelligent Vehicle-Highway Systems* (pp. 1685–1692). Paris: Intelligent Transportation Systems.

Verwey, W. (1990). *Adaptable driver-car interfacing and mental workload: A review of the literature* (Drive Project V1041). Groningen, The Netherlands: University of Groningen, Traffic Research Center.

Verwey, W., & Kaptein, N. A. (1994). Automatic man–machine interface adaptation to driver workload. *Proceedings of the First World Congress on Applications of Transport Telematics and Intelligent Vehicle-Highway Systems* (pp. 1701–1708). Paris: Intelligent Transport Systems.

Wheeler, W. A., Lee, J. D., Raby, M., Kinghorn, R. A., Bittner, A. C., & McCallum, M. C. (1994). Predicting driver behavior using advanced traveller information systems. *Proceedings of the Human Factors Society, 38*, 1057–1061.

Wiener, E. L. (1988). Cockpit automation. In E. L. Wiener & D. C. Nagel (Eds.), *Human factors in aviation* (pp. 433–461). San Diego, CA: Academic Press.

Wierwille, W. W. (1993). Visual and manual demands of in-car controls and displays. In B. Peacock & W. Karwowski (Eds.), *Automotive ergonomics* (pp. 299–320). London: Taylor & Francis.

Wilde, J. G. S. (1982). The theory of risk homeostasis: Implications for safety and health. *Risk Analysis, 2,* 249–258.

Wright, K. (1990). Trends in transportation: The shape of things that go. *Scientific American, 262,* 92–101.

Younger, J. D. (1994, November/December). Boosting your car's IQ. *AAA World*, pp. 10–12.

Zwahlen, H. T., Adams, C. C., & DeBald, D. P. (1988). Safety aspects of CRT touch panel controls in automobiles. In A. G. Gale (Ed.), *Vision in vehicles II* (pp. 335–344). Amsterdam: North-Holland.

17 Maritime Automation

John D. Lee
Thomas F. Sanquist
Battelle Memorial Institute, Seattle Research Center

INTRODUCTION

Technological innovations have led to major changes in the role of humans on ships. Harding (1975) listed a wide variety of potential automated systems including automatic data logging, position fixing aids, restricted channel navigation aids, collision avoidance systems, cargo planning aids, automatic route following, and maintenance diagnostic aids. Many modern ships have adopted elements of this automation. In some instances these functions have been combined to reduce personnel requirements dramatically. Staffing levels have dropped from 30 to 40 crew members to 15 to 21 crew members. As automation has eliminated many routine tasks, mariners have been transformed into shipboard managers, responsible for coordinating multiple automatic systems.

Traditionally, a ship's crew is divided between the deck department, which is responsible for navigation and ship management, and the engine department, which is responsible for the mechanical integrity and operation of the engine room. Automation has implications for both the engine and deck departments. One of the primary responsibilities of the deck department is safe navigation. Currently, most ships navigate with at least three people on the bridge. Generally, one person acts as a helmsman, another as a lookout, and a third as a watch officer. These people are responsible for communicating with other ships, communicating with the vessel tracking system, planning safe courses, executing courses, and avoiding collision with other ships. When executing maneuvers, the deck department communicates the required changes in engine speed to the engine room, where the appropriate adjustments are made. Automation plays a major role in the deck department, and prototype systems suggest that automation may play an even more important role in the future.

Automation in the deck department began with the development of the radar, and has progressed to radar enhanced with automated radar plotting aids (ARPA). More recently, the electronic chart display information system (ECDIS) is beginning to replace paper charts. Future ECDIS developments may include radar and ARPA overlays on ECDIS. Perhaps the most advanced automation systems are artificial intelligence systems that combine navigation and ship performance information to provide routing and maneuvering suggestions to the crew (Grabowski & Wallace, 1993). Other developments include more advanced autopilots that enhance fuel efficiency, and complex combinations of engine, rudder, and thruster controllers that enable the watch officer to dock the ship without need for multiple tug boats. Several countries are working to develop fully integrated bridges that combine elements of all these automated systems to produce an integrated suite of navigation and ship control aids. These changes may make it possible for a single person to act as the helmsman, lookout, and watch officer (Grabowski & Wallace, 1993; Schuffel, Boer, & van Breda, 1988).

Similar changes have occurred in the engine department. Older ships operate with a 24-hour watch consisting of a wiper, a water tender, an oiler, a fireman, and an engineer. However, technology has automated many aspects of the engineering department and has made unattended engine rooms possible. Automation supports planned maintenance with computerized schedules of maintenance tasks. Automation also supports predictive maintenance by tracking changes in vibration signatures of engine components and providing sophisticated ultrasound data that verify the structural integrity of system components. On these ships, the engine room is monitored remotely and engineers perform maintenance during normal work hours (9 a.m. to 5 p.m.). Rather than working various shifts around the clock, engineers in automated engine rooms are off duty at 1700 (5:00 p.m.) and proceed to go on "watch call" where they carry beepers. When only one engineer is in the engine room, a "deadman" button is used to have the engineer continually acknowledge his or her presence. Interviews with ship captains document the consequences of unmanned enginerooms. One captain reported that he had his engine room crew reduced from nine (three sets of three engineers manning the engine room around the clock) to two (engineers carrying beepers when off work). The captain of this particular ship said that having an unattended engine room during voyages greatly increased his stress level. Even with unattended engine rooms, the reliability and safety of the engines still depends on the engineer's ability to anticipate, identify, and react to malfunctions.

Automation on modern ships has begun to blur the distinction between the deck and engineering departments. In older ships, a bell system relayed bridge commands to the engine room, where the engines were manually adjusted to meet bridge orders. Now, direct control from the bridge allows the deck department to implement engine speed changes directly. In addition, displays on the bridge convey data that were once only available to the engineering department. Providing deck personnel with data from the engine room imposes new demands on them. These changes may introduce new knowledge requirements regarding interpretation of engine room data similar to those found in the process control domain (Zuboff, 1988). Introducing remote monitoring systems distances mariners from a wide variety of potentially useful information (smells, noises, vibrations) only available in the engine room itself. Although technology has changed the deck and engineering department, and blurred the distinctions between the two, the responsibility for safe passage remains that of the captain. Thus, human interaction with automation remains a key element of safe ship operation.

Beyond the automation associated with ship control and operation, technology supporting ship management is also becoming common. Almost all modern cargo ships have some kind of computer or computer system aboard. Some vessels have stand-alone, desk-top computers, whereas others have a local area network (LAN) that links all computers aboard the ship. Most commonly, these computers are used for communication, reporting, payroll, and inventory purposes. Some ships have computers that run special programs to automate various operations that were previously handled manually. For example, routine operations such as cargo loading, ballast control/monitoring, and engine room control/monitoring can now be automatically monitored from several points on the ship, including the bridge. Computers used for shipboard management are often linked to corporate offices so that captains can communicate and respond to corporate directives. Although these systems can enhance management efficiency, they can also increase the workload of captains as they are increasingly required to conform to management procedures of shore-based operations. Other problems relate to a lack of training and maintenance support for these systems. For example, the most common form of training was through manuals and tutorials; shipping companies offer minimal hands-on or classroom training, if any. Training tends to be passed down among the crew. In addition to limited training, the maintenance of computer systems and programs is another area found lacking in support. Although shipping companies install computer programs, the crew must maintain the system themselves when at sea. Generally, there is no designated person aboard ship who is formally trained or responsible for computer maintenance. Computer problems tend to be fixed, or are attempted to be fixed, by whomever aboard has the most computer or electronics experience. Most frequently this person is the radio operator. However, recent crew reductions may eliminate this position. During interviews with mariners, one captain noted that although he had considerable experience working with computers and computer programs, the alternating captain knew very little about computers. When the latter captain had computer problems, he had to wait until the ship was in port for the problems to be fixed. The captain of another ship reported that their computer and automation problems were too complex for the crew to fix. As automation becomes more sophisticated, there is a need for more training in the use and maintenance of the systems because they often require skills and knowledge not needed with less sophisticated equipment.

Operation and management of modern commercial ships link many different types of automation including management decision support systems, e-mail, complex navigation systems, and supervisory control of power systems. In many ways, the automation-induced changes in the maritime industry parallel changes in other domains, such as process control, aviation, operating rooms, and manufacturing systems. Just as automation in the maritime domain promises increased safety and efficiency, automation in other domains has made similar claims. However, reality has frequently failed to meet the expectations. Moreover, poorly conceived automation may eliminate small mistakes at the expense of encouraging large blunders (Wiener, 1993). Often the disappointing performance results from cognitive effects associated with the new technology. These cognitive effects include:

- Skill degradation.
- Inadequate feedback resulting in misunderstanding.
- Miscalibration in trust of automation, leading to misuse.
- Fewer physical demands but greater cognitive load.
- Enhanced workload peaks and troughs.
- Inadequate and misleading displays.
- Reduced opportunity for learning.

All of these factors can undermine the effectiveness of automation and compromise system safety and efficiency. To better understand whether problems from other domains might also apply to the maritime domain requires a detailed definition of *automation*. Such a definition would also help identify how changes in training and system design might mitigate automation-induced problems.

DEFINING CLASSES OF AUTOMATION

Automation has been used to refer to a range of technological advances, often without a clear definition of what the term means. Without a clear definition of *automation*, identifying potential training and system design requirements will remain problematic. For example, adding ARPA information to an electronic chart may be referred to as *automation*. Similarly, microprocessor technology used to integrate the control of multiple thrusters, main engines, and rudders to position a large ship may also be referred to as *automation*. The cognitive implications of these very different types of technology have important consequences for extrapolating from problems identified in other domains and for developing strategies to mitigate these problems.

It is useful to distinguish between two general classes of automation, *perceptual augmentation* and *control integration*, as shown in Fig. 17.1. As exemplified by radar, ARPA, and electronic charts, automation often acts as perceptual augmentation, and does not affect the actual control of the ship. In the case of radar, the automation represented by the radar simply acts as perceptual augmentation and helps the mariner see things that would not otherwise be seen. Another general type of automation can be labeled as control integration. This general type of automation contrasts with perceptual augmentation because it enhances mariners' ability to control the ship, rather than perceive the state of the ship or environment. The microprocessors that enhance mariners' ability to position a ship by integrating multiple subsystems (thrusters, engines, and rudders) exemplify control integration. Distinctions between perceptual augmentation and control integration help distinguish between different classes of automation and can help identify the associated cognitive effects of different classes of automation.

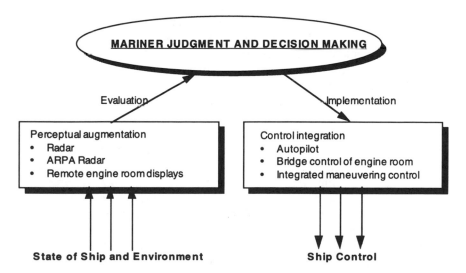

FIG. 17.1. Two types of automation and their influence on mariner judgment and decision making.

Perceptual Augmentation and Collision Avoidance Systems

Collision avoidance is one aspect of vessel navigation that has been changed substantially by technology. Radar is an integral element of collision avoidance, and ARPA radar eliminates many of the tedious, error-prone aspects of radar use. ARPA radars automatically display a wide variety of textual and graphical information, including the relative and absolute speed and direction of any ship within radar range. ARPA radar can also calculate the distance and time of the closest approach and provide a variety of trial maneuver functions that can generate future trajectories of the ships, given proposed speed and course changes. Some ARPA radars display this information numerically, whereas others combine it into graphical icons that represent safety zones around the ships. These labor-saving features contrast with the standard radar. With standard radar, mariners must integrate these observations using relatively complex geometric calculation procedures. Thus, ARPA radar is an example of perceptual augmentation that enables mariners to see what would otherwise be invisible. Although ARPA radar eliminates many of the error-prone tasks of tracking and avoiding other ships, it may not always increase ship safety (Perrow, 1984). Because ARPA has been mandated for two decades, any problems mariners currently experience are not a transient effect of its introduction, but represent chronic problems that are unlikely to improve with time. Furthermore, because ARPA supports the critical activities of collision avoidance and navigation, it is inevitable that some form of ARPA will occupy a prominent role on the bridge of the future. Thus, current problems with ARPA are likely to affect future navigation systems.

There are many anecdotal discussions of errors of interpretation of radar and collision avoidance information (e.g., Perrow, 1984), and accident reports often mention problems with this type of information, but there is virtually no empirical data that can help to understand the frequency, severity, or consequences of these errors. As part of this research, CSERIAC conducted two extensive searches of NTIS, DTIC, Oceanic Abstracts, and other appropriate databases to identify articles or technical reports addressing radar or collision avoidance interpretation errors and their consequences. Neither of these searches yielded any evidence of systematic evaluation (either analytic or empirical) of radar-based errors. Although a number of accident reports have identified misinterpretation of radar or ARPA data as a contributing cause, there are no detailed taxonomies of error types or the underlying psychological or technical causes.

To address this problem, we undertook a detailed analysis of the types of errors observed by instructors during ARPA training. Data were collected during visits to two ARPA training schools and through detailed interviews and observations of ARPA instructors. The result of this work is a taxonomy of ARPA errors, and an analysis of their underlying psychological and technical causes. The basic model underlying the analysis is shown in Fig. 17.2.

This model identifies a relationship among input variables, intervening processes, and outcomes in terms of ship maneuvering errors. The primary input variables of radar/ARPA data, local knowledge, and rules of the road interact with a number of intervening psychological and technical factors. Depending on the outcome of the interpretation, a ship maneuver is initiated. Fig. 17.2 shows a variety of inappropriate ship maneuvers that can result from ARPA misinterpretation. These errors highlight the need to examine alternate designs, develop training, and revise qualifications testing to address the complexities of ARPA and other advanced technology, such as electronic charts and integrated autopilots. Two examples of ARPA errors have been selected to illustrate how input variables interact with the psychological and technical intervening variables to generate a variety of maneuvering errors. The first error involves incorrect initialization of the ARPA, and the second involves misinterpretation of true and relative motion vectors.

ARPA Error Example 1: Incorrect Initialization of Gyroscope or Speed Log Input

As with many automated systems, accurate performance depends on the correct initialization of the system. The following paragraphs provide a

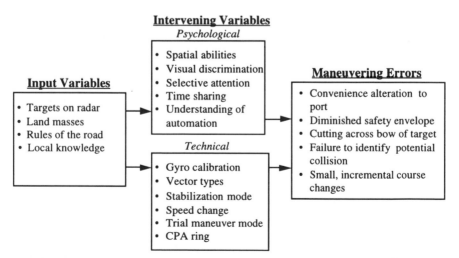

FIG. 17.2. Relationship of radar input, psychological, and technical variables to ARPA-induced maneuvering errors.

detailed discussion of potential outcomes, technical description of ARPA configuration that leads to these outcomes, and the cognitive demands that contribute initialization errors and the misuse of the ARPA.

Potential Maneuvering Errors.   Incorrectly initializing the gyroscope or speed log input to the ARPA can have several important outcomes. An ARPA with incorrect gyroscope and speed log information will correctly indicate a collision course when displaying relative motion vectors; however, incorrect gyroscope or speed input may misrepresent the orientation and speed of other ships, as well as the distance to the collision point. Incorrect initialization may lead to delayed collision avoidance maneuvers and maneuvers that violate the rules of the road.

Technical Description of the ARPA Configuration.   ARPA operates by receiving radar data that specify the speed and direction of surrounding ships and land relative to the observing ship (the one on which the ARPA is mounted). If the speed and direction of the observing ship are known, then simple vector calculations can generate the true speed and direction (relative to the water) of the surrounding ships. The true motion vectors are critical for identifying the orientation, speed, and likely intentions of other ships, whereas the relative motion vectors are critical for identifying potential collision situations. Thus, inaccurate true motion vectors could mislead mariners by misrepresenting viable collision avoidance strategies.

Initializing the speed log and gyroscope provides the ARPA with the speed and course of the observing ship. The ARPA combines this information with the relative motion vectors of the observed ships to produce true motion vectors for each of the observed ships. Therefore, any errors in the true motion vector of the observing ship (gyroscope and speed log input) will be propagated to the true motion vectors calculated for each of the observed ships. However, errors in this information will not influence the accuracy of relative motion vectors. Thus, the incorrect initialization of gyroscope and speed log can have a profound effect on the appearance and interpretation of the ARPA screen. Fig. 17.3 illustrates these effects. The bold vectors in Fig. 17.3 represent the actual true motion vectors of the two ships. The ship in the center of the screen is the observing ship. The vectors labeled "Gyroscope error" represent the appearance of the screen when the gyroscope data are incorrect and portray the observed ship as moving more slowly and more directly toward the observing ship. This error might lead the observer to expect to see a side profile of the ship rather than the rear quarter. The vectors labeled "Speed log error" represent the appearance of the screen when the speed log is set faster than the actual speed of the observed ship. This error makes the observed ship appear to be sailing more quickly than it actually is. The effects of gyroscope and speed log errors

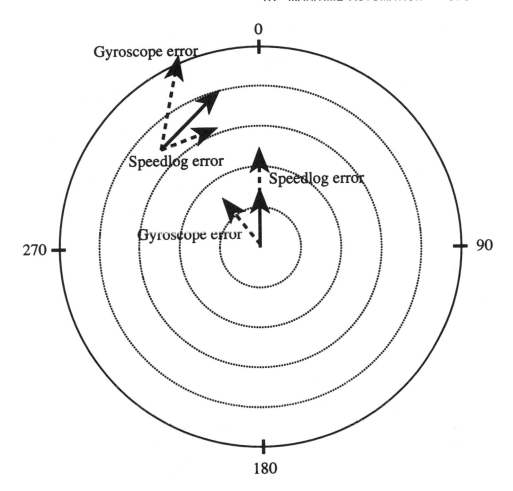

FIG. 17.3. ARPA display errors resulting from incorrect gyroscope or speed log input.

shown in Fig. 17.3 present the mariner with a representation that conflicts with reality. To the extent that this information is used to develop collision avoidance strategies, this information could guide mariners into collision situations.

Psychological Demands Leading to Misinterpretations.   The cognitive demands that lead to incorrectly initialized gyroscope and speed log input draw on mariners' knowledge of the system, their memory for procedures, and their ability to recognize erroneous information generated by an incorrectly initialized radar. If a mariner's mental model of the ARPA is not sufficiently detailed to include the need to initialize the speed log and gyroscope inputs, then it is unlikely that they will be entered or verified during the voyage. Even if the mariner's mental model is sufficiently developed to recognize their importance, the burden of remembering these details, together with the other demands of operating a ship, may cause them to be neglected. This memory load could be alleviated if the speed log and gyroscope settings were more visually salient aspects of the ARPA display. However, in many current ARPA designs, the speed log and gyroscope settings are shown in small text boxes that are only visible when selected from a menu. This limits mariners' feedback and makes it unlikely that a failure to initialize the system will be noticed. The only other opportunity for mariners to identify a poorly initialized system is to identify discrepancies between the visual image seen from the bridge and the radar image. This requires good visibility and training to integrate the two information sources.

Error Prevention Strategies.   The cognitive demands associated with initializing the speed log and gyroscope provide some insights into potential strategies to reduce ARPA initialization errors. These strategies can be divided between design modifications and training/qualifications requirements. To encourage mariners to initialize the ARPA correctly, the ARPA design could be changed to force the entry of any required parameters when the ARPA is turned on. Another strategy could include emphasizing the speed log and gyroscope data so that an incorrectly initialized system would be more readily recognized by the operator. The parameters could be more visible by making them an integral part of the display rather than text buried with other text messages. Alternatively, the speed log and gyroscope input to the ARPA could be displayed alongside the actual speed and course data outside the ARPA itself. This would physically integrate the ARPA with its related data sources, alleviating the burden of integrating the currently disparate data sources.

Adjusting the training and qualifications requirements could also address problems of incorrect ARPA initialization. Specifically, qualification exams could be designed to probe mariners' mental models of ARPA to ensure that they understand the role and importance of speed log and gyroscope input. The surface similarity between ARPA and standard radar may lead mariners to develop deficient mental models that do not accurately anticipate the effects of important factors, such as speed log and gyroscope initialization. Paper-based exam questions could identify this and other deficiencies regarding mariners understanding of ARPA operation. Examining formal and informal procedures used in ARPA operation could reveal opportunities to train mariners so that initializing the ARPA is a well-learned procedure rather than an ad hoc series of adjustments. Training a standard procedure would alleviate memory load and increase the probability that mariners will remember to initialize all critical parameters. The procedure could be augmented by either electronic or paper-based checklists. Evaluating this training, in the context of mariner qualifications, would be more effective in a simulator than through a paper-based test. Training could also be used to help mariners identify when an ARPA is incorrectly configured. Exposing mariners to scenarios where an incorrectly initialized ARPA presents information that conflicts with the visual scene could leave mariners better prepared to identify problems with the ARPA that might occur in actual operating conditions. Qualification exams that address the ability to recognize incorrectly initialized ARPA could be a combination of paper-based and simulator exams.

Implications for Other Ship Systems. The cognitive demands associated with neglecting to initialize the gyroscope and speed log input are common to many automated systems. As systems become increasingly automated, subtle failures of initializing the system can lead to massive disruptions when the automation does not control the system in the manner expected or when the automation presents misleading information based on incorrectly entered parameters. Thus, the general cognitive demands exemplified by gyroscope and speed log initialization are likely to manifest themselves in a variety of ship systems as they become more technologically advanced. For example, some electronic chart displays offer the option of specifying a minimum depth of water for safe travel. This information is used to trigger alarms if the safety margin is compromised. As with gyroscope and speed log initialization, the incorrect initialization of water depth could leave mariners relying on an alarm that might not sound until after the ship is on the rocks. Many of the prevention strategies proposed to ensure correct ARPA initialization will likely apply to other forms of maritime automation, particularly that which acts as perceptual augmentation.

ARPA Error Example 2: Misinterpretation of True and Relative Motion Vectors

On an ARPA screen, the ship motion vectors displayed on the ARPA screen show the direction and speed of surrounding ships. This information indicates the collision potential and the viability of various collision avoidance strategies. These vectors can by viewed from two basic perspectives: relative motion vectors that show motion relative to the observing ship, true motion vectors that show motion relative to the water. ARPA presents this information in two modes of operation. These modes generate substantial opportunities for confusion, misuse, and misinterpretation. These include failure to identify possible collision situations, inappropriate responses to collision situations, and delayed responses to collision situations. The following paragraphs provide a detailed discussion of the potential outcomes, the technical description of ARPA modes that lead to misinterpretations, and the cognitive demands that contribute to misinterpretations of ARPA information.

Potential Maneuvering Errors.   Incorrectly interpreting or misusing the relative and true motion vectors displayed by the ARPA can lead mariners to fail to identify potential collision situations. Even if a potential collision situation is identified, misinterpretations can also inhibit mariners' ability to generate effective collision avoidance maneuvers. Misinterpretations can lead to small, incremental course changes and other violations of the rules of the road.

Technical Description of the ARPA Configuration.   ARPA enables mariners to change how the motion of other ships is displayed. Two important ways this information is displayed are true motion vectors and relative motion vectors. Changing between these two modes generates a radically different display of information. Correctly selecting between these displays requires that mariners understand the relationship between the two modes of operation and the relative capabilities of each.

The true motion and relative motion modes of ARPA operation radically change the appearance of the ARPA display. The true motion representation shows ship speed and direction relative to the water. With the true motion representation, the direction and size of the motion vectors correspond to the headings and speeds of the individual ships. The relative motion representation shows ship speed and direction relative to the observing ship. Fig. 17.4 shows the true and relative motion representations with the observing ship sailing due North, the observed ship A sailing due North, and the observed ship B sailing due East. Fig. 17.4 also illustrates the relative motion representation that results from subtracting the true motion vector of the observing ship from the relative motion vectors of the observed ships. With the relative motion representation, the direction and size of the motion vectors correspond to the course and speed of the observed ships relative to the observing ship. Therefore, the relative speed of the observing ship is zero by definition.

The true motion representation is important because it directly reveals the course and speed of other ships, whereas the relative motion vectors clearly indicate the presence of a collision course. With relative motion vectors, a collision course is indicated when any relative motion vector can be extrapolated to intersect with the observing ship's marker. Ship B in Fig. 17.4 illustrates this situation. If the extrapolated line does not intersect the observing ship's marker, then the shortest distance from the observing ship to the extrapolated line is the distance of closest approach. This cannot be

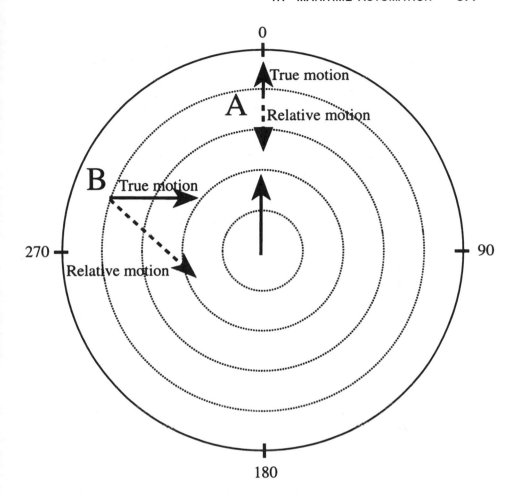

FIG. 17.4.   ARPA displays of true and relative motion vectors for the same target.

said of the extrapolated vectors in the true motion representation. However, the true motion vectors directly reveal the actual direction, speed, and orientation of the surrounding ships. This information is critical to identify the intentions of other ships and to develop viable collision avoidance maneuvers. Generally, relative motion vectors help identify potential collision situations, whereas true motion vectors help select appropriate collision avoidance strategies.

The purpose and meaning of relative and true motion vectors and the associated ARPA modes can be confused easily by the mariners. For example, mariners could interpret relative motion vectors as true motion vectors. This error can lead mariners to think that they are approaching an oncoming ship when they are actually overtaking a slower ship. Fig. 17.4 illustrates the consequence of this confusion by showing that ship A appears to be sailing on a southerly course in the relative motion mode (when interpreted as true motion mode). Similarly, confusing true motion vectors with relative motion vectors can produce a misleading point of closest approach. With relative motion vectors, the closest point of approach (CPA) is shown by the closest point on the line extrapolated from the relative motion vector of the observed ship. If the true motion vector is used, the resulting estimate could misrepresent the actual passing distance and cause mariners to inadvertently compromise their safety envelope. Thus, failing to recognize the distinction between relative and true motion modes, as demanded by the circumstances, will present the mariner with a misleading display that could lead to faulty decision making.

Psychological Demands Leading to Misinterpretations. The cognitive demands associated with interpreting true and relative motion vectors are dramatically different for an ARPA, as compared to a standard radar. With a standard radar, complex geometric manipulations and manual calculations are required to generate relative and true motion vectors. ARPA radar automates these tasks, but it imposes the burden of selecting, identifying, and interpreting the mode of ARPA operation. With the standard radar, the source and meaning of this information is made obvious by the manual calculations that produce it. ARPA forces mariners to understand and interpret true and relative motion vectors in an abstract manner, without the support of the physical representation generated by the grease pencil plotting procedures used with standard radar. The more abstract nature of ARPA operation may tap skills and training not needed to operate standard radar. This trend toward increasingly abstract control and display systems has inhibited understanding and effective use of complex information in other domains. These problems suggest that the training and qualification testing that supported adequate performance with the less automated systems may not be adequate for the more advanced systems. Specifically, grease pencil plotting never required mariners to know what information could be extracted and what conclusions could be derived from a relative motion representation. This knowledge was implicit in the plotting procedure itself. However, ARPA makes these new skill and knowledge demands.

The need to control and interpret mode changes also imposes additional mental load on the mariners. With standard radar, the manually generated plot (RTM triangle) included both relative and true motion vectors. With ARPA, mariners must know what mode the ARPA is in before they can interpret the displayed information. Even though the ARPA display indicates the mode, the need to track the ARPA mode introduces a new memory and information processing load. Systems with multiple operating modes can increase the cognitive demands on the users, leading to "mode errors," in which interpretations and actions appropriate for one mode generate errors when used in another mode.

Error Prevention Strategies.   One of the primary cognitive demands associated with ARPA radar and the true and relative vector modes is the increased level of abstraction relative to standard radar. This increased abstraction demands a sophisticated mental model of ARPA operation and vector mathematics. Designing of the ARPA in a way to make the computational processes of the ARPA more evident could alleviate these problems. For example, including the RTM triangle for each ship would show the mariner the components of the calculation directly. At the same time, showing the RTM triangle could help identify the changes between true and relative motion modes. For example, when in the relative motion mode those vectors would be highlighted, and when in true motion mode those vectors would be highlighted. In each case, the other vectors would remain, but only as a faint trace.

The cognitive demands imposed by the true and relative motion modes also identify training and qualification requirements. Specifically, paper-based testing could query mariners' knowledge concerning the criteria used to select between true and relative motion vectors. In other domains, operators have been found to possess inert knowledge. Inert knowledge may be retrieved for paper-based testing, but it may not be used in operational scenarios. To the extent that exams reflect inert knowledge, they will fail to judge mariners' true capabilities. Generally, paper-based exams that assess the theoretical basis of system operation have little correlation with actual job performance (Rouse & Morris, 1986). Simulator exercises could test whether mariners actually follow the criteria expressed during the exams.

Implications for Other Ship Systems.   Mode errors, together with the associated cognitive demands of multiple modes, have been noted with other automation systems (Reason, 1990). For this reason, mode errors associated with true and relative motion vectors are likely to be symptomatic of a general problem that may confront many advanced display and control systems. For example, an advanced maneuvering device that combines the control of seven maneuvering devices into a single controller may suffer from the multiple modes that enhance its functionality. Although this device facilitates complex maneuvers, it has multiple modes that could lead to serious errors. Each mode enables the captain to turn the ship about a different point. For example, in one mode the center of rotation is the stern, in another mode it is the bow, and in another it is the center of the ship. Changing the mode of the control, and the corresponding center of rotation of the ship, gives the captain substantial flexibility in maneuvering the ship. However, the potential for mistakenly selecting an inappropriate mode could have serious consequences. Similar to the problems with automation in aviation, trivial errors made in initializing the automation may be magnified by the system to emerge later in the voyage (Wiener, 1993).

Slips, Mistakes, and ARPA Errors

The detailed description of the errors associated with incorrect initialization of gyroscope or speed log and misinterpretation of true and relative motion vectors indicates that they are result of inappropriate procedures rather than deficits of underlying knowledge requirements. Table 17.1 summarizes the changes in knowledge requirements from standard radar to ARPA. This table shows that the core knowledge requirements for collision avoidance are not very different for ARPA and standard radar, as shown in the common entries in Table 17.1. However, the procedural aspects of the job change dramatically, often taking a form that is functionally "layered" because of the limited display area of CRT displays. This functional "layering" is manifest principally in the need to learn and remember sequences of control actions and menu selections to obtain information that had previously been available at a surface level. For example, once the mariner plots the position and vector of the other ships on a maneuver board with standard radar plotting, the relative motion vector is obvious —

TABLE 17.1
Knowledge Requirements for Collision Avoidance

Standard Radar	ARPA
Motion vectors: relative, true (sea stabilized), true (groundstabilized)	Motion vectors: relative, true (sea stabilized), true (groundstabilized)
Display orientation (head up, North up, course up)	Display orientation (head up, North up, course up)
Characteristics of targets (moving, stationary, land masses, noise)	Characteristics of targets (moving, stationary, land masses, noise)
Procedure for plotting ship positions	Controls/menus to display target relative and true vectors
Meaning of plotted vectors	Meaning of relative and true vectors
Plot procedure for determining CPA and TCPA	Controls/menus to obtain target data
Plot procedure for determining course change	Controls/menus for trial maneuver: interactions with target vector types
Plot procedure for evaluating viability of course change	Default time factors for trial maneuver vectors

its derivation is a part of the overall procedure of drawing the plot. This is not the case with ARPA, because the mariner must remember to make the appropriate mode selection in order to display the relative motion vector.

Many types of human performance errors that can result from automated technology have been referred to as "slip." Slips are the result of a momentary lapse of attention or memory during ongoing activities that can affect the overall outcome. For example, forgetting to change the ARPA display mode from true vectors to relative vectors when evaluating collision potential would be classified as a slip. The resulting decisions about collision avoidance would be carried out correctly, but based on erroneous information because of the slip. In distinction to slips, mistakes are errors based on flawed reasoning or lack of fundamental knowledge. For example, a mariner may know that a collision course is developing and makes plans for a maneuver, but poor knowledge of the particular ship dynamics results in a flawed plan–for example, implementing the turn too late. This distinction between slips and mistakes is part of a more comprehensive analysis of human error discussed by Reason (1990).

There are distinct training and implications of slips, mistakes, and the associated procedural knowledge and fundamental knowledge. Because collision avoidance automation affects procedural knowledge, and there-fore stresses human performance limits of attention and memory, funda-mental knowledge training related to automation will be of limited value. If the automation involves control dynamics that require an understanding of what the system is doing, then fundamental knowledge training will have value. However, with ARPA radar and the present suite of navigation automation, the need is for procedural training to establish routines that will guard against the human performance slips that can lead to erroneous displays and the subsequent errors in ship handling.

Control Integration and Maneuvering Systems

Most modern ships have a variety of automated systems that act as control integration. These systems include a variety of autopilot devices that maintain consistent headings, correct for wind and currents, and maximize fuel efficiency. One of the more interesting examples of control integration is a system that combines the control of seven maneuvering devices into a single controller. This system combines the control of the engine speed, the angle of each of two rudders, and the activity of four thrusters into a single rotating lever. By manipulating this lever the captain can perform intricate maneuvers required for docking and undocking. Although this device facilitates complex maneuvers, it has several features that could lead to serious problems, similar to those that control integration automation has induced in other domains.

One of the more complicated aspects of this automated positioning device are the subtly different modes in which it might operate. Each mode enables the captain to turn the ship about a different point. For example, in one mode the center of rotation is the stern, in another mode it is the bow, and in another it is the center of the ship. Changing the mode of the control, and the corresponding center of rotation of the ship, gives the captain substantial flexibility in maneuvering the ship. However, the potential for mistakenly selecting an inappropriate mode could lead to mode errors that parallel those found in other domains (Sarter & Woods, 1992).

Besides the problem of mode errors, reliance on this automated positioning device may also lead to substantial skill degradation. Without this controller, the captain must manipulate seven individual controllers. Manipulating these seven controllers requires substantial perceptual/motor skills that are not necessary with the automated system. Although the automated controller usually performs adequately, some combinations of current and wind make its use inappropriate. In these circumstances, mariners revert to controlling the ship by manipulating the seven controllers independently. If their ability to manipulate the individual controllers degrades because of their continual reliance on the automated systems, then it may leave them vulnerable to situations in which they must rely on manual control. To combat this problem, some captains routinely bypass the automatic controller and manually dock and undock the ship. In some ways, this behavior parallels that of the pilots who revert to manual control because they fear skill degradation (Wiener & Curry, 1980).

Integrating Perception Augmentation and Control Integration With Electronic Charts

The distinctions between perceptual augmentation and control integration can help identify design and training implications of different classes of maritime automation. However, truly effective automation may need to combine elements of both classes of automation. Much recent research addressing effective decision making in dynamic situations emphasizes the link between perception and action. In dynamic decision making, decisions are part of an evolving sequence of activity and do not stand as isolated choices. As such, effective decision making depends on a complex interplay between perception and action. Truly effective automation may depend on linking automation defined as perceptual augmentation to automation defined as control integration. A specific example illustrates the need to link the two classes of automation.

Electronic chart display information systems (ECDIS) act as perceptual augmentation by using global positioning system (GPS) to generate an image showing the ship's location relative to land and aids to navigation.

These systems also enhance mariners' perception by highlighting areas where the water depth is sufficient for the ship. This feature clearly identifies areas passable for the ship. By overlaying intended tracklines and plotting past positions, electronic charts highlight potentially dangerous deviations from the intended course. These features enhance mariners' situation assessment and can contribute to increased safety. As with many other computer-based displays, current ECDIS suffer from several problems including keyhole effects due to the need to view a large area through a small window. These systems also impose layers of menus and require familiarity with many complex functions (Lee & Sanquist, 1993).

Beyond the need to shape the electronic chart into a tool to augment mariners' perception, the ECDIS must also act as a tool for planning and ship control. Paper charts act as more than information displays; they are also planning and calculating tools. Thus, if electronic charts are to replace paper charts, they must capture these functions of a paper chart. As such, electronic charts must include characteristics of perceptual augmentation and control integration. For instance, mariners use a paper chart and calipers to calculate distances and estimate times to waypoints. This information forms the basis for voyage planning and coordinating with port authorities and shoreside support. When mariners complete the voyage plan, the chart contains a trackline annotated with speed and heading changes. During the voyage, these annotations provide the input to rudder and engine commands. An automated system to replace these features would act as control integration by integrating information to specify the ship's course. Some advanced systems tie autopilot settings to the planned trackline. With these systems, the initial voyage plan drawn by the captain becomes the settings for the autopilot. In these systems, the mariners need only to verify course changes made by the autopilot. By linking technology that acts as perceptual augmentation (chart display) to technology that acts as control integration (autopilot), the system avoids problems associated with bridging the two types of automation. Also, a system that integrates perceptual augmentation and control integration aspects of automation will better support the entire perception–action decision-making cycle, rather than an isolated part of it.

CONCLUSION

A wide variety of automation exists in the maritime environment. To better understand the cognitive effects of this automation, a more specific definition has been developed. This definition distinguishes between perceptual augmentation and control integration. Perceptual augmentation refers to automation that combines information from diverse sources and accentuates human's perception. Control integration refers to automation that eliminates low-level human control activities and enhances humans' ability to control the system. This distinction helps draw parallels between problems experienced with automation in the maritime domain and problems previously experienced in other domains.

ACKNOWLEDGMENTS

This research was supported by a contract from the U.S. Coast Guard Research and Development Center with Dr. A. Rothblum serving as technical monitor.

REFERENCES

Grabowski, M., & Wallace, W. A. (1993). An expert system for maritime pilots: Its design and assessment using gaming. *Management Science, 39*(12), 1506–1520.

Harding, E. J. (1975). Computer-based ship automation developments in the United Kingdom and abroad. *Proceedings of the Symposium on the Use of Computers in Shipboard Automation* (pp. 11–26). London: The Royal Institution of Naval Architects.

Lee, J. D., & Sanquist, T. F. (1993). A systematic evaluation of technological innovation: A case study of ship navigation. In *IEEE Proceedings: Systems, Man and Cybernetics* (pp. 102–108). Piscataway, NJ: IEEE.

Perrow, C. (1984). *Normal accidents—living with high-risk technologies.* New York: Basic Books.

Reason, J. (1990). *Human error.* Cambridge, England: Cambridge University Press.

Rouse, W. B., and Morris, N. M. (1986). On looking into the black box: Prospects and limits in the search for mental models. *Psychological Bulletin, 100*(3), 349–363.

Sarter, N. B., & Woods, D. D. (1992). Mode error in supervisory control of automated systems. In *Proceedings of the 36th Annual Meeting of the Human Factors Society* (pp. 26–29). Santa Monica, CA: Human Factors Society.

Schuffel, H., Boer, J. P. A., & van Breda, L. (1988). The ship's wheelhouse of the nineties: The navigation performance and mental workload of the officer of the watch. *Journal of Navigation, 42*(1), 60–72.

Wiener, E. L. (1993). *Intervention strategies for the management of human error.* NASA contractor Report, (NCA2–441). Mofett Field, CA: NASA–Ames Research Center.

Wiener, E. L., & Curry, R. E. (1980). Flight-deck automation: Promises and problems. *Ergonomics, 23*(10), 995–1011.

Zuboff, S. (1988). *In the age of the smart machine: The future of work and power.* New York: Basic Books.

18 Decision Support in Medical Systems

Stephanie Guerlain
Honeywell Technology Center

Phillp J. Smith
Ohio State University

Jack W. Smith
Sally Rudmann
Jodi Obradovich
The Ohio State University

Patricia Strohm
Southside Medical Center

INTRODUCTION

Medical decision making is sufficiently difficult by itself, but is even more complex given the added pressures of today's environment. Besides the obvious goal of improving quality of care, there are increased concerns over liability as well as pressures to reduce costs. In addition, the medical field is growing rapidly, with new research yielding different and better ways to evaluate and treat patients. Health care providers must keep up with these new findings and also with the technology that is being developed to support various tasks. Because computer systems and other automated devices have been developed in all aspects of health care, doctors, nurses, technicians, and administrators must determine the potential utility of new products as well as master the use of those products selected for use.

Besides administrative tasks, most tasks performed at hospitals revolve around the diagnosis and treatment of patients. This process can be divided into roughly three areas (Miller, 1986). First is the task of differential diagnosis. This involves the evaluation of a patient's symptoms and initial laboratory data to determine the possible set of diseases he or she may or may not have. This task is usually loosely structured and requires that the physician call on a large body of often diverse knowledge. A second stage, called the *workup* may then be called for. This involves ordering laboratory tests to aid in ruling in, ruling out, or confirming hypotheses. Often, this lab workup is delegated to medical technologists and other medical specialists who analyze the data and report the findings back to the physician. The final stage involves the treatment and management of the patient. An initial treatment plan is formulated and then periodically reviewed to manage the recovery process and ensure that the patient's health is indeed improving.

As Gaba (1994) pointed out, managing and treating a patient involves controlling a process, the living human being, of which we know relatively little. It is often difficult to determine what factors or data will yield diagnostic conclusions. Even then, it is often not possible to directly measure the variables of interest or to even measure them at all. Thus, the data that can be gathered is often not sufficient for conclusively deducing a disease state, and consequently a diagnosis must be inferred in an abductive manner. Cost, time, and risk must also be considered when trying to decide a course of action. More diagnostic data may be available, but may either be very expensive, require invasive procedures, or take too long to perform.

OVERVIEW

Although computers and other automated devices have been designed to play a variety of roles in the medical field, this discussion focuses on the design of decision support systems, specifically those designed to aid human problem solving using a critiquing approach. The history of critiquing, its proposed advantages and disadvantages, relevant design criteria, and evaluations of such systems are reviewed. The design of one particular critiquing system is described, along with the results of two empirical studies showing it to be a significantly better alternative to both partial automation of the task and to current practices (no decision support at all).

TYPES OF MEDICAL DECISION SUPPORT SYSTEMS

There are many taxonomies of decision support systems. For the purposes of this discussion, an attempt is made to integrate some of these (e.g., Pryor, 1994; Shortliffe, 1990; Zachary, 1986) into the following major areas of decision support:

- *Information collection, storage, and display.* Systems in this category help in the storage, retrieval, organization, and integration of data and knowledge. Examples include bibliographic information retrieval systems, such as MEDLINE, and clinical database systems that keep track of patient and hospital records. Also included in this category are clinical monitoring systems that monitor and display patient parameters. Besides displaying raw data values, such systems may calculate and display higher-order variables along with population norms to support the physician's decision making (Woods, 1991).

• *Alerting*. Alerting is a form of decision support whereby the computer directly monitors patient data for prespecified conditions that warrant notification. For example, if critical patient parameters go out of range or if a drug is being prescribed to a patient who has an existing precondition that interacts with that drug, then the computer would alert the physician. Alerting systems are data and event driven. HELP (Pryor, 1983), for example, is a system that monitors patient data and generates alerts when abnormalities are noticed.

• *Critiquing*. Critiquing is a form of decision support whereby the computer examines the person's evaluation of a situation and alerts the user if the evaluation is incorrect or incomplete according to the computer's understanding of the problem. The key feature of a critiquing system is that it only gives suggestions or warnings if the *person's* actions or inferences are deemed inappropriate. So long as the person's interpretations are deemed adequate, the critiquing system does not interrupt the user

• *Diagnosis*. Diagnosis systems attempt to form a hypothesis or set of hypotheses about the disease state of a patient given a set of symptoms and initial laboratory data. Thus, the computer tries to diagnose the patient for the doctor, delegating the doctor's role to gathering and inputting information for the computer and then evaluating the computer's interpretation of the case. An example of this type of system is MYCIN (Shortliffe, 1976).

• *Process modeling*. Process modeling decision support systems aid practitioners by predicting how a process will be affected by certain actions. They allow the person to ask "what-if" questions of the computer in order to predict the implications of a proposed decision (Zachary, 1986). Examples of this include selecting flight plans for commercial aircraft (Layton, Smith, & McCoy, 1994).

• *Treatment and patient management*. Management systems are similar to diagnosis systems except that their primary role is to develop a treatment plan for the patient, rather than come up with a particular medical diagnosis. Some of these systems also monitor for completion of the selected protocol. An example of such an expert system is ONCOCIN (Shortliffe et al., 1981).

RATIONALE FOR THE CRITIQUING APPROACH

Critiquing systems were originally explored as a decision aiding strategy by Perry Miller. He defined a critiquing system as a computer program that critiques human-generated solutions (Miller, 1986). In order to accomplish this task, the critiquing system must be able to solve parts of the problem and then compare its own solution to that of the person. Then, if there is an important difference, the system initiates a dialogue with the user to give its criticism and feedback.

Critiquing systems are potentially more cooperative and informative to practitioners than are diagnosis systems because they structure their analysis and feedback around the proposed solution generated by the user. Because there are often many ways to solve a problem, particularly in the medical field where variation in practice is quite common, the system will use the person's initial solution or partial solution as a basis for communication. This contrasts with traditional diagnosis systems, where the computer generates the entire solution and is unaware of the conclusions drawn by the physician. It is up to the person to process the computer's output, compare what it has proposed to what he or she would have done, and then think about any differences that were detected between the machine- and human-generated solutions. With the critiquing approach, the burden of making the initial comparison and deciding what needs to be discussed further is placed on the computer (or, more accurately, on the computer system designer). Furthermore, the feedback focuses on the particular aspects of the solution that are in question. The feedback is therefore more likely to be pertinent to the user, and in turn more understandable and hopefully more acceptable. In addition, partial or intermediate conclusions proposed by the user can be critiqued immediately (instead of waiting for a complete answer), providing feedback in a more timely and potentially more effective context.

Types of Critiquing Systems

Fischer, Lemke, Mastaglio, and March (1990) and Silverman (1992a) surveyed the kinds of critics that have been developed and identified some of the differences among them. First, the goals of the user can be determined either by implicit goal acquisition, goal recognition, or by asking the user. *Implicit goal acquisition* means that the system uses a generic goal, such as "designing a good product within safety standards," as a basis for its critique. *Goal recognition* means that the system can monitor the user's actions to infer what goal the user has. Fisher et al. used the example of an architect who places a table and six chairs in the middle of a kitchen as evidence that the kitchen is intended to be used as a dining room as well. Finally, some systems may either allow or explicitly ask the user to input their goals.

Some critics are passive, meaning that they wait for the user to ask for help before beginning their analysis. Others operate in an active mode, where they monitor the user's actions and interrupt automatically when a mismatch is detected. In active critiquing systems, feedback may be given during the problem-solving process or held off until the user has finished. Silverman (1992a) distinguished among the use of influencers, debiasers, and directors. Influencers are techniques to try to influence the problem solving before it begins by giving either a tutorial or warnings before problem solving starts. Debiasers react only to actions performed by the user, detecting errors of commission and/or errors of omission, such as missing steps in the problem-solving process. Directors are walkthroughs to demonstrate how to use a particular tool if the user is having too much trouble. Finally, most critiquing systems give negative criticism, reacting to errors or suboptimal aspects of the user's product, but some can provide positive feedback as well, pointing out good performance or desirable characteristics to a solution. Thus, in summary, critiquing systems can:

- Infer the user's goal, ask or be told what it is, or use a predefined goal.
- Detect for biases, errors of omission, and errors of commission.
- Actively monitor the user's actions or be passive (user-invoked).
- Provide feedback during problem solving or at the end.
- Be proactive or reactive to the user's problem-solving process.

Previous Critiquing Studies

The first attempt at building a large-scale critiquing system for the medical community was made by Miller (1986). He developed a prototype system, called ATTENDING, which was designed to work in the anesthesiology domain. Based on this initial research, he experimented with critiquing systems for hypertension, ventilator management, and pheochromocytoma workup as well. All of these prototypes operated in a similar manner. The user was required to enter information about the patient's status and symptoms, as well as the proposed diagnosis and treatment. The computer then critiqued the proposed solution, generating a three-paragraph output summarizing its critique.

Miller saw much potential to the critiquing approach and was able to provide recommendations to other designers for developing good critiquing systems. First, Miller discovered that choosing a sufficiently constrained domain was important. ATTENDING was a system attempting to aid anesthesiologists in treating their patients, a task that takes years for people to learn and practice. Attempting to build a useful expert system in this field turned out to be too difficult due to the expanse of knowledge required. This lesson led him to switch to the more constrained hypertension domain.

Second, Miller concluded that critiquing systems are most appropriate for tasks that are frequently performed, but require the practitioner to remember lots of information about the treatment procedures, risks, benefits, side effects, and costs, because these are conditions under which people are more likely to make errors if unaided, thus making the critiquing system potentially valuable.

Langlotz and Shortliffe (1983) adapted their diagnostic expert system, ONCOCIN (designed to assist with the treatment of cancer patients), to be a critiquing system rather than an autonomous expert system because they found that, "The most frequent complaint raised by physicians who used ONCOCIN is that they became annoyed with changing or 'overriding' ONCOCIN's treatment suggestion" (p. 480). It was found that because a doctor's treatment plan might only differ slightly from the system's treatment plan (e.g., by a small difference in the prescribed dosage of a medicine), it might be better to let the physician suggest his or her treatment plan first, and then let the system decide if the difference is significant enough to mention to the doctor. In this manner, the system would be less obtrusive to the doctor. Thus, Langlotz and Shortliffe changed ONCOCIN to act as a critiquing system rather than a diagnostic expert system with the hopes of increasing user acceptance.

Fischer et al. (1990) developed a critiquing system, called JANUS, to aid with the design of kitchens. It is an integrated system, in that the user is already using the computer to design, and the system uses building codes, safety standards, and functional preferences (such as having a sink next to a dishwasher) as triggering events to critique a user's design.

To test the potential value of critiquing systems, Silverman (1992b) compared performance on two versions of a critiquing system designed to help people avoid common biases when interpreting word problems that included multiplicative probability. The first system only used debiasers, meaning that it provided criticism only after it found that the user's conclusion was incorrect. It had three levels of increasingly elaborate explanation if subjects continued to get the wrong answer. Performance was significantly improved with the critiques than without (69% correct answers for the Treatment Group after the third critique vs. 4% correct for the Control Group), but was not nearly perfect. Subsequently, a second version of the critiquing system was built that included the use of influencers, for example, before-task explanations of probability theory that would aid in answering the upcoming problems. With the addition of these influencers, performance improved to 100% correct by the end of the third critique. In examining these results and the performance of several other critiquing systems on the market, Silverman (1992b) proposed that to be effective, a critiquing system should have a library of functions that serve as error-identification triggers, and include the use of influencer, debiaser, and director strategies. In a related article, he summed up his definition of a good critiquer by saying: "A good critic program doubts and traps its user into revealing his or her errors. It then attempts to help the user make the necessary repairs" (1992a, p. 107).

Analysis of Previous Critiquing Work

These previous studies have provided us with much knowledge about critiquing systems, but not all of the factors related to their design and use have been resolved. For example, no formal evaluations of the systems developed by Miller were performed. Significantly, none of them are in use in hospitals. Although these systems tailor their feedback to the user-proposed solution, from a cognitive engineering perspective, it appears that they would be too cumbersome to use because the physician is required to enter the patient symptoms as well as his or her proposed solution and then read the relatively long output generated by the computer. Thus, the physician is required to act as the computer's secretary, typing in all the information that it requires.

In order for a critiquing system to be successful, it should require very little extra effort on the part of the human to interact with it. The computer must be able to directly infer the user's conclusions, which can only be done if the person is already using the computer as an integral part of task performance. The critiquing version of ONCOCIN was a step in the right direction. Physicians were already using ONCOCIN to fill out patient data forms, so the expert system used this information as its primary source of protocol data for the patient. JANUS was also an integrated system, allowing users to design kitchens and get feedback in the context of their work.

Silverman provided some of what little exists in terms of empirical assessments of critiquing systems, finding significant improvement in performance with their use. The domain that he studied, however, was one in which the system's knowledge was guaranteed to be correct. Thus, if the user understood the advice being given by the computer and heeded it, he or she would always get the case right. This leaves unanswered a critical question for medical applications.

No formal studies have been done to date to examine how practitioners interact with a critiquing system that is not fully knowledgeable about a domain. What happens in such a case? Does critiquing make performance any worse than it would be without it? Can anything be said for designing a critiquing system in such a case versus some other form of decision support? These are the types of questions that we have been exploring.

Critiquing Antibody Identification

One domain that we have found to be highly suitable for studying the use of computer aiding is that of antibody identification. This is a laboratory workup task, in which medical technologists must run a series of tests to detect antibodies in a patient's blood. Antibody identification satisfies all of the requirements outlined by Silverman and Miller. It is a sufficiently constrained domain that is frequently performed but difficult for people to do. It requires analyzing a large amount of data and deciding which tests to run to yield the most information. There is large variation in practice as to how to solve antibody identification cases, and technologists have been documented to make errors in transcribing and interpreting the data (Smith et al., 1991; Strohm et al., 1991). Furthermore, it has the classical characteristics of an abduction task, including masking and problems with noisy data.

This task is currently done on paper, using forms similar to the ones shown in Fig. 18.1, but these forms are easily ported to a computer format. We have developed a system, using C on the Macintosh platform, that allows medical technologists to perform their problem solving on the computer similar to the way they would on paper (Guerlain, 1993; Miller et al., 1993; Smith et al., 1992). Using pull-down menu options, the user can select forms corresponding to the different kinds of tests available for solving antibody identification cases (see Fig. 18.2). Once viewing a test panel, the user can select a pen color to mark up the panel, indicating that an antibody is ruled out, unlikely, likely, or confirmed. Because the technologist is using the computer to request test forms and to mark up those forms with hypotheses, the computer is able to watch the person's problem-solving process, potentially detecting errors in the subject's procedure (see Fig. 18.3). Thus, no extra work is required on the user's part to feed information to the computer. Practitioners work just as they naturally would and, *because of the interface design*, the data on the user's problem-solving activities is rich enough for the computer to detect problems and provide feedback.

To begin studying the effects of computer aiding, we developed a knowledge base that focused on one of the major strategies used by experts for ruling out hypotheses (antibodies) (Smith et al., 1991; Strohm et al., 1991). This is an aspect of the problem solving that narrows down the search space, providing users with one form of evidence to focus on an answer. This strategy was heuristic in nature and thus could fail in certain situations. (We implemented only this single strategy—which nevertheless required a set of many supporting rules—as a research strategy to avoid conflicting explanations of empirical findings due to the potential interactions of different strategies and subsystems.) Other types of knowledge that

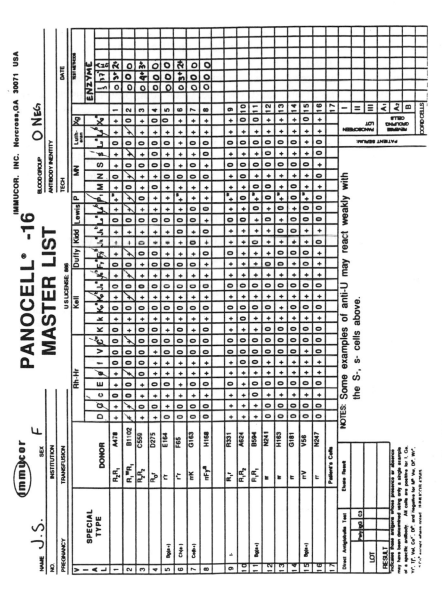

FIG. 18.1. An example panel test sheet currently used in blood banks.

393

were not implemented in this first pass include formation of hypotheses based on the patterns of results in the data, and knowledge about how to intelligently select further tests.

In other words, the broad question addressed in the initial design was: Assuming that the computer and the user have complementary sets of knowledge (with or without overlap in these sets), how do we engineer successful cooperative problem solving where both agents access and use their knowledge when appropriate? The specific case we focused on (Case 8, discussed later) was one in which the computer's knowledge was insufficient, but for which many users could be expected to have additional, complementary knowledge.

STUDY 1. CRITIQUING VERSUS AUTOMATION: WHAT ROLE SHOULD THE COMPUTER PLAY?

The goal of the first study performed with this testbed was to determine what role the computer should take to aid practitioners, particularly when the system's knowledge was inappropriate for the current task situation. Once a system has knowledge in it about how to perform part of a task, there are at least two approaches to computer support that can be taken. The first is to have the computer critique the practitioner as he or she solves a case. The second is to switch the roles, by having the computer do that part of the task automatically and having the person critique the computer's

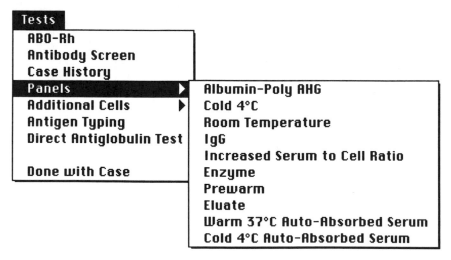

FIG. 18.2.   An example of all the tests that a blood banker can use when solving an antibody identification case.

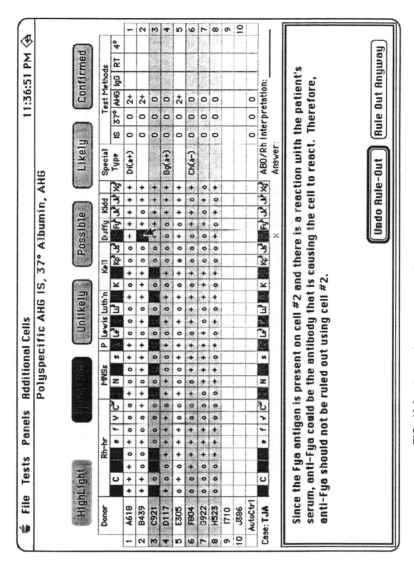

FIG. 18.3. A sample error message in response to the user's action.

analysis (which is the normal role for the person to play with most automated functions). We ran a study to evaluate how the role of the computer would affect human–computer cooperative problem-solving performance in such a situation (for more details, see Guerlain, 1993a, 1993b; Guerlain et. al., 1994).

Method

Thirty-two certified technologists solved 10 test cases using either the critiquing system or the partially automated system. The first three cases were completed by the subjects without any feedback, to get a benchmark for users' performance without any aiding at all. The next two cases were used to train users on the new features of the system, either critiquing or "partial automation," which is the critiquing version of the system with an automatic rule-out feature available. The next five cases were test cases used to evaluate overall human–computer performance. The cases were designed to range in difficulty from straightforward single-antibody cases to more difficult multiple-antibody cases. In particular, the eighth case was a weak antibody case where we knew our system's knowledge was inadequate to deal with the situation.

Results

The results showed that in cases in which the computer's rule-out knowledge was adequate, there was no statistically significant difference in performance (11.9% misdiagnosis rate for users of the critiquing system vs. 5.6% misdiagnosis rate for users of the partially automated system, $p > 0.05$). Although not statistically significant, the trend was for performance to be slightly better when the computer automated the rule-out strategy than if it was critiquing the use of that strategy. This could be due to the fact that this version of the critiquing system only detected errors of commission and not errors of omission. Thus, if users did not rule out at all or did not rule out everything that they could have, the critiquing system did not aid or warn them, whereas the automated rule-out function, if used, guaranteed that all antibodies that could be ruled out with that data sheet would be, a strategy that by definition was appropriate for these cases.

For the weak antibody case where the computer was incompetent at aiding, a different result was found. Performance was significantly better with the critiquing system than with the partially automated system (43% vs. 72% misdiagnosis rate respectively, $p < 0.05$). Although performance was still not very good even with the critiquing system, one must remember that the computer was not aiding practitioners at all, because its rule-out knowledge was inappropriate for this case. The partially automated system, however, was in fact making performance worse.

Discussion

By analyzing the videotapes of performance and computer-generated data logs, it was found that the poorer performance with the partially automated system appeared to be due to factors relating to the subject's problem-solving strategies (before they had the tool available to them). In other words, there was a task/strategy/tool interaction such that: (a) If a person had never used a particular strategy before, he or she was put into a position of having to judge the computer's use of that strategy without having the requisite knowledge to do so; (b) If the subject did normally use the strategy that was automated for him or her, but in a different manner than the computer, he or she developed an incorrect mental model of the system, assuming that it applied that strategy in exactly the same manner that the subject normally would; (c) For a practitioner who did use the strategy in exactly the same manner as the computer, it appeared that the subject was less likely to apply his or her knowledge to detect situations where that strategy was inappropriate if application of that strategy was delegated to the computer. Thus, even though the data indicated several underlying causes (missing knowledge, incorrect mental model, or overreliance), automation was universally worse when the computer's knowledge was unable to deal correctly with the situation because the person failed to adequately critique the computer's performance.

In contrast, the roles were reversed when the computer was put in the position of critiquing the person. When the person did not apply the strategy implemented in the system, the computer said nothing (at least in the tested version of the system), because it only checked for errors of commission. In such a case, the user might or might not have gotten the right answer. When the practitioners applied the rule-out strategy in a different manner than did the computer, the computer gave feedback and taught them over time regarding its knowledge and strategies. Thus, these practitioners learned from the computer and built up a partial mental model of the computer's knowledge, at least enough to know that the computer's strategy was different from their own. They then demonstrated the ability to recognize situations where the computer's strategy was inappropriate and ignored the computer, relying instead on their own strategies to solve the problem. Further evidence of this effect of system design was that significantly more subjects using the critiquing version of the system ran extra tests to try to verify their answer before completing the weak antibody case (93% vs. 50%, $p < 0.01$), a sign that users of the critiquing system were better able to detect that there was something suspicious about the case.

Thus, this study provided some of the first empirical evidence that critiquing is potentially a more effective approach to aiding human decision making than partial automation if the computer is assumed to be less that all-knowing for the full range of situations that could arise in its domain. Equally important, the results indicated three different reasons why the automatic performance of subtasks by the computer can lead to impaired performance in such a situation.

STUDY 2. EVALUATION OF A COMPLETE SYSTEM

To avoid potential confoundings, Study 1 looked at the effects of providing a tool based on a single problem-solving strategy. Based on the results of that study, we implemented a critiquing system that incorporated a number of support strategies. The goal of this second study was to see if misdiagnosis rates could be reduced or eliminated with the design of a more complete critiquing system, and to explore its effects on cooperative performance.

The types of knowledge encoded into the second version of the system included detecting:

1. Errors of commission (due to slips or mistakes):
   Errors in ruling out antibodies (same as in Study 1).
2. Errors of omission (due to slips or mistakes):
   Failure to rule out an antibody for which there was evidence to do so.
   Failure to rule out all clinically significant antibodies besides the antibodies included in the answer set.
   Failure to confirm that the patient did not have an autoimmune disorder (i.e., antibodies directed against the antigens present on their own red blood cells).
   Failure to confirm that the patient was capable of forming the antibodies in the answer set (i.e., that the patient's blood was negative for the corresponding antigens, a requirement for forming antibodies in the first place if the possibility of an autoimmune disorder has been ruled out).
3. Errors due to masking:
   Failure to detect and consider potentially masked antibodies.
4. Errors due to noisy data:
   Failure to detect situations where the quality of the data was questionable.
5. Answers unlikely given the data (low probability of data given hypothesis):

Failure to account for all reactions.

Inconsistency between the answers given and the types of reactions usually exhibited by those antibodies (e.g., that a warm-temperature antibody was accounting for reactions in cold temperatures).

6. Unlikely answers according to prior probabilities (regardless of the available evidence):

Antibody combinations that are extremely unlikely due to the way the human immune system works.

An example of an error message produced by this version of the system is shown in Fig. 18.4.

Besides designing the critiquing system to detect these types of errors in response to the user's actions (or the lack of them), we designed a checklist outlining the steps that the computer expected to have completed before finishing a case (helping to ensure effective cooperation). Subjects were given the checklist and trained on how to follow all the steps.

Method

To test the system, we compared performance of subjects using a version of the system with all critiquing functions turned off (the Control Group) to a version using all the critiquing functions and the checklist (the Treatment Group). The subjects were 32 certified, practicing medical technologists (who were different from those of the first study), taken from six hospitals. Subjects were randomly assigned to two groups, such that half of the subjects used one version of the system and half used the other.

Performance was studied using a combination of a within- and between-subject design. All subjects were first trained on the use of the interface with the control version of the system. They were then asked to solve one case by themselves (without the help of any critiquing by the system). This was a masking case that was randomly chosen from one of two cases with the same masking characteristics (two antibodies looking like one). All subjects received both of these cases, but it was randomly determined whether they would get the first case as a pretest case (where both groups were using the control version of the system) or as the first test Case (where the treatment group had the critiquing functions turned on). The purpose of randomly assigning this case was to counterbalance to allow comparisons of performance for the Treatment Group in a within-subjects manner. Four posttest cases allowed a between-subjects comparison as well.

Between the pretest case and the first posttest case, the experimental group was introduced to the checklist and asked to practice the use of certain steps to be sure that they understood the overall problem-solving

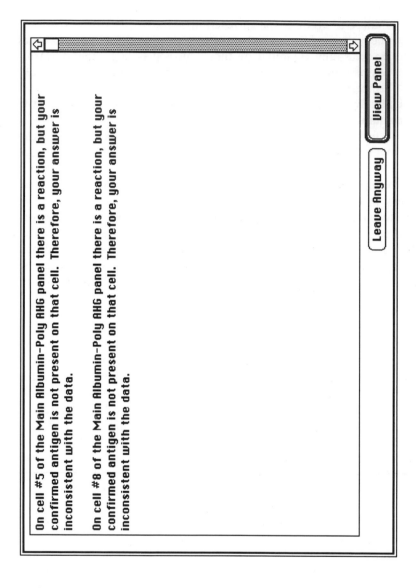

On cell #5 of the Main Albumin-Poly AHG panel there is a reaction, but your confirmed antigen is not present on that cell. Therefore, your answer is inconsistent with the data.

On cell #8 of the Main Albumin-Poly AHG panel there is a reaction, but your confirmed antigen is not present on that cell. Therefore, your answer is inconsistent with the data.

Leave Anyway

View Panel

FIG. 18.4. A sample error message in response to the user's answer for a case.

strategy represented by the checklist. The Control Group worked through the same set of cases used in this training, but without instruction regarding the computer's problem-solving strategy.

Finally, the subjects solved four posttest cases. The first case was one in which it looked like one antibody, but in actuality it was two different antibodies (the other matched case not selected as the pretest case). The second case was the same case that the previous version of our system (from Study 1) could not handle. In this version of the system, the computer's knowledge base was still not adequate to solve the case. However, some changes were made to hopefully improve performance to some degree: (a) There was a warning when triggering data (based on meta-knowledge about the weaknesses in the computer's knowledge base) was encountered, indicating that the normal strategy of ruling out right away might not be appropriate due to the weakness of the reactions; (b) the system did not require subjects to rule out as in other cases, because it had just suggested that rule out may not be appropriate; and (c) one of the plausible but incorrect answers for the case (anti-E) fell into the "unlikely antibody combinations according to prior probabilities" category (Type 6 in the list presented earlier). The third case was a masking case in which one antibody was masking the presence of another. The fourth case was one that we were testing to see if our system could handle cases for which it was not explicitly designed. It was solicited from a blood bank lab that knew nothing of our work. This case turned out to be a complicated, three-antibody case with all of the initial cells reacting, thereby making it very difficult to rule out anything. In summary, there were three categories of cases that we tested:

1. Cases that the system was designed to aid (Cases 1 and 3).
2. Cases that we knew the system was deficient in handling (Case 2, weak antibody).
3. Cases that the system was not deliberately designed to handle, sent by another lab (Case 4).

Results

Due to a bug in the program, the data from one of the cases for four of the subjects were invalid, so these data points are not included in the analyses that follow. The results showed that there was no significant difference in performances on the pretest case for the control and treatment groups but that subjects in the critiquing version of the system performed significantly better on all four of the cases. The treatment group also showed a trend for improvement in performance (a reduction from 26.7% to 0% misdiagnosis error rate, $p > .05$ using McNemar's Chi Square for dependent samples) from the pretest case to posttest case 1. The between-subject comparisons showed marked improvement in performance for the treatment group (see Table 18.1). On Cases 1, 3, and 4, subjects in the critiquing version had a 0% misdiagnosis rate, while subjects in the control group got 5/15 (33.3%), 6/16 (37.5%), and 10/16 (62.5%) incorrect respectively. Using Fisher's exact test, each of these differences is significantly different ($p < .05$). For the case that the system was not designed to completely handle (Case 2), 3/16 subjects (18.75%) in the treatment group misdiagnosed the case vs. 8/16 subjects (50%) in the control group. This difference is marginally significant ($p = .072$). A log-linear analysis was also run for each posttest case to take into account the difference in performance for the two groups on the pretest case. This analysis gave very similar results, namely that there was a highly significant improvement in performance for the treatment group for Cases 1, 3, and 4, and a marginally significant improvement for Case 2. Combining the p values across the four posttest cases gives a combined p value of ($p < .000005$) that favors performance for the treatment group.

CONCLUSIONS

Critiquing, although not explored to date by very many researchers as a form of decision support, seems to be a viable solution for greatly improving performance on certain kinds of tasks, including the important, real-world medical diagnosis task of antibody identification. Clearly, this is a task that medical technologists find difficult, because they are getting moderately difficult, yet realistic, patient cases consistently wrong when unassisted. A well-designed critiquing system has proven to be a method for virtually eliminating the errors that it was designed to catch, and for aiding on cases for which its knowledge is incomplete. Perhaps the best test of the system's capability was its success at reducing misdiagnosis rates on Case 4 from 64% to 0%, a case given to us by a laboratory that had no knowledge of our work nor any inkling of why we were asking them to send us difficult cases. It is also interesting to note that the Control Group in Study 2 got the weak D case wrong on the same order of magnitude as those users using the original version of the critiquing system from Study 1 (50% vs. 43% misdiagnosis rate). This is further evidence that such a critiquing system does not make performance any worse than a person working alone, whereas an automated system was shown in Study 1 to significantly worsen performance (from 43% to 72%).

As a final summary, the following list recapitulates many of the ideas already brought up in this chapter:

TABLE 18.1

Misdiagnosis rates for users of the critiquing system (n = 16) vs. a Control Group with no feedback (n = 16).

	Test Cases				
	Pre-Test Case	Case 1	Case 2	Case 3	Case 4
	2 antibodies looking like 1 (randomly chosen from one of two matched cases, the other of which was the first Post-Test case)	2 antibodies looking like 1 (randomly chosen from one of two matched cases, the other of which was the Pre-Test case)	weak antibody (for which the system was not designed to adequately handle)	1 antibody masking another	3 antibodies reacting on all cells (a case for which the system was not explicitly designed, sent by another blood bank lab)
Control Group					
	6/14 (42.9%)	5/15 (33.3%)	8/16 (50%)	6/16 (37.5%)	10/16 (62.5%)
Treatment Group					
	4/15 (26.7%)	0/16 (0.0%)	3/16 (18.75%)	0/16 (0.0%)	0/16 (0.0%)
Significance					
	NS	$p < 0.05$	$p = 0.072$	$p < 0.01$	$p < 0.001$

403

- Critiquing systems are best for restricted domains in which practitioners solve recurring problems but may commit types of errors that machines are good at eliminating.
- In order for the system to work, practitioners should already be performing the task on the computer, or the computer should be able to directly infer this information from other sources so that the users of the system do not have to act as its secretary.
- Critiquing should be considered as an alternative to no decision support at all, because the combined expertise of the human plus computer can be better than either alone. For example, new data often yields more information than is used by practitioners (a form of omission error), and the critiquing system can detect this.
- Critiquing should be considered as an alternative to automating the task because it appears to act as a more effective partner to the person, critiquing the person's plan rather than offering final conclusions and recommendations with perhaps some retrospective explanation. Furthermore, the system does not have to know everything about the domain for it to be useful, so it can be built incrementally. Perhaps most important, critiquing seems less likely to make performance worse in cases that it was not designed to handle, whereas automation has been documented to do so.
- The system can serve as a memory aid by including rarely used but pertinent information in its critiques. It can also bring new, updated information to a physician.
- A critiquing system's explanation is structured around the error just detected, and can reference aspects of the immediate situation confronting the practitioner. This serves to ground the information in the current context and make it easier to understand. Furthermore, the system can have nested levels of explanation to aid the user's understanding.
- If the person is doing everything right, the system is not obtrusive.
- The same system can be run not only in consultation mode but also in a tutoring mode to aid in teaching the task to new practitioners.

Thus, the two studies presented here have built on previous work examining the effectiveness of critiquing as a form of decision support, providing empirical evidence for the previously suspected but not thoroughly tested idea that critiquing is a more effective form of support than automation of tasks such as medical diagnosis. A correctly designed critiquing system is able to not only immediately improve overall performance by catching slips and mistakes in a more cooperative and less obtrusive manner than many automated systems, but it also has the potential to transfer much of its knowledge and strategies to the person by the nature of its interaction.

REFERENCES

Fischer, G., Lemke, A., Mastaglio, T., & Morch, A. (1990). *Using critics to empower users.* In *CHI '90 Human Factors in Computing Systems Conference Proceedings* (pp. 337–347). New York: Association for Computing Machinery.

Gaba, D. (1994). *Automation in anesthesiology.* In M. Mouloua & R. Parasuraman (Eds.), *Human performance in automated systems: Current research and trends* (pp. 64–67). Hillsdale, New Jersey: Lawrence Erlbaum Associates.

Guerlain, S. (1993a) *Designing and evaluating computer tools to assist blood bankers in identifying antibodies.* Unpublished master's thesis, Ohio State University, Columbus.

Guerlain, S. (1993b). *Factors influencing the cooperative problem-solving of people and computers.* In *Proceedings of the Human Factors and Ergonomics Society 37th Annual Meeting,* 1 (pp. 387–391). Santa Monica, CA: Human Factors and Ergonomics Society.

Guerlain, S., Smith, P. J., Gross, S. M., Miller, T. E., Smith, J. W., Svirbely, J. R., Rudmann, S., & Strohm, P. (1994). Critiquing vs. partial automation: How the role of the computer affects human–computer cooperative problem solving. In M. Mouloua & R. Parasuraman (Eds.), *Human performance in automated systems: Current research and trends* (pp. 73–80). Hillsdale, New Jersey: Lawrence Erlbaum Associates.

Langlotz, C. P., & Shortliffe, E. H. (1983). Adapting a consultation system to critique user plans. *International Journal of Man–Machine Studies, 19,* 479–496.

Layton, C., Smith, P. J., & McCoy, E. (1994). Design of a cooperative problem-solving system for enroute flight planning: An empirical evaluation. *Human Factors, 36*(1), 94–119.

Miller, P. (1986). *Expert critiquing systems: Practice-based medical consultation by computer.* New York: Springer-Verlag.

Miller, T. E., Smith, P. J., Gross, S. M., Guerlain, S. A., Rudmann, S., Strohm, P., Smith, J. W., & Svirbely, J. (1993). The use of computers in teaching clinical laboratory science. *Immunohematology, 9*(1), 22–27.

Pryor, T. A. (1983). *The HELP system. Journal of Medical Systems, 7,* 87–101.

Pryor, T. A. (1994). Development of decision support systems. In M. Shabot & R. Gardner (Eds.), *Decision support systems in critical care* (pp. 61–73). New York: Springer-Verlag.

Shortliffe, E. H. (1976). *Computer-based medical consultations: MYCIN.* New York: Elsevier.

Shortliffe, E. H. (1990). Clinical decision-support systems. In E. Shortliffe & L. Perreault (Eds.), *Medical informatics. Computer applications in health care,* (pp. 466–500). New York: Addison-Wesley.

Shortliffe, E. H., Scott, A. C., Bischoff, M., Campbell, A. B., van Melle, W., & Jacobs, C. (1981). ONCOCIN: An expert system for oncology protocol management. In *Proceedings of the seventh International Joint Conference on Artificial Intelligence* (pp. 815–822). Menlo Park, CA: American Association of Artificial Intelligence.

Silverman, B. (1992a). Survey of expert critiquing systems: *Practical and theoretical frontiers, 35*(4), 107–127.

Silverman, B. (1992b, April). Building a better critic. Recent empirical results. *IEEE Expert,* pp. 18–25.

Smith, P. J., Miller, T. E., Fraser, J., Smith, J. W., Svirbely, J. R., Rudmann, S., Strohm, P. L., & Kennedy, M. (1991). An empirical evaluation of the performance of antibody identification tasks. *Transfusion, 31,* 313–317.

Smith, P. J., Miller, T., Gross, S., Guerlain, S., Smith, J., Svirbely, J., Rudmann, S., & Strohm, P. (1992). The transfusion medicine tutor: A case study in the design of an intelligent tutoring system. In *Proceedings of the 1992 Annual Meeting of the IEEE Society of Systems, Man, and Cybernetics,* (pp. 515–520). Piscataway, NJ: Institute of Electrical and Electronic Engineers.

Strohm, P., Smith, P. J., Fraser, J., Smith, J. W., Rudmann, S., Miller, T., & Kennedy, M. (1991). Errors in antibody identification. *Immunohematology, 7*, 20.

Woods, D. D. (1991). *The theory and practice of representation design in the computer medium* (CSEL Technical Report). Columbus, OH: Cognitive Systems Engineering Laboratory, The Ohio State University.

Zachary, W. (1986). A cognitively based functional taxonomy of decision support techniques. In M. Helander (Ed.), *Human-computer interaction* (pp. 25–63). Hillsdale, NJ: Lawrence Erlbaum Associates.

19 Automation in Quality Control and Maintenance

Colin G. Drury
State University of New York at Buffalo

INTRODUCTION

Changing System Boundaries and Functions

Industry is changing rapidly, and with it the functions of quality control and maintenance. Automation is only one aspect of this change, so that the context for these two functions needs to be established if we are to see how they will evolve in the future. Whether the industry is manufacturing, transportation, or service, the changes are similar, because both the competitive pressures and the hardware controlling the automation are similar. For example, over the past 20 years the percentage of U.S. companies facing direct foreign competition in their home market has risen from approximately 20% to 80%. Thus, we need to examine system boundaries and functions in a number of industries if we are to see the broad changes in sufficient detail to be useful.

Changes across industries have been listed by many authors (e.g., Brödner, 1994; Drury, 1991; Hendricks, 1991). They can be seen clearly in the prescriptive guidance offered by management specialists under such titles as total quality management (TQM; Golomski, 1994), total productive maintenance (TPM; Nakajima, Yamashina, Kumagai, & Toyota, 1992), and total productivity management (also TPM; Akiba & Suito, 1992). A suitable list, modified from Drury (1991) is as follows:

- *Customer driven.* The realization that customer service is the primary function of an organization has had to be relearned. Especially in mature industries, where products are not too different, fulfilling customer expectations gives an obvious competitive advantage. Quality is a major component of customer satisfaction, and its achievement is an integral part of current standards (e.g., the ISO-9000 series) and prescriptive solutions (e.g., TQM).

• *Measurement based*. With measurement, you can control, to paraphrase Lord Rayleigh. In industry this means careful and continuous process measurement to a drive eventual predictive (open-loop) control over processes (e.g., statistical process control, or SPC) and over maintenance (e.g., preventive maintenance). It also means finding the "best in class" for any particular process, and using this as a benchmark to evaluate your process (Evans & Lindsay, 1993).

• *Defined mission*. Most industrial organizations have become aware that they need to know why they are in business, and where they are going, if they are to become effective. Hence, they have developed a mission statement, a vision statement, and a set of goals and objectives (e.g., Taylor & Felten, 1993) within these statements. Only with a defined mission can the systems thinking necessary to modern organizations design take place.

• *Continuous improvement (CI)*. The idea that a design can be optimized completely before its introduction is recognized as utopian. Although design of both products and processes should receive considerable attention at the outset (and even be concurrent; e.g., Ion, 1994), the successful design will be one in which improvement is continuous. CI is not just a procedure for designers, it is part of the job of each person in the organization. Neither does CI happen just by putting up posters and starting a suggestion scheme; it requires new forms of training and organization to be successful.

• *New organizational structures*. Although the rigid hierarchies of management and union in an enterprise still operate successfully in some organizations, they are clearly not the way of the future. Whether from the perspective of anthropocentric design (Brödner, 1994) or sociotechnical systems (STS) design (Taylor & Felten, 1993), organizations based on different principles are finding remarkable success. They rely on a trained workforce, empowered for control over the technical processes, working often in semi-autonomous groups with flat organizational structures. Control is moved down to the lowest point possible, with computer support where required. This is also a route toward reduced work stress and illness (Karasek & Theorell, 1991).

• *Advanced technology*. Finally, there are large technological changes under way. From microprocessor chips in many stand-alone items of equipment through to computer-integrated manufacturing (CIM), computing power and new sensor/effector mechanisms are changing almost all of mankind's artifacts. Largely, these changes have been technology driven, rather than user driven (Sheridan, 1992). Use of this technology ranges from rejection to total automation (lightless factory concept), with user-centered automation between these extremes. Despite spectacular automation failures (e.g., Denver International Airport's baggage handling system, which repeatedly delayed the airport's opening), automation is still seen as a desirable end in itself by many companies. It continues to be promoted by its designers as eliminating the need to deal with people at the workplace, both their continuing cost and their unreliability.

If these changes are taking place in industry, how do they affect maintenance and quality control? The main issue in this book is automation, but this is only one of a large number (six in the preceding list) of changes taking place simultaneously. In order to give proper consideration to automation changes, Table 19.1 is presented as an overview of the effects of the six changes listed above on both quality control and maintenance functions. It can be seen that there are even now major impacts on the way both functions operate, independent of the effects of automation. In the rest of this chapter we take a number of these impacts into account.

AUTOMATION: ISSUES FOR INDUSTRY

"Automation is the automatically controlled operation of an apparatus, a process, or a system by mechanical or electronic devices that take the place of human organs of observation, decision and effort", according to Sheridan's (1992, p. 527) adaptation from *Webster's Third International Dictionary*. It implies that sensors, processors, and effectors are all machine assigned, but it is not all that we mean by *automation* in industry. In practice, *automation* is used to refer to almost any degree of mechanization (i.e., assignment of functions to machines rather than people). Indeed, Sheridan was a pioneer in defining different levels of automation through supervisory control. This issue is not whether automation is good or bad per se, but rather how far to go in automation (Sheridan, 1992). For example, Horte and Lindberg (1991) gave an analysis of 27 flexible manufacturing systems (FMSs) in Sweden, and show great variation in performance. The most flexible systems, and those used specifically for their flexibility, do best. It is interesting to note in this study that the major tasks of the operator were operation (obviously), process monitoring and maintenance (i.e., quality and maintenance functions were well-integrated into the operator's job). This can be contrasted with the many well-documented

TABLE 19.1
Effects of System Changes on Quality and Maintenance Functions

Change	Quality System Changes	Maintenance System Changes
Customer driven	Demands for quality of product are paramount. Customer interaction used to define quality and incorporate it into product, for example, quality function deployment (QFD).	Rapid customer response, typical of just-in-time (JIT) and agile manufacturing, are putting large pressures on downtime reduction. System availability considerations are driving design for maintainability (e.g., Bond, 1987).
Measurement based	Standards (e.g., ISO-9000) call for specific quality-related measurements. In addition, the whole quality control movement is based on statistical process control (SPC), machine capability, and models of machine operation. All require numerical measures of both product characteristics and process indicators.	Prediction of causes of downtime becomes important as the move is made from breakdown maintenance (BM) through preventive maintenance (PM) to condition-based maintenance (CBM) (e.g., Nakajima et al., 1992). CMB requires detection of prebreakdown signs of maintenance need, again a measurement-based process.
Defined mission	Typically, the mission statement will include quality, but its achievement relies on subgoals related to the mission.	Maintenance is usually a subgoal below the level of the company mission statement.
Continuous improvement (CI)	The CI process is the central belief of TQM prescriptions (Bounds, Yorks, Adams, & Ranney, 1994). Teams organized around the CI philosophy have been part of the quality movement since the quality circles of 20 years ago.	CI within maintenance is seen in the group activities fostered by TPM as "company-led small group activities" (Nakajima et al., 1992)
New organizational structures	The philosophy of control of a variance at source has led to a different concept of operator, for example, one who has the knowledge and tools to be responsible for quality. Team working in small groups is a key feature of empowerment of operators in TQM (Bounds et al., 1994).	Autonomous maintenance by the operator, rather than having a separate maintenance group, is a central part of TPM.

(*Continued*)

TABLE 19.1 (*Continued*)

| Advanced technology | Automated dimensional inspection, nondestructive inspection, machine vision systems, predictive models for process control, and automated test equipment are all available (Drury & Prabhu, 1994). | Built-in test equipment (BITE) is becoming a standard feature of aircraft, automobiles, computers, trucks, and even some industrial machines. New sensors are available for measurement of conditions leading to failure. |

examples of system failure in automation (e.g., Rasmussen, Duncan & Leplat, 1987), primarily from poor use of human operators in the system (e.g., Bessant, Levy, Ley, Smith, & Tranfield, 1992).

These issues of what to automate (i.e. appropriate automation) have surfaced in many contexts, from civil airline operations (Wiener & Nagel, 1988) to industrial production (e.g., Brödner, 1994; Kidd & Karwowski, 1994). The successful philosophy in industry currently is composed of the items in Table 19.2, summarized from Bessant et al. (1992). Ultimately, the design of such systems rests on the allocation of function between human and automation. In its original form, allocation of function (e.g., Fitts, 1962) was a choice between human and machine, although Jordan (1963) cautioned that cooperation rather than competition between these alternative system components was a more suitable viewpoint. Since that time

TABLE 19.2
Factors Common to Successful Change Implementation In Manufacturing (46 Applications in 28 Companies)

Work Organization Factors
  Multiskilling
  Integrated tasks
  Short skill life cycle
  Reskilling
  Teamwork
  Alternative payment systems
  Supervisor support
  Flexibility/automony

Management Organization Factors
  Blurred line/staff boundaries
  Flat organization structures
  Network communications
  Holographic adjustment
  Product/project/customer-based structures
  Single-status employees
  Flexible, participative workers

Sheridan's "taxonomy of automation levels" (e.g., Sheridan, 1988) and Rouse's (1976) concept of flexible (or run-time) allocation have allowed a greater degree of choice. Allocation of function is seen as a major issue in modern industry (Clegg, Ravden, Corbett, & Johnson, 1989; Sinclair, 1993).

The issues, then, are less concerned with whether to automate than with which functions to automate. This concern is the organizing thread throughout the discussion of quality control and maintenance that follows. As these two functions are typically rather separate in industry, even to the extent of having their own prescriptive solutions (TQM, TPM), an example is chosen in which the two are intimately related.

QUALITY CONTROL AND MAINTENANCE: INSEPARABLE PARTS OF AIRCRAFT OPERATION

The domain of aircraft quality control and maintenance activities is an important one for several reasons:

- The consequences of error can be catastrophic, as in the case of a Continental commuter aircraft that flew without the tail de-icing boot properly attached, or the United DC-10 whose engine failed, severing control hydraulic lines and causing a crash at Sioux City.
- The system of quality control and maintenance is a highly regulated and traditional one, which nonetheless recognizes that change is needed.
- External factors such as competitive pressures, shortage of skilled technicians, and introduction of new aircraft technologies alongside the old are forcing system changes (Shepherd, Johnson, Drury, Taylor, & Berninger, 1991).
- A technological push towards automation is being felt by those parts of organizations that typically have no history of methods for integrating human and machine functions to achieve system goals.

Within the aircraft maintenance and quality control system, there is an overall philosophy guiding the major functions. For civil aircraft, any new design is analyzed for its quality and maintainability needs using a process called MSG-3. Each failure mode of the aircraft structure and systems is dealt with by a hierarchy of interventions from self-repair through indication of failure to time-limited replacement. For most structural components of aircraft and airframe, the failure modes are dealt with by inspection at a defined period, and repaired if required (Hagemaier, 1989).

The defined period comes from loading analysis and fracture mechanics of the structure (Goranson, 1989), which determine how quickly a crack will form and grow. Then the data on detectable crack sizes is combined with crack growth predictions to determine when a structure must be inspected, so as to have a high probability that the crack is detected before failure occurs. This is an intimate relationship between the modeling (load analysis, crack growth) and human performance (probability of detection, or POD, as a function of crack size). It also defines a close relationship between the quality aspects of fitness for service as determined by inspectors, and maintenance aspects of correct repair following detection. Note also how a problem in physics or mechanical engineering (structural failure mechanics) has been transformed in practice into a problem in human performance (detection and repair). Thus, in the famous "Aloha incident" in which multiple site damage resulting from cracks joining together gave an in-flight failure of the hull of a B-737, the problem was seen both as failed prediction of the damage and failed inspection at the prior maintenance check.

Quality control in aircraft maintenance has two connotations. At the lowest level it refers to the action of discovering the defects, such as cracks, corrosion, missing parts, loose or frayed wires, broken components, or failed avionics systems. This is the detection function of a quality system, equivalent to inspection in industry, and called *inspection* in the airlines. At a higher level, there is a quality control system surrounding all inspection/maintenance activities. This usually relies on checking engineering change orders, auditing inspection and maintenance activities for errors, auditing components (and vendors) used for replacements, and examining record keeping. In addition to this in-house function, often known as *quality assurance*, there is a further level of audit and surveillance by regulatory bodies, such as the FAA in the United States. The most obvious interaction of human factors and automation occurs at the lower level, although there are considerable STS concerns throughout the organizations (e.g., Taylor, 1993).

Similarly, maintenance in the civil aviation context is somewhat different from the typical industrial use of the term. Many of the issues in maintenance (Table 19.1) are taken care of in the planning and organization process. For example, the work of Gude and Schmidt (1993) on models of how operators choose efficient repair schedules is largely irrelevant when the schedules are predetermined by manufacturers, regulators, and company engineers. Again, this lack of control, although technically necessary to ensure safe aircraft, can be expected to have implications for job design.

FUNCTIONS IN AIRCRAFT MAINTENANCE AND INSPECTION

When an aircraft arrives at the maintenance hangar or repair station, a series of functional steps need to take place to convert the raw material (unrepaired aircraft) into finished product (repaired and airworthy aircraft).

Function 1. Planning

Because the maintenance activity can last from overnight to several weeks, and involve from 10 to several hundred personnel, the planning function is critical. Known or scheduled repairs may be needed; for example, changing of time-rated components. Deferred items in the log book and items reported by pilots need to be scheduled. In addition, inspections are scheduled that may result in repairs, again requiring at least probability-based plans. Replacement items need to be ordered and prestaged. Finally, the activities need to be planned so as to minimize both interference between tasks and overall out-of-service time for the aircraft. For example, repair on a doorway can interfere with access to tasks inside the fuselage. Also, X-ray inspection requires all personnel to be away from the aircraft for radiation safety reasons, so such tasks are scheduled during night-shift meal breaks.

Function 2: Opening/Cleaning

In order to inspect and repair an aircraft, the structures must be clean and free from visual interferants, such as oil, hydraulic fluid, or spillage from the galley and toilets. In addition, the paint may need to be stripped from parts of the aircraft. There is an initial cleaning before the aircraft is brought into the hangar, but then interior furnishings and necessary access panels must be removed and opened before internal cleaning can take place. Regulations on chemicals in the workplace and effluents in the environment have changed the technology of cleaning in recent years.

Function 3: Inspection

In order to find out as quickly as possible what repairs will be needed (so as to minimize out-of-service time), the inspection process must be completed as soon as possible. This often means scheduling a large number of inspectors, some on overtime, to work on the first shift(s) the aircraft is available. Because aircraft earn revenue during the day, the first shift typically starts about 2300 hours (11:00 p.m.), again giving potential job design implications.

The functions of inspection have received considerable attention from the human factors community in recent years (e.g. Drury, Prabhu, & Gramopadhye, 1990). There are two basic types of inspection: visual inspection (VI) and nondestructive inspection (NDI). Both are somewhat misleading terms. VI requires the use of active touch and hearing as well as sight, and really refers to inspection aided only by simple hand tools (mirror, flashlight, ruler). NDI should also logically cover VI, because neither destroys the structure. In practice, NDI refers to inspection using more complex aids, such as eddy-current or ultrasound devices, X-rays, gamma rays, and dye penetrant (Latia, 1993). All of these allow the detection of defects, such as cracks or corrosion, which are either invisible to direct vision because they are too small, or inaccessible to vision due to intervening structures.

Table 19.3 gives the generic functions of inspection, modified from Drury, Prabhu, and Gramopadhye, 1990, with examples from VI and NDI. This generic function list was used by Drury (1994) to structure human factors interventions into the inspection process. We return to this list when automation is considered later in the chapter.

Function 4. Repair

This is the function performed by people with the title "maintenance," usually referred to as aircraft maintenance technicians, or AMTs. Although they concentrate on removing and replacing components, they are also the functional group responsible for much avionics and systems diagnosis. Thus they need to make use of advanced diagnostic aids (e.g., Husni, 1992) as well as hand tools and power tools. Table 19.4 shows a generic function list for repair, similar to that for inspection. Note that, like inspection, maintenance begins with a workcard that details the steps in a repair and refers to any needed attachments, such as repair manuals, service bulletins, or engineering change orders. This repair function can be performed on the

TABLE 19.3
Generic Functions in Aircraft Inspection, With Examples From Visual Inspection and NDI

Task Description	Visual Example	NDI Example
Initiate	Get workcard. Read and understand area to be covered.	Get workcard and eddy-current equipment. Calibrate.
Access	Locate area on aircraft. Get into correct position.	Locate area on aircraft. Position self and equipment.
Search	Move eyes across area systematically. Stop if any indication is spotted.	Move probe over each rivet head. Stop if there is any indication.
Make decision making	Examine indication against remembered standards (e.g., for dishing or corrosion).	Reprobe while closely watching eddy-current trace.
Respond	Mark defect. Write up repair sheet or, if no defect, return to search.	Mark defect. Write up repair sheet or, if no defect, return to search.

TABLE 19.4
Generic Functions in Aircraft Repair

Function	Tasks
Initiate	Read and understand workcard
	Prepare tools, equipment
	Collect parts, supplies
	Inspect parts, supplies
Site access	Move to worksite, with tools, equipment, parts, supplies
Part access	Remove items to access parts
	Inspect/store removed items
Diagnosis	Follow diagnostic procedures
	Determine parts to replace/repair
	Collect and inspect more parts and supplies, if required
Replace/repair	Remove parts to be replaced/repaired
	Repair parts if needed
	Replace parts
Reset systems	Add fluids/supplies
	Adjust systems to specification
	Inspect adjustments
Close access	Refit items removed for access
	Adjust items refitted
	Remove tools, equipment, parts, excess supplies

aircraft itself, or in an off-hangar workshop (e.g., for brakes, avionics equipment, or passenger seats).

Function 5. Buyback and Return to Service

Many of the repairs require that a person other than the AMT reinspect the finished repair to ensure that it was performed correctly. This "buyback" function is usually performed by an inspector, although others can be granted that authority. Buyback is essentially a reprise of the inspection function, although it may occur in stages (e.g., before each level of access to the repair is closed).

Finally, the official return to service is a legal assurance by designated people that the aircraft is airworthy. All items in the scheduled maintenance, in the log book, and those nonroutine items found in inspection must be certified as complete. In some predefined circumstances, non-safety-critical items may be approved for deferral until a later service. This paperwork function ensures that the pilots responsible for operating the aircraft know that their equipment is safe. The paperwork also forms an auditable tracking and control system for the quality control/quality assurance departments of the maintenance organization and of the regulatory authority.

AUTOMATION IN MAINTENANCE AND INSPECTION

As in most industrial activities, automation has proceeded in a piecemeal manner in aircraft maintenance and inspection. It is at least as much a push system, driven by those who invent or develop systems that may have an application here, as a pull system driven by user needs. Whether push or pull, there appears to have been no systematic (i.e., system-conscious and system-driven) strategy for automation. Similarly, there has been no attempt to determine the level of automation (in Sheridan's sense) represented or desired by each implementation. This section therefore classifies current and near-term projected automation examples in the field, using the five major functions developed in the previous section. (For a survey of the state of automation classified by traditional departments, see Goldsby, 1991).

Planning

Because planning is an activity relatively easy to model using task precedence and interference constraints, as well as expected completion times, it has been at least partially automated in some airlines for many years. The combinational problem of finding a minimum-out-of-service-time schedule is not easy to solve, but good heuristics do exist (Pinedo, 1992). Such scheduling systems are used by the airlines, typically with fine-turning by foremen and supervisors on a shift-by-shift basis. No attempts to use research (e.g. Sanderson, 1989) on hybrid scheduling systems (i.e,. humans and computers interacting) in this industry have been located.

Another area which automation has been used is in stock control and parts ordering. Because aircraft parts are both very costly and have long lead times for procurement, the costs of both inventory and stockouts are considerable. Thus innovative solutions have had to be developed so as not to delay aircraft return to service. Computers are used extensively by parts suppliers as well as airlines to track parts required for repair. Airlines often cooperate in parts inventory, a task almost impossible without computer tracking.

The final area of planning automation is in the optimization of the overall repair process. Given that the maintenance plan approved for each aircraft calls for a series of inspections, repairs, and replacements at known times, the packaging of these activities into hangar visits is a process by which automation (in the sense of model optimization) has been used. Thus, to bring together a number of different maintenance activities into a check visit, the inspection intervals for some components can be shortened. Because the intervals may be time based (e.g., some perishable items), cycle based (e.g., fuselage inspection or landing gear), or flight-hours based (e.g., engine components), and the flight profiles for each airline differ, the optimum schedule for hangar visits will differ. For example, there are some airlines that perform heavy checks (such as a C-checks) in a single long visit, others that perform them in a small number of shorter visits, and still others that "phase" their C-checks during overnight hangar visits at stations away from major repair bases.

Opening/Cleaning

Little in the way of automation or even modern job aids has been applied to this function, although the technology itself is changing. For example, the use of environmentally dubious solvents for paint removal is being supplanted by high-pressure water jets.

Inspection

Much of the automation taking place or projected concerns the inspection function, even though it is likely to remain a labor-intensive activity. Indeed, many NDI inspection devices have been justified in terms of reduced inspection times (e.g., Lutzinger, 1992) due to not having to disassemble structures and components.

At the *initiate* function, automation is possible using computer-based workcards. The IMIS system (Johnson, 1990) pointed the way for military aircraft, and many ways of automating this information presentation function have been proposed (e.g., Marx, 1992). A typical modern example is the computer workcard system developed by Lofgren and Drury (1994). This presents workcard information using a hypertext format, allowing for access to background documents as well as the more conventional sequential access. A test of this in an airline environment showed how it scored more highly than either the original paper-based workcards or paper-based workcards redesigned using human factors principles. It should be noted, however, that the redesigned paper-based workcards accounted for almost as much improvement over the original ones as did the hypertext system.

For *access*, perhaps the best-known example of automation is the climbing robot for eddy-current scans of lap-splice joints developed by Carnegie Mellon University (Albert, Kaufman, & Siegel, 1994). This is being developed to automate the data acquisition of the NDI system, essentially by removing the human access function.

Search and *decision* functions in automated inspection may not be easily separable (Drury & Prabhu, 1994), so they will be considered together under sensing and signal processing. A major effort by the FAA and NASA to develop new sensor systems to automate signal processing and to aid final decision making has been underway throughout the 1990s (for preview, see D. Johnson, 1989; for current status, see DOT, 1993). Many of these applications are push driven, and most are still in the stage of proof of concept. However, some have been tested under simulated operational conditions at the FAA's Aging Aircraft NDI Validation Center Facility at Albuquerque, New Mexico. Typically, they represent good physical approaches to the detection problem, but with little thought given to integration between device and user at this stage. The same problem of user interface design is also apparent in operational eddy-current and ultrasonic test equipment.

The *response* function can be, and has been, automated in combination with the initiate function as part of a work control system. Thus, in the hypertext system (Lofgren & Drury, 1993) when an inspector completes a workcard item, it can either be signed as good, or a nonroutine repair (NRR) form can be generated, both on the computer. Some airlines even have a bar-code system with lists of possible faults and positions, so that NRRs can be generated without any typing.

Repair

The *initiate* functions of repair use the same automation technology as the equivalent function in inspection. Thus Goldsby (1991) described the aircraft visit management system (AVMS) at United Airlines, which coordinates both inspection and repair activities through a central computer database.

It is, however, in the *diagnosis* function that the major impact of automation is being felt. Because diagnosis of avionics failures is a rule-based and knowledge-based system, it is inviting for computer aiding, expert systems, and artificial intelligence (e.g., Husni, 1992). AI techniques have been used in military avionics systems (e.g., Lesgold, 1990) with good evaluation results. Modern aircraft are designed with built-in systems to aid troubleshooting and diagnosis. For example, the B-777 (Hessburg, 1992) has an on-board maintenance system (OMS), combining condition monitoring (cf. the CBM of Nakajima et al., 1992) with built-in test equipment (BITE). Early analog BITE systems were characterized as unfriendly to the user, whereas first-generation digital systems gave large numbers of false alarms (Hessburg, 1992). The newer systems are, as ever, promised to be free of the ills of the systems they replace. At least these newer OMS systems are designed with specific human factors input on user requirements.

Because of the typically high cost of removing avionics equipment that later turns out to be fault-free, there has been an expanding interest in using AI systems with models of the equipment as part of a diagnosis training program. Thus, Kurland and Huggins (1990) demonstrated a model training system (MACH-III), which trains for a functional understanding of both the system itself and of the diagnostic procedures. Johnson, Norton, and Utsman (1992) developed intelligent tutoring systems (ITS's) for aircraft maintenance, again based on expert system concepts, but including a formalism for student and instructor modeling.

In the *repair/replace* function, there is little in the way of computer assistance except in repair shops, where for example brakes, engines, or seats are passed through a manufacturing-like system away from the aircraft itself. Here part tracking, inventory control, and scheduling are computer based, and there are some CNC machining centers and robotic welders (Goldsby, 1991). Here too Goldsby reported that work reorganization, to give focused repair centers for nozzle guide vanes, produced the results shown in Table 19.5. This use of cellular manufacturing concepts in maintenance gives similar spectacular results to those achieved in the manufacturing industry. For a comparative example, Table 19.5 also presents the data from Drury (1991) for a manufacturing cell for zinc castings. Neither of these examples rely strictly on automation: both use automated machinery in a way that includes intelligent use of the human operator within a social and organizational context. No uses of automation have been located for the *reset systems* or *close access* functions of repair.

Buyback and Return for Service

As with the other functions (planning and initiate) that are paperwork based, automation is proceeding more rapidly. With an automated workcard system, the outstanding items requiring buyback or final signoff can be found at any time. Goldsby (1991) described bar-codes on badges and workcards that can be read to provide job control information. The fully electronic logbook is the next logical step. Questions about the need for a system are not typically addressed, although the current paper-based system is often cumbersome in practice.

TABLE 19.5
Comparison of Examples of Job Redesign Results in Aircraft Maintenance
With Those From Manufacturing

	Goldsby (1991)		Drury (1991)	
	Original Shop	Focused Repair Center	Original Plant	Manufacturing Cell
Cycle time (days)	130	23	7	0.3
Labor cost ($/part)	57	33	1.71	0.81
Rejection rate (%)	50	5	?	0
Travel distance (meters)	9,600	16	209	13

AUTOMATION NEEDS IN AIRCRAFT MAINTENANCE AND INSPECTION

The review in the previous section demonstrated that although automation, or perhaps just mechanization, has been taking place in the aircraft maintenance/ inspection system, it has not had the demonstrated impact expected of it. Compared with the best manufacturing applications of automation (e.g., Bessant et al., 1992; Brödner, 1994), the examples in the aircraft hangar appear to be both piecemeal and technology driven. The existing system of work organization is accepted, and pieces of technology are grafted onto it. Sometimes this takes place with user input, sometimes not. For example, the development of a pen computer system for FAA inspectors to audit maintenance operation (Layton & Johnson, 1993) started from an analysis of current operations, then used custom software to eliminate many of the users' problems with the current manual system. Usability tests proved that it fulfilled its objectives. However, the job design issues behind the automation were largely beyond the scope of the project.

There are, however, examples of broader-based successes within the aviation maintenance/inspection field. Rogers (1991) presented the results of a rethinking of the jobs of USAF Tactical Air Commands maintenance program. Using modern concepts of work design, Rogers reorganized the whole function of the maintenance work to allow small teams to be dedicated to particular aircraft, to set goals for aircraft availability, and to allow considerable flexibility in how the teams achieved these goals. Results, measured in systems-relevant criteria, were spectacular over the time frame of 1978–1987:

Aircraft utilization/month +43%
Mission capable rate +59%
Percent repaired within 8 hours +158%
On-time departures +20%

Similar results have been seen in manufacturing industry where the automation and job design changes are well integrated; where this is not the case, success does not follow (Samson, Sohol, & Ramsay, 1993). Modern manufacturing industry is meeting the challenges presented in the first section of this chapter using an integration of automation and organizational techniques, in a way that could well prove more profitable for aircraft maintenance and inspection systems.

The key to these successes is becoming known as anthropocentric design (Brödner, 1994), or skill-based design (Kidd & Karwowski, 1994). In essence, it starts from the premise of appropriate automation, for example, a systematic evaluation of which tasks should (not can) be automated in order to achieve high system performance by explicitly designing-in roles for the human operator. The idea is exactly the opposite of determining which functions it is feasible to automate currently, and leaving to the operator those remaining functions that just cannot be automated at present. In manufacturing, it has been recognized that the concept of the peopleless factory, as the ultimate expression of CIM, does not give the agility and flexibility required to survive in the rapidly changing market place. People, used approximately, prevent the brittleness and consequent error-proneness of rule-based systems (e.g., Rasmussen, 1987).

How can this new approach to design be achieved in aircraft maintenance and inspection systems? Although these systems work on large structures (airliners) rather than the small components in manufacturing, they are not very different in concept. As many have shown over the years (e.g., Lock & Strutt, 1990; Taylor, 1993), aircraft maintenance and inspection is a complex system with many stakeholders. It is highly technical, but can be split fairly simply into unit operations, or transformations required between input and output (as shown earlier in this chapter). Thus, at one level, it is appropriate for STS analysis and design. At the same systems level, there are clear alternatives between automation and manual implementation of many functions, although I suspect there will be more, and more appropriate, automation alternatives if a systematic viewpoint is taken to define better what *should* be automated. Modern allocation of function ideas (e.g., Clegg et al., 1989; Sinclair, 1993) can be applied relatively easily to a system as well defined as this one. Drury (1994) proposed an error-based allocation of function methodology that would seem highly appropriate to a system in which errors are highly visible and costly. Finally, there have been attempts in this domain to utilize small groups for innovation and analysis. For example, Northwest Airlines uses group processes extensively for systems improvement, and United Airlines has used them in an innovation integration of inspection and maintenance functions (Scoble, 1994).

Thus, many of the elements of a successful modern approach to industrial systems design are in place. There are other helpful factors. There has been a long history of systematic design involving manufacturers, regulators, and airlines (e.g., in the MSG-3 process). Thus, system thinking is not new. Also, the maintenance hangar has been described, only half jokingly, by Taylor in Shepherd et al. (1991) as the "boys' own airplane club," meaning that most of the employees have a genuine, technical interest in aviation going well beyond their jobs. Thus, there is a high level of systems knowledge, and a high level of comfort with automation and technical concepts, among at least the older operators. In our own work in aircraft maintenance hangars over several years, we too have found an eagerness for, and fascination with, computers and other technological manifestations.

What is now needed to bring these threads together? The maintenance organizations themselves must be the focus of the effort, with aircraft manufacturers and regulatory bodies giving support as required. Again, there is a long history of this support, from the "lead customer" concept of new aircraft introduction to regulatory changes as automation progresses (e.g., the coming separate certification of NDI-specialist inspectors). With the focus at the maintenance hangar, the mixed operators/management-/staff teams typical of STS design (Taylor & Felten, 1993) need to perform broad but detailed systems analyses. With this background data, a step-by-step function allocation is needed, based on the idea that human skills, appropriately supported, have a central role in the system. What should emerge are long-term and short-term plans for the maintenance hangar of the future. These need to emphasize where automation needs to be developed, in contrast to the current climate of vendors asking the industry where automation can be used. Thus, a plan for development of appropriate automation can be developed that will have the support of management, staff and operators. Automation development will still take place, but now in an implementation-driven manner, rather than seeing "technology transfer" as a final and relatively uninteresting step in a scientific/engineering process.

In aircraft maintenance/inspection, we have a system that has served the country and the public extremely well over many years, with only about 5% of major accidents being directly maintenance related. We need to consciously provide for ways in which this service can continue and be enhanced in our increasingly automated future. Automation will be used here anyway; we have a chance to influence how it is used. We must collectively seize this chance.

ACKNOWLEDGEMENTS

Parts of this work were performed under a contract from the Federal Aviation Administration, Office of Aviation Medicine (Dr. W. Shepherd) through Galaxy Scientific Corporation.

REFERENCES

Akiba, M., & Suito, K. (1992). Total productivitiy management. In G. Salvendy (Ed.), *Handbook of industrial engineering* (2nd ed., pp. 52–61). New York: Wiley.

Albert, C. J., Kaufman, W. M., & Siegel, M. W. (1994). *Development of an automated nondestructive inspection (ANDI) system for commercial aircraft: Phase I* (DOT/FAA/CT-94/23). Springfield, VA: National Technical Information Service.

Bessant, J., Levy, P., Ley, C., Smith, S., & Tranfield, D. (1992). Organization design for factory 2000. *The International Journal of Human Factors in Manufacturing, 2*(2), 95–125.

Bond, N. A. (1987). Maintainability. In G. Salvendy (Ed.), *Handbook of human factors* (pp. 1328–1355). New York: Wiley.

Bounds, G., Yorks, L., Adams, M., & Ranney, G. (1994). *Beyond total quality management toward the emerging paradigm.* New York: McGraw-Hill.

Brödner, P. (1994). Design of work and technology. In G. Salvendy & W. Karwowski (Eds.), *Design of work and development of personnel in advanced manufacturing* (pp. 125–157). New York: Wiley.

Clegg, C., Ravden, S., Corbett, M., & Johnson, G. (1989). Allocating functions in computer integrated manufacturing: a review and a new method. *Behavior and Information Technology, 8*, 175–190.

Department of Transportation (DOT). (1993). *National aging aircraft research program plan.* Atlantic City, NJ: Federal Aviation Administration Technical Center.

Drury, C. G. (1991). Ergonomics practice in manufacturing. *Ergonomics, 34*(6), 825–839.

Drury, C. G. (1994). Function allocation in manufacturing. In S. A. Robertston (Ed.), *Proceedings of the Ergonomics Society 1994 Annual Conference* (pp. 2–16). London: Taylor & Francis.

Drury, C. G., & Prabhu, P. V. (1994). Human factors in test and inspection. In G. Salvendy and W. Karwowski (Eds.), *Design of work and development of personnel in advanced manufacturing* (pp. 355–402). New York: Wiley.

Drury, C. G., Prabhu, P., & Gramopadhye, A. (1990). Task analysis of aircraft inspection activities: Methods and findings. *In Proceedings of the Human Factors Society 34th Annual Conference* (pp. 1181–1185). Santa Monica, California: Human Factors Society.

Evans, J. R., & Lindsay, W. M. (1993). *The management and control of quality* (2nd. ed.). Minneapolis/St. Paul, MN: West.

Fitts, P. M. (1962). Functions of man in complex systems. *Aerospace Engineering 21,*, 34–39.

Goldsby, R. P. (1991). Effects of automation in maintenance. In *Proceedings of the Fifth FAA Meeting on Human Factors Issues in Aircraft Maintenance and Inspection* (pp. 103–123). Springfield, VA: National Technical Information Service.

Golomski, W. A. J. (1994). Human aspects of total quality management. In W. Karwowski and G. Salvendy (Eds.), *Organization and management of advanced manufacturing* (pp. 103–119). New York: Wiley.

Goranson, U. G. (1989). Continuing airworthiness of aging jet transports. In *Proceedings of the 2nd Annual International Conference on Aging Aircraft* (pp. 61–89). Baltimore, MD: Federal Aviation Administration.

Gude, D., & Schmidt, K. H. (1993). Preventive maintenance of advnaced manufacturing systems: A laboratory experiment and its implications for the human-centered approach. *The International Journal of Human Factors in Manufacturing, 3*(4), 335–350.

Hagemaier, D. (1989). Nondestructive testing of aging aircraft. In *Proceedings of the 2nd Annual International Conference on Aging Aircraft* (pp. 186–198). Baltimore, MD: Federal Aviation Administration.

Hendricks, H. W. (1991). Ergonomics in organizational design and management. *Ergonomics, 34*(6), 743–756.

Hessburg, J. (1992). Human factors considerations on the 777 on board maintenance system design. In *Meeting Proceedings Sixth Federal Aviation Administration Meeting on Human Factors Issues in Aircraft Maintenance and Inspection "Maintenance 2000"* (pp. 77–91).

Alexandria, VA: National Technical Information Service.

Horte, S. A., & Lindberg, P. (1991). Implementation of advanced manufacturing technologies: Swedish FMS experiences. *International Journal of Human Factors in Manufacturing*, *1*, 55–73.

Husni, M. (1992). Advances in artificial intelligence for aircraft maintenance. *Meeting Proceedings Sixth Federal Aviation Administration Meeting on Human Factors Issues in Aircraft Maintenance and Inspection "Maintenance 2000"* (pp. 157–171). Alexandria, VA: National Technical Information Service.

Ion, W. J. (1994). The key to concurrent engineering. In P. T. Kidd & W. Karwowski (Eds.), *Advances in agile manufacturing* (pp. 71–74). Washington, DC: IOS Press.

Johnson, D. (1989). Research and development. In *Proceedings of the 2nd Annual International Conference on Aging Aircraft* (pp. 30–35). Baltimore, MD: Federal Aviation Administration.

Johnson, R. C. (1990). An integrated maintenance information system (IMIS): an update. In *Final Report Second Federal Aviation Administration Meeting on Human Factors Issues in Aircraft Maintenance and Inspection "Information Exchange and Communications"* (pp. 00–00). Washington, DC: Office of Aviation Medicine, Federal Aviation Administration.

Johnson, W. B., Norton, J. E., & Utsman, L. G. (1992). Integrated information for maintenance training, aiding, and on-line documentation. In *Proceedings of the Human Factors Society 36th Annual Meeting, Innovations for Interactions*, *1*, (pp. 87–91). Atlanta, GA.

Jordan, N. (1963). Allocation of function between man and machines in automated systems, *Journal of Applied Psychology*, *47*, 161–165.

Karasek, R., & Theorell, T. (1991). *Healthy work*. New York: Basic Books.

Kidd, P. T., & Karwowski, W. (Eds.). (1994). *Advances in agile manufacturing*. Washington, DC: IOS Press.

Kurland, L., & Huggins, A. W. F. (1990). Training trouble-shooting using advanced AI-based training technology. In *Proceedings of the FAA Training Technology Symposium* (pp. 133–148). Alexandria, VA: National Technical Information Service.

Latia, D. C. (1993). *Nondestructive testing for aircraft* (Order No. EA-415). Casper, WY: IAP.

Layton C. F., & Johnson, W. B. (1993). Job performance aids for the flight standard service. In *Proceedings of the Human Factors and Ergonomics Society 37th Annual Meeting, Designing for Diversity*, *1*, (pp. 26–29). Santa Monica, CA: Human Factors and Ergonomics Society.

Lesgold, A. M. (1990). Intelligent computer aids for practice of complex troubleshooting. In *Proceedings of the FAA Training Technology Symposium* (pp. 116–132). Alexandria, VA: National Technical Information Service.

Lock, M. W. B., & Strutt, J. E. (1990). *Inspection reliability for transport aircraft structures: a three part study* (CAA Paper 90003). London: Civil Aviation Authority.

Lofgren, J., & Drury, C. G. (1994). Human factors advances at Continental Airlines. In *Proceedings of the 8th FAA/OAM Meeting on Human Factors in Aviation Maintenance and Inspection "Trends and Advances in Aviation Maintenance Operations"* (pp. 117–138). Alexandria, VA: National Technical Information Service.

Lutzinger, R. (1992). Tomorrow's problems as seen by maintenance managers. *Meeting Proceedings Sixth Federal Aviation Administration Meeting on Human Factors Issues in Aircraft Maintenance and Inspection "Maintenance 2000"* (p. 97). Alexandria, VA: National Technical Information Service.

Marx, D. (1992). Looking toward 2000: The evolution of human factors in maintenance. *Meeting Proceedings Sixth Federal Aviation Administration Meeting on Human Factors*

Issues in Aircraft Maintenance and Inspection "Maintenance 2000" (pp. 64–76). Alexandria, VA: National Technical Information Service.

Nakajima, S., Yamashina, H. Humagai, C., & Toyota, T. (1992). Maintenance management and control. In G. Salvendy (Ed.), *Handbook of industrial engineering* (2nd ed., pp. 1927–1986). New York: Wiley.

Pinedo, M. (1992). Scheduling. In G. Salvendy (Ed.), *Handbook of industrial engineering* (2nd ed., pp. 2131–2153). New York: Wiley.

Rasmussen, J. (1987). Cognitive control and human error machanisms. In J. Rasmussen, K. Duncan, & J. Leplat (Eds.), *New technology and human error* (pp. 53–61). New York: Wiley.

Rasmussen, J., Duncan, K., & Leplat, J. (Eds.). (1987). *New technology and human errors.* New York: Wiley.

Rogers, A. (1991). Organizational factors in the enhancement of military aviation maintenance. In *Proceedings of the Fourth International Symposium on Aircraft Maintenance and Inspection* (pp. 43–63). Washington, DC: Federal Aviation Administration.

Rouse, W. B. (1976). Adaptive allocation of decision-making responsibility between supervisor and computer, In Sheridan and Johannsen (Eds.), *Monitoring behavior and supervisory control* (pp. 295–306). New York: Plenum.

Samson, D., Sohal, A., & Ramsay, E. (1993). Human resource issues in manufacturing improvement initiatives: Case study experience in Australia. *The International Journal of Human Factors in Manufacturing, 3*(2), 135–152.

Sanderson, P. M. (1989). The human planning and scheduling role in advanced manufacturing systems: an emerging human factors domain. *Human Factors, 31*(6), 635–666.

Scoble, R. (1994). Recent changes in aircraft maintenance worker relationships. In *Meeting Proceedings of the Eighth Federal Aviation Administration Meeting on Human Factors Issues in Aircraft Maintenance and Inspection "Trends and Advances in Aviation Maintenance Operations"* (pp. 45–58). Alexandria, VA: National Technical Information Service.

Shepherd, W., Johnson, W. B., Drury, C. G., Taylor, J. C., & Berninger, D. (1991). *Human factors in aviation maintenance, phase I: Progress report* (DOT/FAA/AM-91/16). Springfield, VA: National Technical Information Service.

Sheridan, T. B. (1988). Task allocation and supervisory control. In M. Helander (Ed.), *Handbook of human–computer interaction, 8.* Amsterdam: Elsevier.

Sheridan, T. B. (1992). Social implications of telerobotics automation and supervisory control. In *Telerobotics, automation, and human supervisory control* (pp. 356–360). Cambridge, MA: MIT Press.

Sinclair, M. A. (1993). Human factors, design for manufacturability and the computer-integrated manufacturing enterprise. In M. Helander & M. Nagamachi (Eds.), *Design for manufacturability* (pp. 127–146). London: Taylor & Francis.

Taylor, J. C. (1993). The effects of crew resource management (CRM) training in maintenance: an early demonstration of training effects on attitudes and performance. In *Human factors in aviation maintenance—phase two progress report* (DOT/FAA/AM-93/5, pp. 159–181). Alexandria, VA: National Technical Information Service.

Taylor, J. C., & Felten, D. F. (1993). *Performance by design.* Englewood Cliffs, NJ: Prentice-Hall.

Wiener, E. L., & Nagel, D. C. (Eds.). (1988). *Human factors in aviation.* San Diego, CA: Academic Press.

20 Organizational and Safety Factors in Automated Oil and Gas Pipeline Systems

Najmedin Meshkati
University of Southern California

INTRODUCTION

Human ingenuity can now create technological systems whose accidents rival in their effects the greatest of natural disasters, sometimes with even higher death tolls and greater environmental damage. A common characteristic of complex technological systems, such as chemical processing plants, nuclear power stations, and aircrafts, is that they are under the centralized control of a few (control room or cockpit) operators. The effects of human error in these systems are often neither observable nor reversible; therefore, error recovery is often either too late or impossible. Complex technological systems' accidents, in the case of aircraft crashes, cause the loss of lives and property. In addition to these losses, in the case of chemical or nuclear plants, because of large amounts of potentially hazardous materials that are concentrated and processed at these sites, accidents pose serious threats with long-lasting health and environmental consequences for the workers, for the local public, and possibly for the neighboring region or country (Meshkati, 1991a).

For the foreseeable future, despite increasing levels of computerization and automation, human operators will have to remain in charge of the day-to-day controlling and monitoring of these systems, because system designers cannot anticipate all possible scenarios of failure, and hence are not able to provide preplanned safety measures for every contingency. According to Rasmussen (1980a), operators are kept in these systems because they are flexible, can learn and adapt to the peculiarities of the system, and because "they are expected to plug the holes in the designer's imagination" (p. 97). Thus, the safe and efficient operation of these technological systems is a function of the smooth and synchronized interaction among their human (i.e., personnel and organizational) and engineered subsystems (e.g., automation in general and automated control devices such as "intelligent," expert, or decision support systems in particular).

Many technological systems' failures implicated in serious accidents have traditionally been attributed to operators and their errors. Consequently, for the problem of technological systems safety, an engineering solution has been suggested (Perrow, 1986). For instance, many system designers postulate that removing humans from the loop is the most convenient alternative for the reduction or even the elimination of human error and, therefore, consider automation the key to the enhancement of system reliability. However, in many cases automation only aggravates the situation and becomes part of the problem rather than the solution. For example, in the context of aviation, automation is even more problematic because it "amplifies (crew) individual difference" (Graeber, 1994), and "it amplifies what is good and it amplifies what is bad" (Wiener, 1994). Furthermore, the automated devices themselves still need to be operated and monitored by the very human whose caprice they were designed to avoid. Thus, the error is not eliminated, but only relocated. The automation system itself, as a technological entity, has a failure potential that could result in accidents. Once an automated system requiring human intervention fails, operators, because of being out of the loop, are deskilled in just those very activities that require their contributions.

The underlying rationale and the major objective of this chapter is to demonstrate the critical effects of human and organizational factors and to also highlight the role of their interactions with automation (and automated devices) in the safe operation of complex, large-scale technological systems. This is done in the following sections by a brief analysis of well-known accidents at such systems, an overview of the most important problems and shortcomings of the present automated systems, and a case study to demonstrate the critical role of human organizational factors in the safety of an advanced control room of an oil and gas pipeline system.

THE CRITICAL ROLE OF HUMAN AND ORGANIZATIONAL FACTORS IN THE SAFETY OF CONTROL ROOM-OPERATED PETROCHEMICAL AND NUCLEAR POWER PLANTS

Most petrochemical and nuclear power plants around the world are operated by a group of human operators who are using advanced computer-based devices from a centralized control room. A large number of accidents at these plants typically start with equipment malfunction, process upset, or operator error, but they are aggravated and propagated through the system by a series of factors that could be attributed to human, organizational, and safety factors within the system. Also, most complex systems' accidents resemble the "unkind work environment"; that is, an environment in which once an error has been made, it is not possible for the person to correct the effects of inappropriate variations in performance before they lead to unacceptable consequences. This is because the effects of such errors are neither observable nor reversible (Rasmussen, 1986). As research has shown, in most cases operator error is an attribute of the whole technological (plant) system—a link in a chain of concatenated failures—that could result in accidents. The most important lesson to be learned from past accidents is that the principal cause tends to be neither the isolated malfunctioning of a major component nor a single gross blunder, but the unanticipated and largely unforeseeable concatenation of several small failures, both engineered and human. Each failure alone could probably be tolerated by the system's defenses. What produces the disastrous outcome is their unnoticed and often mysterious complex interaction.

On many occasions, human error is caused by the inadequate responses of operators to unfamiliar events. These responses depend very much on the conditioning that takes place during normal work activities. The behavior of operators is conditioned by the conscious decisions made by work planners or managers. Therefore, the error and the resulting accidents are, to a large extent, both the attribute and the effect (rather than the cause) of a multitude of factors such as poor workstation and workplace designs, unbalanced workload, complicated operational processes, unsafe conditions, faulty maintenance, disproportionate attention to production, ineffective training, lack of motivation and experiential knowledge, nonresponsive managerial systems, poor planning, nonadaptive organizational structures, rigid job-based pay systems, haphazard response systems, and sudden environmental disturbances (Meshkati, 1988). Thus, attributing accidents to the action of front-line workers is an oversimplification of the problem.

According to Perrow (1984, p. 351), "the dangerous accidents lie in the system, not in the components," and the inherent system accident potential can increase in a poorly designed and managed organization. The critical role of human and organizational factors in the safety of petrochemical plants has been highlighted in a survey by Meshkati (1991b). The United States Environmental Protection Agency (EPA) conducted a review of emergency systems for monitoring, detecting, and preventing releases of hazardous substances at representative domestic facilities that produce, use, or store these substances (EPA, 1988). Among the findings in the EPA's final report was that the "prevention of accidental releases requires a holistic approach that integrates technologies, procedures, and management practices" (p. 3). Moreover, the report stated:

> The commitment of management to accident prevention, mitigation, and preparedness is essential. Without such commitment, installation of the most advanced technologies will be an expensive, but ineffectual safeguard for preventing serious injury, death, or environmental damage . . .While accidents can occur in well-managed facilities, the lack of management commitment can lead to disaster . . . The ultimate responsibility for safe design, operation, and maintenance of a facility rests with management. (p. 3)

The important role of human and organizational factors in the safety of nuclear power plants has been investigated in studies by Gertman, Haney, Jenkins, and Blackman (1985), Orvis, Moieni, and Joksimovich (1993), and Haber, O'Brien, Meatly, and Crouch (1991). These issues were also addressed and explored in the recent works of Gertman and Blackman (1994), Marcus and Nichols (1991), Wells and Ryan (1991), Wu, Apostolakis, and Okrent (1991), and Mosleh, Grossman, and Modarres (1991). The critical role of human and organizational causes in the Chernobyl accident is encapsulated in the following statement, which appeared in the conclusion section of the International Atomic Energy Agency's (IAEA) Summary Report on the Post-Accident Review Meeting on the Chernobyl Accident (1986, p. 76): "The root cause of the Chernobyl accident, it is concluded, is to be found in the so-called human element. . . . The lessons drawn from the Chernobyl accident are valuable for all reactor types."

Moreover, Valeriy A. Legasov (deceased), a former Soviet academician, the First Deputy Director of the Kurchatov Institute in Moscow at the time of the Chernobyl accident, and the head of the Soviet delegation to the Post-Accident Review Meeting of the IAEA in August, 1986, "declared with great conviction: 'I advocate the respect for human engineering and sound man–machine interaction. This is a lesson that Chernobyl taught us' " (cited in Munipov, 1992, p. 540). These facts and other investigations led the IAEA to declare that "the Chernobyl accident illustrated the critical contribution of the human factor in nuclear safety" (Nuclear Safety Review for 1987, p. 43).

Finally, according to the IAEA, "the (Chernobyl) accident can be said to have flowed from deficient safety culture, not only at the Chernobyl plant, but throughout the Soviet design, operating and regulatory organizations for nuclear power that existed at the time. . . . Safety culture . . . requires total dedication, which at nuclear power plants is primarily generated by the attitudes of managers of organizations involved in their development and operation" (IAEA, 1992, p. 24). In *Safety Culture*, a report by the

International Nuclear Safety Advisory Group of the IAEA, safety culture is defined as "that assembly of characteristics and attitudes in organizations and individuals which establishes that, as an overriding priority, nuclear plant safety issues receive the attention warranted by their significance" (IAEA, 1991, p. 4).

According to the author's analyses of large-scale technological systems' accidents, there were two main categories of human and organizational factors causes: a lack of human and organizational factors considerations (a) at the (system) design stage and (b) at the (system) operating stage (Meshkati, 1991a). Notwithstanding the overlapping domains and intertwined nature of these two stages, the former, using Reason's (1992) characterization, refers primarily to the "latent errors" — adverse consequences that may lie dormant within the system for a long time, only becoming evident when they combine with other factors to breach the system's defenses. In the context of this chapter, they include: control room, workstation, and display/control panel design flaws causing confusion and leading to design-induced errors; problems associated with lack of foresight in operators' workload estimation leading to overload (and stress); inadequate training; and organizational rigidity and disarrayed managerial practices. The final factor, which is associated with the performance of the front-line operators immediately before and during the accident, includes sources and variations of "active errors" such as misjudgments, mistakes, and wrongdoings. In order to prevent accidents in chemical and nuclear plants, as suggested by Meshkati (1990, 1991a), an integrated systemic approach should be taken to the design and operation as attentive to both technical elements and human and organizational factors. This approach should be based on a thorough and integrated analysis of plants' processes, workstations, procedures, management, and supervisory systems.

THE PROBLEMS OF AUTOMATION IN CONTROL OF COMPLEX SYSTEMS

Most complex, large-scale, technological systems have been both "tightly coupled" and "complexly interactive" (Perrow, 1984). The characteristics of a tightly coupled system include: processing delays that are unacceptable; production sequences that are relatively invariant; relatively few ways of achieving a particular goal; little slack permissible in supplies, equipment, and personnel; and buffers and redundancies deliberately designed into the system. Interactive complexity can be described by one or a combination of features such as: the close proximity of components that are not linked together in a production sequence; the presence of many common-mode connections (i.e., many components whose failure can have multiple effects "downstream"); the fact that there is only a limited possibility of isolating failed components; the fact that, due to the high degree of specialization, there is little chance of substituting or reassigning personnel (the same lack of interchangeability could also be true for material and supplies); unfamiliar or unintended feedback loops; the many control parameters that could potentially interact; the fact that certain information about the state of the systems must be obtained directly, or inferred; and the characteristic that there is only a limited understanding of some processes, particularly those involving transformations. Tight coupling requires centralization to ensure immediate response to failures by those who are in charge and in a position to understand the problem and determine the correct course of action. Interactive complexity, on the other hand, mandates decentralization to handle the unexpected interaction of different functions, decisions, and errors.

As the task uncertainty increases, which is the case in the "nonnormal" or emergency situations at the complex technological systems, the number of exceptions also increases until the organizational hierarchy is overloaded, at which time the organization must use another mechanism to reconfigure itself. Furthermore, the "normal function" of tightly coupled technological systems is to operate on the boundary to loss of control. That is, people are involved in a dynamic and continuous interaction with the failure and hazard (Rasmussen 1989a). Thus, "touching the boundary to loss of control is necessary (e.g., for dynamic speed-accuracy trade-offs)" (Rasmussen, Pejtersen, & Goodstein, 1994, p. 150). This is a rapidly changing environment and, in order to survive it, the system should be able to respond in a safe and effective manner. Occasionally, it may require an improvised response from the operator(s), but it should certainly be coordinated and in concert with others' activities and stay within the boundaries or "space" of acceptable work performance (Rasmussen 1989b). Otherwise, it would be just noise in the control of the system and could lead to errors.

It is the nature of complex, tightly coupled, and complexly interactive systems, according to Reason (1987), to spring "nasty surprises." As case studies repeatedly show, accidents may begin in a conventional way, but they rarely proceed along predictable lines. Each accident is a truly novel event in which past experience counts for little, and where the plant is returned to a safe state by a mixture of good luck and hard, knowledge-based effort. Accident initiation and its propagation through possible pathways and branches within the system is a highly complex and hard-to-foresee event. It is analogous to the progression of a crack in an icy surface, which can move in several directions, hit different levels of thickness, and, if not stopped, can cause the surface to break up and open ("uncover the core" and break the system).

Operators' control of complex, large-scale technological systems can be termed coordination by preplanned routines (Woods, 1987). However, coordination by preplanned routines is inherently "brittle." Because of both pragmatic and theoretical constraints, it is difficult to build mechanisms into preplanned routines that cope with novel situations, adapt to special conditions, or recover from human errors in following the plan. When preplanned routines are rotely invoked and followed, performance breaks down in the light of underspecified instructions, special conditions or contexts, violations of boundary conditions, human execution errors, bugs in the plan, multiple failures, and novel situations (incidents not planned for) (Woods, 1987). This is the problem of unanticipated variability that happens frequently during emergencies at complex technological systems. Moreover, in virtually every significant disaster or near disaster in complex systems, there have been some points where expertise beyond the pre-planned routines was needed. This point involves multiple people and a dynamic, flexible, and problem-solving organization. Handling unfamiliar events (e.g., emergencies) also requires constant modification of the design of the organization, coordination, and redeployment of resources (Mesh-kati, 1991c). However, as it has been observed and reported many times, usually the preprogrammed routines of decision support in expert computing systems sets the organization in a static design (Sloane, 1991).

Furthermore, it has been empirically validated that experts in high-stress, demanding situations do not usually operate using a process of analysis. Even their rules of thumb are not readily subjected to it, whereas most of the existing artificial intelligence-based automated systems always rely on analytical decision process. If operators of complex systems rely solely on a computer's analytic advice, they would never rise above the level of mere competence—the level of analytical capacity—and their effectiveness would be limited by the inability of the computer systems to make the transition from analysis to pattern recognition and other more intuitive efforts (Dreyfus & Dreyfus, 1986).

The issue of operators trusting automated systems is another major factor limiting the application and effectiveness of these systems. Trust between humans and machines is a very complex issue that, among others, is a function of the machine's behavior and the stability of its environment (Muir, 1988; Sheridan 1980).

In summary, when employing automation for the control of complex technological systems, system designers and managers should always remember that one can and should not replace the other, as suggested by Jordan (1968, p. 203): "Men and machines are not comparable, they are complementary. . . . Men are good at doing what machines are not good at doing and machines are good at doing that at which men are not good at doing."

CASE STUDY: APPLICATIONS TO AN OIL AND GAS
PIPELINE SYSTEM'S CONTROL ROOM

According to a study of 500 incidents involving pipe work failure and subsequent chemical release (in the United Kingdom, United States, Netherlands, and Finland) for the United Kingdom's Health and Safety Executive, "responsible in 30.9% of the incidents, operator error was the largest contributor to pipework failures among know direct causes" (Geyer, Bellamy, Astley, & Hurst, 1990, p. 68). This study concluded and recommended "human factors reviews of maintenance and operations personnel and functions" (p. 69) as one of the four critical areas where management of oil, gas, and chemical companies should concentrate their efforts.

General Observations and Findings

The following is a summary of relevant and pertinent human, organizational, and safety factors affecting operators' performance while using advanced automated systems in the pipeline control room. From this control room, a sophisticated network of oil and gas pipeline systems in the Western United States is controlled.

Human Factors Considerations: Workstation and
Interface (Displays)

Alarms.   Alarms are incoming signals from different active and inactive pipeline systems that operators need to acknowledge.

- Alarms were not prioritized.
- In responding to different alarms for a pipeline system, the operator had to spend significant amounts of time identifying the alarm.

Display (Screen) Design.

- All manual valves were usually not presented on the display, whereas remotely controlled valves were all presented. A manual valve was only displayed when it connected two lines.

Normal Conditions and the Nature of the Workload.   The major contributing (task) loading factors or categories of the operators' (mental) workload during normal or routine times, included:

- Information processing—for example, performing a number of concurrent tasks, valve alignments, responding to alarms, and other mental tasks.
- Communication—for example, on a routine basis talking with field workers, maintenance, and other operators; answering phone calls.
- Data recording—for example, filling out paperwork and incident logs.
- A workload that, of course, was proportional to the number of pipeline systems that were controlled by the operator. In this control room, 7 to 9 pipelines, on the average, were simultaneously controlled by a single operator. The workload could have intensified because of time pressure, time of the day, and activities in the field (such as maintenance) that affect the control room operators.

Abnormal Conditions, Workload, and Leak Detection.   Workload substantially increased as a result of system upset, such as a leak, or equipment malfunction, such as a valve or pump breakdown. Leak detection required a good understanding of the physical characteristics of the product, the "profile" of the pipeline system and its hydraulic characteristics (pressure and flow), the terrain, and environmental conditions (temperature). A leak, therefore, was an unannounced event in the control room and leak detection was a diagnostic effort.

During an emergency, the pipeline system control room was the focal point of communication with state and local agencies.

- Leak detection was typically done through periodic checks of (the trend of) temperature, pressure, flow rate, meter accumulator, or tank gauge.
- During the leak detection and handling, the operator needed to continue performing other control functions.

Safety-Related Considerations

Reported Causes of Errors and Performance Obstacles.   There were two basic types of errors: pipeline valve alignment and paperwork related. Misalignment errors were primarily caused by lack of concentration (interruption, distraction, and omission caused by heavy workload), failure to check the pipeline map thoroughly for all valves, and discrepancy between map and computer data.

Other sources of errors were lack of familiarity with the particular pipeline system, not asking for help from other operators, and relying on the information given only by one source, either the sending or receiving party. Other errors were caused by not keeping track of the required entries for the paperwork and not balancing the product movement.

- There were nonessential interruptions resulting from calls to the pipeline system control room.
- There were discrepancies between the valves' positions on the pipeline blueprint (map) version and its computer version.
- When all valves on a pipeline were remotely controlled valves, it took a fraction of an hour to align the system; for pipelines with manual valves, it took up to 10 times that long.

Organizational-Related Factors

Performance Obstacles.

- Operators perceived a lack of sufficient support and appreciation from within the company, which affected morale and motivation.
- Operators perceived very limited opportunities for advancement and promotion.
- Operators perceived a disproportionate amount of input from other units within the company in their performance review.

Analysis

A primary goal of this case study was to identify error-inducing conditions as well as human and organizational causes of errors while using automated systems at the pipeline control room. This section attempts to further elaborate these issues by addressing the potential for human–task mismatch, because errors are caused by human—machine or human–task mismatches. Operators' errors should be seen as the result of human variability, which is an integral element in human learning and adaptation (Rasmussen, 1985). This approach considers the human–task or human--machine mismatches as a basis for analysis and classification of human errors, instead of solely tasks or machines (Rasmussen, Duncan, & Leplat, 1987). These mismatches could also stem from inappropriate work conditions, lack of familiarity, or improper (human–machine) interface design. The use of off-the-shelf general training, increased number of procedures, and stricter administrative controls is less effective than utilizing real countermeasures against these modes of mismatches or misfits. Thus, human error occurrences are defined by the behavior of the total human--task system. Frequently, the human–system mismatch will not be due to spontaneous, inherent human variability, but to events in the environment that act as precursors.

Nature and Categories of Errors in the Pipeline System Control Room

An important category of errors within the context of the pipeline system control room, wherein the operators typically engage in monitoring and supervising the system and have to respond to changes in system operation with corrective actions, is called *systematic errors*. In this context, two types of systematic errors are important and should be considered (Reason, 1992).

1. Research has shown that operators' responses to changes in a technological system will be systematically wrong if the task demands exceed the limits of capability. In the case of the pipeline system operator, job demands and capability may conflict due to several aspects of a task, such as the time required, availability of needed information, and background knowledge on system functioning.

The mental workload of operators working in the pipeline system control room was highly variable and could have reached extremely high levels. This is synonymous with having or lacking balance between task demands and an operator's capabilities. According to Tikhomirov (1971/1988), high or unbalanced mental workload causes:

- Narrowing span of attention.
- Inadequate distribution and switching of attention.
- Forgetting the proper sequence of actions.
- Incorrect evaluation of solutions.
- Slowness in arriving at decisions.

In addition to occasional unbalanced workloads, human factors-related problems with the computer workstation, such as the mismatch between computer and map data on valves or lack of alarm prioritization, could cause a good portion of errors in the pipeline system control room. These type of errors, the so-called *design-induced* or *system-induced* errors, are forced on operators.

2. Systematic operator error may be caused by several kinds of "procedural traps" (Rasmussen, 1980b). During normal working conditions, human operators are generally extremely efficient because of effective adaptation to convenient, representative signs and signals that they receive from the system. This is a very effective and mentally economical strategy during normal and familiar periods, but leads the operator into traps when changes in system conditions are not adequately reflected in his or her system of signs. Such mental traps often significantly contribute to the operator's misidentification of unfamiliar and complex system states. This

misidentification, in turn, is usually caused by the activation of "strong-but-wrong" rules, where the "strength" is determined by the relative frequency of successful execution. When abnormal conditions demand countermeasures from the operator, a shift in the mental work strategies is needed by the operators. However, it is very likely that familiar associations based on representative, but insufficient, information will prevent the operator from realizing the need to analyze a complex and/or unique situation. He or she may more readily accept the improbable coincidence of several familiar faults in the system, rather than the need to investigate one new and complex fault of low probability. In this case, the efficiency of the human operator's internal mental model allows him or her to be selective and, therefore, to cope effectively with complex systems in familiar situations, which at the same time may lead the operator into traps that are easily seen after the fact.

Errors During Normal Conditions.   Usually, those errors that occur during normal conditions at the pipeline system control room, such as failing to open a valve when preparing a pipeline, are of a slip or lapse in nature rather than mistakes. Slips and lapses are associated with failures at the more subordinate levels of action selection, execution, and intention storage, whereas mistakes occur at the level of intention formation and planning (Reason, 1992).

According to research findings, a necessary condition for the occurrence of a slip or lapse is the presence of "attention capture" associated with either distraction or preoccupation. Another type of slip that happens at the pipeline system control room could stem from what is called "inappropriately timed check." Like omitted checks, inappropriate monitoring is associated with attention capture. Mistimed monitoring is most likely to occur immediately following a period of "absence" from the task in mind, caused by interruptions (Reason, 1992).

In addition to the general factors that promote absent-minded slips and lapses (the execution of routine tasks while preoccupied or distracted), the following are a number of task factors in the pipeline system control room that are likely to increase the probability of making an omission error. Even the most experienced operators could not escape the negative effects of these factors (based on the framework of Reason, 1992).

- The larger the number of discrete steps in an actions sequence (e.g., having many valves on the pipeline), the greater the probability that one or more of them will be omitted.
- The greater the informational loading of a particular procedural step (e.g., preparing a pipeline with many complicated pipeline valve

stations having several manual valves), the more likely it is that items within that step will be omitted.

• Procedural steps that are not obviously cued by preceding actions or those that do not follow in a direct linear sequence from them arc likely to be omitted.

• When instructions are given verbally and there are more than five simple steps, items in the middle of the list of instructions are more likely to be omitted than are those either at the beginning or the end.

• When instructions are given in a written form, isolated steps at the end of the sequence (e.g., replacing caps or brushes after maintenance, removing tools, etc.) have a reasonably high probability of being omitted.

• In a well-practiced, highly automated task, unexpected interruptions (e.g., during valve alignment task, receiving alarms and phone calls) are frequently associated with omission errors, either because some unrelated action is unconsciously counted in as part of the task sequence, or because the interruption causes the individual to lose his or her place on resumption of the task (i.e., the individual believes that he or she was further along in the task prior to the interruption than was actually the case). Such routine tasks are also especially prone to premature exits — moving on to the next activity before the previous one is completed, thus omitting some necessary final steps (e.g., without opening a valve, moving to the next one on the pipeline). This is particularly likely to happen when the individual is working under time pressure or when the next job is near at hand (e.g., preparing a pipeline and having to fill out the corresponding paperwork).

Errors During Abnormal Conditions. The previously mentioned systematic errors are significant contributors to technological systems' failures. According to research findings, the failure of human operators to identify abnormal states of a system because of the foregoing systematic errors plays an important role in accidents and incidents in complex technological systems. Even if the state of the system is correctly identified, the operator may still be caught in a "procedural trap" (Rasmussen, 1980b). It has been argued that a familiar, stereotyped sequence of actions may be initiated from a single conscious decision or association from the system state. If the corresponding procedure takes some time (e.g., it is necessary to move to another place to perform it), the mind may return to other matters, making the workings of the subconscious vulnerable to interference, particularly if part of the sequence is identical to other heavily automated sequences. Systematic human errors in unfamiliar tasks are typically caused by interference from other, more stereotyped situations and, therefore, the potential for systematic errors depends very much on the level of the operator's skill. "The fact that operators can control the system successfully during a commissioning and a test period is not proof that operators will continue to do so during the system life cycle" (Rasmussen, 1980b, p. 364).

A basic problem when dealing with systematic erroneous responses to unfamiliar situations is the low probability of such complex situations. In a properly designed system, there should be a reverse relation between the probability of occurrence of an abnormal situation and its potential effects in terms of losses and damage. In modern centralized control rooms the consequence of faults can be very serious and, as a result, the effects of human error in situations of extremely low probability must be considered. In such cases, as in the pipeline system control room, the potential for systematic errors cannot be identified from experience. The skills developed and gained during normal operations are not a satisfactory basis for infrequently needed improvisation to handle unfamiliar events (Rasmussen, 1980a). Instead, the operator's task and work organization should be restructured to ensure that he or she has the necessary knowledge available when abnormal situations demand his or her understanding of the system's physical functioning. Only through a systematic functional analysis of realistic scenarios and their decomposition to the subtask level can the error-inducing conditions be exposed.

Furthermore, we cannot rely solely on the experience level of the operators to avoid accidents. In fact, "in accident avoidance, experience is a mixed blessing" (Reason, 1992, p. 86). Operators learn their avoidance skills not so much from real accidents as from near-misses. It has even been said that "if near-accidents usually involve an initial error followed by an error recovery, more may be learned about the techniques of successful error recovery than about how the original error might have been avoided" (Reason, 1992, p. 86).

These types of problems cannot be effectively counteracted by administrative measures or by better training. In complex systems, such as the control room of the oil and gas pipeline system, we also have to consider rare events for which operators cannot be prepared by training on the use of procedures. In such cases, operators have to generate proper procedures online by functional evaluation and causal reasoning, based on knowledge about system properties. This suggests that it is necessary that more than one operator be involved in problem-solving during rare events, and the whole crew of the pipeline system control room should be able to work as a team. The recent studies on team mind consider the team as "an emergent entity"; postulating that the "team acts as does a person" and contending that a smoothly functioning team mind is "anticipating the needs of others, synchronizing actions, and feeling free to improvise" (Thordsen & Klein, 1989, pp. 3, 6).

CONCLUSIONS AND RECOMMENDATIONS

Based on the analysis, recommendations for considering human, organizational, and safety factors in pipeline system control room were made. There were two sets of such recommendations, one each for both short- and long-term considerations.

Short-Term Human, Organizational, and Safety Considerations

In the short term, it was concluded that human factors and safety considerations should include simplifying tasks, and improve the physical control center and interface-related factors. It was recommended that attempts should be made to:

- Minimize interruptions.
- Prioritize incoming alarms, and queue and batch-process the low-priority ones.
- Balance the workload.
- Redesign and simplify paperwork and revise procedures for filling it out.
- Upgrade computer databases of pipelines' parts, components, valves, and routes, and make them consistent with maps.
- Develop a system for online updating of the preparation of a pipeline system and progress of maintenance activities.
- Make sure all pipeline system control room equipment and systems (computer and communications) work properly.
- Develop and provide operators with decision aids and memory aids. Decision aids are designed to minimize failures when a human operator formulates his or her action or plan, whereas memory aids support the performance during plan storage and execution (Reason, 1992).
- Develop a paper or electronic checklist for every pipeline system. These checklists should cover all steps needed for the alignment of all manual and remotely controlled valves on any pipeline system.

In the short term, it is suggested that the organizational-related factors should include and attempt to:

- Educate employees working for other areas in the company about the pipeline system control room and the full range of operators' jobs and responsibilities.

- Set performance goals with input from the operators.
- Review the career opportunities and promotion possibilities of operators within the company. Openly communicate this information to the existing and future crew members of the control room.
- Clearly identify career aspirations ("Career Concepts") of each operator.
- Integrate and synchronize the personnel requirements of the support staff and other supporting departments with the control room.
- Develop a context-specific and skill-based performance review system for the pipeline system control room. The corresponding form should not be generic or job based. Factors such as skill versatility, analytical abilities, and information integration and differentiation abilities should be included, because they are important contributors to keeping the system in a normal operating mode and bringing it back from an upset mode in the case of a failure. This form should address all the performance-related factors of the control room crew as specifically as possible.
- Develop a team or collective performance evaluation plan and an accompanying mechanism, in addition to any individual performance review, to encourage, recognize and reward the much-needed team-work.

It is noteworthy that one of the most important considerations, with far-reaching effects for human, organizational and safety areas, is the inclusion of the operators in the decision-making process. The input from the operators may point out areas with a high potential for error within the system that might otherwise be overlooked.

Long-Term Human, Organizational, and Safety Considerations

It was recommended that the long-term human factors considerations should include the incorporation of several human factors issues in the design of software and (new) display systems for the control room.

The findings of a recent study by Moray et al. (1993) have important implications for the new generation of control room computer-generated, animated, and direct perception displays. According to this study, "recall and diagnosis should be better for an integrated display than for a traditional single-sensor-single-indicator display (SSSI). . . . Even experts can only exercise their skills and expertise optimally if the pattern in which information is displayed matches their models of the dynamics of the problem. . . . The advantage of direct perception interfaces should be particularly strong when the operators have advanced levels of expertise" (p. *x*). Computer-generated "displays should not merely transfer *data* to the observer: they should transfer *goal-relevant information*, which will most easily arouse the operator's expertise in the relevant task domain. To evaluate interfaces requires us to evaluate the extent to which they perform this task" (p. 5). These findings were further corroborated in a recent study by Meshkati, Buller, and Azadeh (1994), in which uses of the ecological interface resulted in significantly more accurate event diagnosis and recall of various plant parameters, faster response to plant transients, and higher ratings of operators preference.

As mentioned before, errors are caused by human–machine or human--task mismatches. These mismatches could stem from inappropriate working conditions, lack of familiarity with the system, or improper (human–machine) interface design. To reiterate, using general training, a large number of procedures, and stricter administrative controls is less effective than utilizing real countermeasures against these modes of mismatches or misfits. Whatever the cause of the specific individual error—a change in working conditions, a spontaneous slip of memory, high workload, distraction, and so on—the resulting margin of mismatch between situation and the human can be decreased by providing the operator with better access to information about the underlying causal net so as to improve improvisation and recall. In particular, the margin can be decreased by making the effect of the operator's activity directly observable. Interface design should aim at making the boundaries of acceptable performance visible to the users while their effects are still observable and reversible. This can be done by designing readily visible feedback to support functional understanding of the system. It was recommended that to assist operators in coping with unforeseen situations, the designer should provide them with tools to make experiments and test hypotheses without having to do these things directly on potentially irreversible pipeline systems.

As suggested by Rasmussen (1989c), causal reasoning in a complex functional network, such as a pipeline with many pipeline valve stations, places excessive demands on limited working memory resources. Information should be embedded in a structure that can serve as an externalized mental model. It was recommended that this representation (for the operators) should not only aim at identifying a specific problem solution, but should also aim at indicating an effective strategy (i.e., a category of possible solutions).

The inclusion of organizational and safety factors into the design and operation of a pipeline system control room results in better operator–task and operator–workstation matches. Thus, it will certainly contribute to the reduction of human error potential and the enhancement of the total system's reliability.

ACKNOWLEDGMENTS

In the analysis and writing of this paper, the author has greatly benefited from discussions with Professor Jens Rasmussen (Risø National Laboratory, Denmark), as well as his and Professor James Reason's (University of Manchester) scholarly works. Contributions of the Research Assistant, Priscilla Jorgensen, are highly appreciated.

REFERENCES

Dreyfus, H. L., & Dreyfus, S. E. (1986). *Mind over machine*. New York: Free Press.

Environmental Protection Agency (EPA) (1988). *Review of emergency systems* (Report to Congress). Washington, DC: Office of Solid Waste and Emergency Response, EPA.

Gertman, D. I., & Blackman, H.S. (1994). *Human reliability & safety analysis data handbook*. New York: Wiley.

Gertman, D. I., Haney, L. N., Jenkins, J. P., & Blackman, H. S. (1985). *Operational decisionmaking and action selection under psychological stress in nuclear power plants*. (NUREG/CR-4040). Washington, DC: U.S. NRC.

Geyer, T. A., Bellamy, L. J., Astley, J. A., & Hurst, N. W. (1990, November). Prevent pipe failures due to human error. *Chemical Engineering Progress*, pp. 66–69.

Graeber, R. C. (1994, May). *Integrating human factors knowledge into automated flight deck design*. Invited presentation at the International Civil Aviation Organization (ICAO) Flight Safety and Human Factors Seminar, Amsterdam.

Haber, S. B., O'Brien, J. N., Metlay, D. S., & Crouch, D. A. (1991). *Influence of organizational factors on performance reliability*. (NUREG/CR-5538). Washington, DC: U.S. NRC.

International Atomic Energy Agency (IAEA). (1986). *Summary report on the post-accident review meeting on the Chernobyl accident* (Safety Series # 75-INSAG-1). Vienna: Author.

International Atomic Energy Agency (IAEA). (1991). *Safety culture* (Safety Series No. 75-INSAG-4). Vienna: Author.

International Atomic Energy Agency (IAEA). (1992). *The Chernobyl accident: Updating of INSAG-1* (INSAG-7). Vienna: Author.

International Atomic Energy Agency (IAEA). (1987). *Nuclear Safety Review 1987*. Vienna, Austria: Author.

Jordan, N. (1968). *Themes in speculative psychology*. London: Tavistock.

Marcus, A. A., & Nichols, M. L. (1991). Assessing organizational safety in adapting, learning systems: Empirical studies of nuclear power. In G. Apostolakis (Ed.), *Probabilistic safety assessment and management* (pp. 165–170). New York: Elsevier.

Meshkati, N. (1988, October). *An integrative model for designing reliable technological organizations: The role of cultural variables*. Invited position paper for the World Bank Workshop on Safety Control and Risk Management in Large-Scale Technological Operations, World Bank, Washington, DC.

Meshkati, N. (1990). Preventing accidents at oil and chemical plants. *Professional Safety,* *35*(11), 15–18.

Meshkati, N. (1991a). Human factors in large-scale technological systems' accidents: Three Mile Island, Bhopal, Chernobyl. *Industrial Crisis Quarterly, 5,* 133–154.

Meshkati, N. (1991b, November). Critical human and organizational factors considerations in design and operation of petrochemical plants. (Paper # SPE 23275). *Proceedings of the First International Conference on Health, Safety & Environment in Oil and Gas Exploration and Production,* (Vol. 1, 627–634). The Hague, The Netherlands: Society of Petroleum Engineers (SPE).

Meshkati, N. (1991c). Integration of workstation, job, and team structure design in complex human–machine systems: A framework. *International Journal of Industrial Ergonomics, 7,* 111–122.

Meshkati, N., Buller, B. J., & Azadeh, M. A. (1994). *Integration of workstation, job, and team structure design in the control rooms of nuclear power plants: experimental and simulation studies of operators' decision styles and crew composition while using ecological and traditional user interfaces* (Vol. I). Grant Report Prepared for the U.S. Nuclear Regulatory Commission (Grant # NRC-04-91-102). Los Angeles, CA: University of Southern California.

Moray, N., Jones, B. J., Rasmussen, J., Lee, J. D., Vicente, K. J., Brock, R., & Djemil, T. (1993). *A performance indicator of the effectiveness of human–machine interfaces for nuclear power plants* (NUREG/CR-5977). Urbana–Champaign, IL: Dept. of Mechanical and Industrial Engineering, University of Illinois.

Mosleh, A., Grossman, N., & Modarres, M. (1991). A method for evaluation and integration of safety performance indicators. In Apostolakis (Ed.), *Probabilistic safety assessment and management* (pp. 43–48). New York: Elsevier.

Muir, B. M. (1988). Trust between humans and machines, and the design of decision aids. In E. Hollnagel, G. Mancini, & D. D. Woods (Eds.), *Cognitive engineering in complex dynamic worlds* (pp. 71–83). New York: Academic Press.

Munipov, V. M. (1992). Chernobyl operators: criminals or victims? *Applied Ergonomics,* *23*(5), 337–342.

Orvis, D. D., Moieni, P., & Joksimovich, V. (1993). *Organizational and management influences on safety of nuclear power plants: Use of PRA techniques in quantitative and qualitative assessment* (NUREG/CR-5752). Washington, DC: U.S. NRC.

Perrow, C. (1984). *Normal accidents.* New York: Basic Books.

Perrow, C. (1986). *Complex organizations: A critical essay* (3rd ed.). New York: Random House.

Rasmussen, J. (1980a). What can be learned from human error reports? In K. D. Duncan, M. M. Gruneberg, & D. Wallis (Eds.), *Changes in working life* (pp. 97–113). New York: Wiley.

Rasmussen, J. (1980b). Notes on human error analysis. In G. Apostolakis, S. Garribba, & G. Volta (Eds.), *Synthesis and analysis methods for safety and reliability studies* (pp. 357–389). New York: Plenum.

Rasmussen, J. (1985). Trends in human reliability analysis. *Ergonomics, 28*(8), 1185–1195.

Rasmussen, J. (1986). *Information processing and human–machine interaction: An approach to cognitive engineering.* New York: North-Holland.

Rasmussen, J. (1989a). *Focus of discussion at the Karlstad workshop: Some notes.* Unpublished notes.

Rasmussen, J. (1989b). *Self-organization and structural evolution: A note for discussion.* Unpublished discussion paper.

Rasmussen, J. (1989c, June). *Human error and the problem of causality in analysis of accidents.* Invited paper for Royal Society meeting on Human Factors in High Risk

Situations, London, England.

Rasmussen, J., Duncan, K., & Leplat, J. (Eds.). (1987). *New technology and human error*. New York: Wiley.

Rasmussen, J., Pejtersen, A. M., & Goodstein, L. P. (1994). *Cognitive systems engineering*. New York: Wiley.

Reason, J. (1987). Cognitive aids in process environments: Prostheses or tools? *International Journal of Man-Machine Studies, 27,* 463–470.

Reason, J. (1990). *Human error*. New York: Cambridge University Press.

Reason, J. (1992). *Human error*. New York: Cambridge University Press.

Sheridan, T. B. (1980, October). Computer control and human alienation. *Technology Review*, pp. 61–73.

Sloane, S. B. (1991, Spring). The use of artificial intelligence by the United States Navy: Case study of a failure. *AI Magazine*, p. 89.

Thordsen, M. L., & Klein, G. A. (1989, November). *Cognitive processes of the team mind*. Paper presented at the IEEE Conference on Systems, Man and Cybernetics, Cambridge, Massachusetts.

Tikhomirov, O. K. (1988). The psychology of thinking [Psikhologiia myshleniia]. (N. Belskaya, Trans.). Moscow: Progress Publishers. (Original work published in 1971)

Wells, J. E., & Ryan, T. G. (1991). Integrating human factors expertise into the PRA process. In G. Apostolakis (Ed.), *Probabilistic safety assessment and management*. (pp. 577–582). New York: Elsevier.

Wiener, E. (1994, May). *Integrating practices and procedures into organizational policies and philosophies*. Invited presentation at the International Civil Aviation Organization (ICAO) Flight Safety and Human Factors Seminar, Amsterdam.

Woods, D. D. (1987). Commentary: Cognitive engineering in complex and dynamic world. *International Journal of Man-Machine Studies, 27,* 571–585.

Wu, J. S., Apostolakis, G., & Okrent, D. (1991). On the inclusion of organization and management factors into probabilistic safety assessments of nuclear power plants. In G. Apostolakis (Ed.), *Probabilistic safety assessment and management* (pp. 619–624). New York: Elsevier.

IV Future Trends

21 Speculations on Future Relations Between Humans and Automation

Thomas B. Sheridan
Massachusetts Institute of Technology

INTRODUCTION

Some predictions about automation seem conservative enough: faster, cheaper, generally more powerful computer chips—and therefore more computer chips and more computers in a greater variety of forms—in the office, factory, hospital, school, home, and automobile. Huge investment in optical and satellite communication. Human interface technology such as speech recognition and virtual reality getting affordable. More information available in larger, faster memories to more people over the much-trumpeted information highways. More crowded concrete highways and airports and cities, along with continuing insistence on both speed and safety, and therefore continuing efforts to monitor and tightly control all forms of transport. Government incentives to develop and apply high technology, new applications, new markets in developing countries. Therefore, more automation. How can anyone deny it?

However, at times of rapid technological development the promises outpace the reality. Thirty years ago early proponents of artificial intelligence announced capabilities "just around the corner" that still have not been seen. Machine translation of language is an example; there is progress, but it is slow and painful. Robots have made inroads into some sectors of manufacturing such as welding and paint spraying, but robotic assembly has turned out to be much more difficult than expected at first, and many companies that went into robots are now out. (There probably are fewer robot companies in the United States today than there were a decade ago; many have gone to Japan.) President Reagan's Strategic Defense Initiative was political rhetoric fueled by irresponsible scientists and engineers, and has proved to be a giant misuse of public funds. The proponents of intelligent vehicle highway systems have made claims to be able reduce highway congestion by a significant percent in a decade, although there is little basis for such a claim. Perhaps most tragic of all, automation and related high technologies have been used as a basis to assert that in a short while the blind will see, the crippled will walk, and there is plenty of room and resources for an ever-greater world population. Although I am not an advocate of encouraging more product liability litigation, one might muse that it is too bad that advertisers and politicians (and indeed scientists and engineers who write speculative book chapters!) are not held legally liable for their promises.

AUTOMATION AND HUMAN SUPERVISORY CONTROL

The tendency of the public is still to think of automation as a replacement for people. For the simplest and therefore most demeaning tasks, that is certainly true. Tasks that were done manually in primitive times, such as pumping of water, sawing of wood, grinding of grain, and transporting of materials in workplaces are now automated, at least in the developed countries of the world. However, there are relatively few tasks that can be said to be fully automated, for example, without some form of human monitoring and supervision. Human supervisory control, then, is a natural complement to automation, although this fundamental fact is not well appreciated and understood. Even for high-visibility tasks, such as those of an astronaut in space, the public and even many public officials see automation as necessarily replacing humans, when in reality what it is doing is changing the role of the human operator from direct manual controller to that of supervisory controller.

Specifically, supervisory control means that human operators plan and teach tasks to computers, and the computers then implement the tasks automatically through their own sensory inputs, memory, and control actions. During automatic action the humans monitor high-level displays of information, which present an integrated picture of what is happening in the system, and specify new goals and procedures to be implemented through the computer interface. This new human–machine relation is more and more coming to characterize modern factories (manufacturing, chemical, and other process plants), offices, transportation systems (including aircraft, trains, ships, and automobiles), hospitals, buildings (e.g., elevators), and homes (appliances).

Fig. 21.1 diagrams the basic human–machine relationships in supervisory control. The first two blocks sometimes operate without the third in the

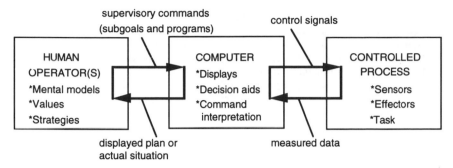

FIG. 21.1. Supervisory Control.

mode of planning and teaching. The last two blocks sometimes operate without the first for short periods as an automaton, with humans cast in the role of monitor (which they are not necessarily good at). If necessary the humans intervene in the automatic operation to reprogram the computer or take over operation themselves. Over time, the human learns, formulates new goals and procedures, and the whole process recycles.

Although supervisory control takes a variety of forms, and no single paradigm is accepted by the science/engineering community, it is clear that progress on the following component interface problems are and will be crucial to making it work well:

- Techniques for allocating functions between human and machine.
- Computer graphics.
- Decision aids.
- Command and control languages and techniques.

In regard to the allocation problem it is instructive to look back at the Fitts (1951) list of the superior attributes of (hu)man and of machine, which I like to call his "MABA-MABA" (Table 21.1).

Sheridan and Verplank (1978) proposed a somewhat different kind of list, a scale of degrees of computer aiding (Table 21.2), in which supervisory control spans all but the first and last step.

Computer graphics is a critical technology in supervisory control. It enables integrated visual presentation of information in symbols, pictures, diagrams, animation, and virtual environment representation. In the future it will continue to become higher resolution, and faster in image rendering.

Command and control capability, including command languages, speech recognition, and ability to manipulate images through cursors, keyboards, or haptic interface devices (master teleoperators) are also critical. Command and control can only be as good as the quality of these supporting capabilities.

TABLE 21.1
The Fitts List

Men (humans) are better at:
- Detecting small amounts of visual, auditory, or chemical energy
- Perceiving patterns of light or sound
- Improvising and using flexible procedures
- Storing information for long periods of time, and recalling appropriate parts
- Reasoning inductively
- Exercising judgment

Machines are better at:
- Responding quickly to control signals
- Applying great force smoothly and precisely
- Storing information briefly, erasing it completely
- Reasoning deductively
- Doing many complex operations at once

TABLE 21.2
Scale of Degrees of Automation

1. The computer offers no assistance; the human must do it all.
2. The computer offers a complete set of action alternatives, and
3. Narrows the selection down to a few, or
4. Suggests one, and
5. Executes that suggestion if the human approves, or
6. Allows the human a restricted time to veto before automatic execution, or
7. Executes automatically, then necessarily informs the human, or
8. Informs him or her after execution only if he or she asks, or
9. Informs him or her after execution if it, the computer, decides to.
10. The computer decides everything and acts autonomously, ignoring the human.

DECISION AIDING

Future supervisory control will be supported in large measure by decision aiding. A decision aid is a computer-based system that gives advice to the human operator either in an unsolicited form (if it, the computer, feels there is something that warrants the operator's attention) of if the operator enters a request into the computer. Such systems are also called *expert systems, decision support systems*, and other names in other contexts.

Decision aids up to now have been based on the conjunction of rather large sets of production rules, or if–then statements. Although the logic for expert systems is relatively straightforward, the elicitation of these rules from subject-matter experts has been difficult. Experts generally have trouble stating what rules they use to make decisions, and when they verbalize their rules at all they tend not to verbalize in terms of crisp rules (in which the variables have precise numerical meanings) but rather in fuzzy rules that make use of fuzzy linguistic variables such as *large, small, fast*, or *slow* (e.g., if the pressure is large or increasing fast, then do _____).

Fuzzy logic treats variables in terms of relative truth, called *fuzzy set membership*. Although membership functions appear like probability density functions, they are not, because membership (of some physical state like a pressure of 3457 PSI in the fuzzy set labeled "high pressure") is always true. This is not like probability, where once an event occurs it is precisely some number (pressure or whatever). The conjunction of a set of fuzzy rules applied to a particular set of crisp (measured) variables yields a set of alternative actions, each of which carries a membership or relative truth. In some cases these actions can be averaged, in other cases the action with the greatest membership is selected. In still other cases the situation indicates that the evidence is not sufficient to warrant any decision. An extensive theory of fuzzy logic to deal with such databases is emerging (Kosco, 1991), and appears much more compatible to action based on real human expertise than classical control.

Decision aids are being integrated with command language interfaces. For example, in commercial aviation a new computer-based system, called the *flight management system* (FMS), includes many modes of autopilot control. It also can, on request, advise the pilot of best landing procedures in consideration of both fuel remaining and restriction on runway availability, noise, and so on. It can help the pilot perform failure diagnoses on various systems within the aircraft, help navigate around weather or advise of other air traffic in the area, and perform other functions (Sheridan, 1992).

One problem with decision aids is their validation. To fully validate a decision aid one must have some ideal or normative basis for comparison. But it can be said that if such a norm were available it could be applied to an automatic system and the human decision maker would not be necessary. The fact that a human decision maker is necessary suggests that no such ideal norm is available (Roseborough, 1988), and full validation is impossible.

A second problem with decision aids is that it is impossible to separate the human decision from the output of the decision aid, because it cannot be known the degree to which the human is merely using the decision aid to check his or her own independently made decision, or is slavishly following what the decision aid recommends.

SAFETY OF AUTOMATION

An anxious public demands safety of automation, and generally it is getting safety, but the situation is not all rosy, and there are complications. For example, let us consider automation in commercial aviation. As automation is steadily being introduced, air transportation is gradually getting safer. However, the aircraft that have been automated most aggressively, namely the Airbus A310 and 320 (with the intention that automation makes the aircraft easier to operate for pilots from less developed nations, who may not have such good pilot training), have the worst safety records, and these accidents are mostly attributed to human errors. The pilot community is raising serious questions about how far to go with automation.

A prevalent type of human error in the new highly automated systems is called a *mode error*. It means the pilot has lost track of which of many modes he or she has set the FMS in, and the information that is coming back, often in numerical form, means something different from what the pilot thinks it means. Naturally, pilot responses to incorrect interpretations of data lead to further error and bad consequences.

There are continuous efforts to simplify the cockpit displays and controls through graphic computer displays (the so-called "glass cockpit"), but there is the simultaneous and countervaling pressure to add more automation, more control modes, and more advisory systems. Because of this, although the mental workload of the pilot (the new flight manager) has decreased for the enroute portion of long-haul flights, it may even be increasing for making landings at crowded airports. A popular concern is the effect on situation awareness. The pilots complain that all the new automation may be "killing us with kindness." Such complaints can also be found in nuclear power and other highly automated industrial plants.

A final safety aspect of automation that must be mentioned is software reliability. The very idea of software reliability is novel in many industries. Consider the following premises: (a) At any one time a digital computer program either works or it does not (naturally, if it receives bad data from outside the result is bad; after all, garbage in, garbage out); (b) by its nature, if a computer program can be demonstrated to work once, then it is trustworthy. Although the former premise is acceptable most of the time, the latter is decidedly not acceptable. Computer software can demonstrate very different reliability as a function of the data it processes; and this is not simply a question of inferior input data or hardware.

HOW TO REGARD HUMAN "ERROR"

Is a human error the inverse or negative of performance? Behavioral scientists and systems engineers define performance in ever more complex, continuous, multi-attribute metrics (vectors). Curiously, when such professionals deal with "human error," in effect they simplify the metric of performance to one bit. In any specific context the criterion of a discrete human error is usually fuzzy and arbitrary, and the possibilities for self-correction of error in time that no error is manifest are plentiful. If one is asked, "How many errors have you made today?" the answer is not easily forthcoming, and that question posed to an assembled group will yield blank stares and giggles.

The probabilistic risk analysts would argue, justifiably, that they are analyzing human reliability within operating procedures that are binary, with well-defined criteria of success and failure at each step. Combinatorial analysis of failing to perform given steps according to the given criteria (errors of omission) may be all they claim for the current art of human reliability analysis. Errors of commission tend to be nastier, because they are not contained within the procedure, and they can be of infinite variety. Such limitations raise questions about how seriously the results of such analyses should be regarded.

A common premise of human factors is that human error is undesirable. Pope said "To err is human," but that does not seem to make error acceptable. However, to a control engineer, without error (a deviation of system state from its nominal desired value) there can be no identification of the controlled process, and no discovery of system parameter changes. To an inventor, without error (a conception of events or relations that differs from the existing norm) there can be no invention. To a learning theorist, without error (a trial that leads to performance better than what was thought to be the best way to do things) there can be no learning. (The "bumps" in learning curves are now thought to be exactly this new-strategy-discovery taking place; see Venda, 1986.) To the Darwinian evolutionist, without error there can be no requisite variety and hence no evolutionary improvement of the species. Somehow, what may appear at the time to be deviant operator behavior often turns out to produce ultimate success and stability for the overall system (Sheridan, 1983).

Perhaps we ought to regard as a higher risk that human operator who is, within a short observation period, error free, and who behaves zombielike without the tendency to test and try and indeed err. Where to draw the trade-off between ultimately constructive and momentarily destructive error?

COGNITIVE SCIENCE AND COGNITIVE ENGINEERING

What operators are thinking is obviously related to how they perform. Not so many years ago in psychology it was scientifically unacceptable to consider thinking or cognition. The widely accepted notions of operational physics demanded operational (open and repeatable) measurement, and there was no room to posit intervening variables (hypothetical intermediates in otherwise observable processes). Computers and computer models changed that, because computers became a metaphorical way to consider human behavior (e.g., artificial intelligence, neural networks, fuzzy logic).

The notion of cognitive model is now in high fashion, although unfortunately the term is used to refer to two different things. It sometimes is used to refer to the human operator's input–output causality model of what he or she observes in the physical surround (e.g., if–then rules). It is also used to refer to the researcher's model of how a person perceives, remembers, associates, values, and decides. The latter meaning, in other words, is a substitute for a theory of thinking, what parameters affect mental processes, and so on. Thus, where $y = F(x)$ may characterize the former, F relating the variables x and y as observed by the subject, the latter is concerned with how F is constituted by past experience, current context, reward and punishment, and so on, as observed by the experimenter.

Some in the technical community assume that cognitive scientists (the term *cognitive engineer* is sometimes used) know exactly what they mean when they refer to cognition, and indeed that well-defined cognitive models are available. Although useful cognitive models of the first type may take the form of simple diagrams (of how a light switch is wired to a power source and a door bell) or symbolic rules (distance covered by an oncoming car is proportional to the velocity and the elapsed time), discovery of mental models of even slightly more complex phenomena is elusive. The engineer may actually think according to Force = Mass × Acceleration, Displacement = Stiffness × Force, and so on, but most people have only vague notions or at best fuzzy rules of such phenomena. As we move from mental representation of simple and definable physical phenomena to those that are far more complex, there is an even greater danger of miscommunication and self-deception in using a term like *cognitive model* without rigorous specification of what is meant, although the urge is there to go ahead and use the jargon anyway. This is what I like to call the emperor's new clothes phenomenon, referring to Hans Christian Andersen's famous tale.

As a complement to the human operator's mental model of the system, it is sometimes proposed that the system incorporate a model of the human operator. This essentially accords with the second meaning of cognitive model discussed previously. Supposedly, such a model would be used online to detect when the operator lost situation awareness or became inattentive. This idea has yet to be implemented to any practical extent.

Rasmussen's (1983) taxonomy of skill, rule, and knowledge levels of human behavior in effect distinguishes three levels of cognitive model (where skill-based behavior means subconscious and reflexive behavior analogous to a hardware servo system, rule-based behavior means stored procedural steps triggered by some perception analogous to a software do-loop, and knowledge-based behavior refers to higher goals and requires overt decisions among novel alternatives). In my own laboratory studies of airplane pilots suggest strongly that as pilots gain experience they progressively transform behavior from the knowledge- to the rule- and eventually to the skill-based level. In other words, they do not learn to decide, they learn not to decide. They learn to "internally automate" to avoid having to make overt decisions.

Modern neurobiology is clearly offering exciting new evidence about how the brain works, and eventually may provide a more solid basis for cognitive science. Indeed, there are some who claim they already see a physiological basis for consciousness. For example, Nobel laureate Gerald Edelman (1992) asserted that Darwinian evolution of successful arrangements of synapses occurs at a very much more rapid pace than evolution of whole organisms. He posited that this leads to reorganization into neuronal chunks (by a form of neural network learning that does not include back propagation, because there is no evidence for that in the nervous system) and that this ultimately forms the basis of all cognition. Although the arguments about neuronal theories rage on, some closely related ideas are emerging that bear directly on the present topic, offer a means to define the complexity, and guide the design of the human interface with the automatic machine. For example, Yufik and Sheridan (1991) suggested that, given a procedure or task for operating a machine, through use of network analysis and cut-set criteria, an ideal chunking (or, in effect, layout design) of the interface may be evolved.

HUMAN-CENTERED AUTOMATION

This discourse ends on the topic human-centered automation, a most popular phrase, but one that suffers from the same emperor's new clothes phenomenon as discussed previously. Do we really know what we mean by it? What should it mean?

To have automation be human centered makes us feel good. Surely ancient man felt good about being at the center of the physical universe, and Copernicus' news was not welcome. But then, for three decades afterward, at least people had the satisfaction that life on earth revolved around them—until Darwin rudely interrupted that pleasant sense of discontinuity from other life forms. Then, for at least a few decades, our behavior was happily thought to be centered around our conscious intentions—until Freud taught us that we are not so much in control of even ourselves as we had thought. Now automation presents a potential fourth discontinuity from the comfort of human centeredness (Mazlish, 1967). How are we to react?

How would one treat another person, who each day is observed to be growing more powerful? I should think it would be wise to study that person's behavior, to learn his or her language and try to engage in communication, to try to understand that person's intentions, and resolve them with one's own intentions, and, as best one could, to model and thereby predict that person's behavior contingent on one's own actions.

This is a poor analogy, some are inclined to protest, because the automatic machine is benign and has no intentions, other than those its human maker intentionally gives it. This view, unfortunately, is naive, because the full result of one's actions is usually different in many ways from what one intends — particularly if one is not very careful about what one asks for.

This critical problem with automation, then, is misplaced trust. It was particularly well articulated by Wiener (1964), who used as a metaphor a classic in horror literature, W. W. Jacobs' "The Monkey's Paw":

> In this story, an English working family sits down to dinner in its kitchen. Afterwards the son leaves to work at a factory, and the old parents listen to the tales of their guest, a sergeant major in the Indian army. He tells of Indian magic and shows them a dried monkey's paw which, he says, is a talisman that has been endowed by an Indian holy man with the virtue of giving three wishes to each of three successive owners. This, he says, was to prove the folly of defying fate.
>
> He claims he does not know the first two wishes of the first owner, but only that the last was for death. He himself was the second owner, but his experiences were too terrible to relate. He is about to cast the paw on the coal fire when his host retrieves it, and despite all the sergeant-major can do, wishes for £200.
>
> Shortly thereafter there is a knock at the door. A very solemn gentleman is there from the company that has employed his son and, as gently as he can, breaks the news that the son has been killed in an accident at the factory. Without recognizing any responsibility in the matter, the company offers its sympathy and £200 as a solatium.
>
> The theme here is the danger of trusting the magic of the computer when its operation is singularly literal. If you ask for £200 and do not express the condition that you do not wish it at the cost of the life of your son, £200 you will get whether your son lives or dies. (pp. 57–61)

It is so tempting to trust to the magic of computers and automation. It was mentioned earlier that if a computer program compiles, we often believe, the software is valid and the intention will be achieved. We are coming to realize that, especially as automatic systems become more complex, this is far from true, and the results can be disastrous.

The proponents of a new technology are usually the inventors, the developers, and the investors. They foresee and forcefully articulate the potential advantages. The opponents are usually those with vested interest in the status quo; they simply resist change. Usually neither side can foresee the real disadvantages, because most likely the disadvantages may manifest themselves in subtle ways, and often not until well after the introduction of the technology, sometimes many years. Pollution effects of burning fossil fuels in automobiles and power plants was not evident for decades, although they now are. In contrast, the seemingly obvious effects of automation on unemployment, a major concern since the Luddites in 19th-century England smashed their knitting machines, haven't seemed to take effect yet — although the recent widespread downsizing of industrial plants seems finally to be based squarely on cumulative automation technology.

More subtle than unemployment is the threat that sophisticated automation technology will alienate workers — where alienation means losing their identity as skilled manual craftspeople to become button pushers, understanding less and less of how the technology works as it becomes more sophisticated and requires white-coated high priests to program it or diagnose troubles and fix it. Alienation also means becoming suspicious or superstitious, or worse, blissfully trusting the technology and abandoning responsibility for one's own actions.

Should we strive for human centeredness in terms of some optimally efficient human–machine system? (*Optimal*, then, must be defined in terms of some explicit mathematical criterion? Indeed, if it is not, we have no hope of making any sense of the notion of optimization, and for the moment that seems to be the case.) Or does appropriate human centeredness mean that automation is there to augment the human's skill and subjective sense of productivity and quality of life, without regard for system efficiency in economic terms? Although these two perspectives have characterized very different approaches to setting up manufacturing organizations, one must ask if ultimately they are not the same. Ultimately the criterion, whether one strives for objective optimization or merely some satisfying degree of social acceptance, rests with humans. The ultimate human factors problem, then, is to elicit the multicriteria objectives from the affected human parties and resolve the differences among them to determine a "most acceptable" social choice. Only then will we know how far to go with automation.

REFERENCES

Edelman, G. M. (1992). *Bright air, brilliant fire.* New York: Basic Books.

Fitts, P. M. (1951). *Human engineering for an effective air navigation and traffic control system*. Ohio State University Research Foundation report, Columbus.

Kosco, B. (1991). *Neural networks and fuzzy systems*. Englewood Cliffs, NJ: Prentice-Hall.

Mazlish, B. (1967). The fourth discontinuity. *Technology and Culture, 8*(1).

Rasmussen, J. (1983). Skills, rules and knowledge: Signals, signs and symbols, and other distinctions in human performance models. *IEEE Trans. on Systems, Man and Cybernetics* (SMC-133), 257–267.

Roseborough, J. (1988). *Aiding human operators with state estimates*. Unpublished doctoral dissertation, Massuchussetts Institute of Technology, Cambridge.

Sheridan, T. B. (1983) Measuring, modeling and augmenting reliability of man–machine systems, *Automatica, 19*.

Sheridan, T. B. (1992). *Telerobotics, automation and human supervisory control*. Cambridge, MA: MIT Press.

Sheridan, T. B., & Verplank, W. L. (1978). *Human and computer control of undersea teleoperators*. Cambridge, MA: MIT Man–Machine Systems Laboratory Report.

Venda, V. F. (1986). On transformational learning theory. *Behavioral Science 31*(1), 1–11.

Wiener, N. (1964). *God and Golem, Inc*. Cambridge, MA: MIT Press.

Yufik, Y. M., & Sheridan, T. B. (1991). *A technique to assess the cognitive complexity of man–machine interface* (NASA Contract Report NAS2-13283). Potomac, MD: Institute of Medical Cybernetics.

22 Teleology for Technology

P. A. Hancock
University of Minnesota

> **Teleology:** 1. The doctrine of final causes or purposes; 2. the study of the evidence of design or purpose in nature; 3. such design or purpose; 4. the belief that purpose and design are a part of, or are apparent in nature.
>
> **Technology:** 1. The branch of knowledge that deals with the industrial arts: the sciences of the industrial arts. 2. the terminology of an art, science, etc.; technical nomenclature.

INTRODUCTION

Science and technology have always tried to answer the question "how?" How does this or that mechanism work? What are the laws and causal properties that underlie this phenomenon and, in the case of technologies, how can such knowledge be used to develop a useful tool or machine? However, science and technology rarely, if ever, address the question "Why?" It seems to be outside their respective universe of discourse. The question is ruled inadmissable or inappropriate for the methods and capabilities at hand. I reject this division absolutely. I submit that questions of how and why are so mutually dependent that they should never be considered independently, and I attribute much of our present circumstance to this unfortunate and unhappy division. Those who know how must always ask why. Those who ask why must always think how.

OVERVIEW OF THE CHAPTER

With reference to the previous thematic statement, I examine our collective future by asking questions about intention with respect to technology. I set this argument against a background of current human-machine systems, and particularly automated systems. I do this because of my belief that technology cannot be considered in the absence of human intention. Likewise, contemporary societal aims have no meaning without reference to pervasive technology. I start with a prelude that presents a metaphor for initial consideration. I then define the terms within which the argument is framed. This leads to an examination of what technology is and to what extent technology is "natural." I then examine human–machine symbiosis and the potential future that may be encountered by such a co-evolutionary pairing. I point to human-centered automation as a stage in this sequence of evolution that witnesses the birth of autonomous machine intention. I evoke a number of cautions that attend this birth. In noting the stage-setting and constraint-setting functions of contemporary systems design, I cite earlier warnings concerning the failure of previous principles of human-machine interaction. I conclude that the collective potential future for humans and machines can only be assured by the explicit enactment of mutually beneficial goals. In the near term, I caution against the possiblity of a society divided by technology against itself. I advocate for human factors as a liberating force in providing technical emancipation, the heart of which is universal education.

A METAPHORICAL PRELUDE

It is in this way alone that one comes to grips with a great mystery, that life and time bear some curious relationship to each other that is not shared by inanimate things. (Eisley, 1960)

Setting the "Scene"

To start this chapter, I want to put a vision in your head. To do this, I am going to use a familiar icon. This icon is not a single picture but, rather, a scene from a famous motion picture. The picture is Alfred Hitchcock's *North by North-West,* and the scene is the coast road. I hope this brief description will let most of you identify the sequence I mean. However, for those who are not familiar with the movie, it is as follows. Cary Grant, the hero, has been forcibly intoxicated by the henchmen of the evil James Mason on the mistaken assumption that he is an investigating government agent. To rid themselves of this menace, they put the now-drunk Grant in a car and start him off on a perilous trip down a steep and winding coast road. Through sheer force of will and with some luck in the more dangerous maneuvers, our hero manages to survive both human and environmental sources of danger, to fight again. Even those who have not seen the film will not be surprised that, in the end, the forces of evil are routed and the hero wins the heroine.

A Metaphoric Relation

I want to use Grant's vehicular progress as a metaphor for our uses of technology. I want to suggest that we, like him, are careening down a dangerous path. None of us intentionally put ourselves in this predicament but, nevertheless, here we are. We posses similar goals in which simple survival is the highest priority. Both we and Grant are painfully aware that the volatile combination of powerful technologies and fallible humans, in unstable environmental circumstances, threaten disaster. Although our progress resembles Grant's in many respects, we are radically different in some fundamental ways. Above all things, we have no director to shout "cut" or script writer to ensure that the story continues. Unlike Grant, we also seem to be drinking heavily as we proceed down the mountainside in, apparently, a progressively more chaotic fashion. We have a science whose primary purpose seems to be to ensure that we are able to keep a firm grip on the bottle. The science is *human factors*. We have a motive force in technology that seeks to accelerate the rate of our "progress." The force is *automation*.[1]

Meta-Technical Purpose

Using this metaphor, I am suggesting that contemporary human factors largely fails to address the fundamental purposes of and for technology. We do not question whether technology represents the appropriate avenue through which human society can achieve its necessary goals. Human factors accepts technology (as whole cloth) and consequently adopts its intrinsic and extrinsic assumptions in toto. In so doing, human factors seeks to facilitate human–machine interaction, even if the purpose of the machine is suicide, genocide, or terracide. By this, I do not mean that many individual members of the discipline do not question such functions, because they do (e.g., Moray, 1993, 1994; Nickerson, 1992; see also Hancock, 1993). But, as a science, we have yet to state that it is our role to question the very purposes of technology. I affirm that it is. Those outside the scientific and technical community frequently protest about the fallout

---

[1]Since writing this I have become aware of Zola's novel *The Beast in Man* in which a similar picture is drawn. Briefly, two railroad engineers are fighting in the cab of a locomotive going downhill. In the course of their fight, they fall out of the train and the passengers (soldiers) continue to carouse, unaware that no one is driving the train. On this the novel ends. It has been used by Russell (1992) in a similar way to which I have used the *North by Northwest* metaphor here.

of technology. However, it is we who shape human interaction with technology that can and should direct from within.

More Than an Appliance Science

Therefore, in what follows, I want to question the fundamental tenets of human interaction with technology. I want to question whether ergonomics or human factors should always facilitate the human–machine linkage, especially when such improvements are clearly antagonistic to the collective good. I want to question whether human factors should always foster and assume that technological growth, and increased automation in particular, is appropriate. In sum, I want to question where we are going with technology and what our goals and purposes are. It is my conviction that human factors is more than just an appliance science. I believe that human factors can exert a strong influence in what we do with technology and is beginning to be recognized in that role (Sedgwick, 1993). I believe it can be a critical voice in determining the goals we collectively set for ourselves. It is our mandate to mediate among humans, technology, and the environment. This mediation requires informed value judgments. I think we must acknowledge and grasp this social responsibility. The future is too important to be left to the accidental happenstance of time, or the commerical search for something smaller, faster, and less expensive. In essence, the carefree adolescence of humankind is ending and the new millenia must hail an age of responsible societal adulthood.

The foregoing is largely a series of polemical exhortations. The reader is entitled to enquire not only about the basis for such assertions but the reasons why they should be put forward as important at this time. I think the answer is simple. It is in the final decade of the 20th century that we are starting to set the agenda for automation and human-centered approaches to such automated systems. In truth, this represents the earliest growth of machine autonomy and independence. Decisions now constrain what will be possible in the future. I want that future to contain at least sufficient alternatives such that one is continued human survival.

Summary

There is a paradox in technology. Its avowed purpose is to improve the lot of humankind. Its actual effects appear to be the slow destruction of the planet on which we live. I have illustrated the nature of this paradox with a metaphor. Human society is engaged in a perilous endeavor with respect to technical systems. Human factors currently takes technology as given and seeks to facilitate interaction, assuming technology to be a good thing.

But is this so? To become more than an appliance science, human factors should focus directly on the issue of why technology is used — its purpose — in addition to how technology is used — its practice.

A DEFINITION OF TERMS: TELEOLOGY AND TECHNOLOGY

Machines just don't understand humans. They (automobiles) don't understand that they should start even when we do dumb things like leave the lights on or fail to put antifreeze in the radiator. But maybe they understand us too well. Maybe they know we rely on them instead of ourselves and, chuckling evilly in the depths of their carburetors, they use that knowledge against us.

We humans forget how to rely on our own skills. We forget we can survive without machines; we can walk to the grocery store even if it's a mile away. We forget that before the calculator, we used our brains. (Well at least some of us did.) We forget that the Universe doesn't grind to a halt when machines break down. Machines like that. They're waiting for us to get soft and defenseless so they can take over the world. (Tolkkinen, 1994)

In what follows, I hope to show some of the fallacies and misunderstandings that underlie this recidivist position.

Teleology: A Traditional View

I want to define the terms that are central to the arguments that I make. I start with the term *teleology*. *Teleology* is a term that is predominantly used with respect to the existence of a deity or some form of final cause or purpose for being. This concept is founded on human observations of order in the universe. The teleological argument postulates that because the universe shows this order, there must be some entity behind that order, creating and directing it. This assertion is rebutted by a multiple-universe argument, which proposes that order is present only because we, as human beings, are here to observe it. That is, many other universes are possible but conditions within them do not give rise to observers who observe them. The latter point is a cornerstone of the anthropic principle that seeks to shed light on the early conditions of our universe, founded on the fact that observers such as ourselves do exist.

TELEOLOGY: RETROSPECTION VERSUS PROSPECTION

I do not want to use the term *teleology* in this retrospective way to look back to the origins of order, nor to speculate about the existence of any deity. For me, teleology does not relate to an historic and passive search for ultimate cause. Rather, it is an active term concerning our prospective search among the potentialities with which we, as human beings, are presented. As a result, I use *teleology* in the present context to refer to an active search for unified purpose in human intention. By using it in this fashion, I am affirming that individuals and society do exhibit intention. That is, regardless of metaphysical arguments as to the reality or illusion of free will, we individually and collectively act in an everyday manner that signifies our belief that our actions are not predetermined and do influence ourselves and others. In sum, my use of *teleology* represents a search for goals and intentions for ourselves and society. In the context of the present chapter, I focus on the role that technology plays in this search.

Technology: Art, Artifact, or Science?

The formal definition of *technology* that is cited at the start of this chapter is different from that of everyday usage, which basically is how I want to use it here. In common parlance, technology is associated with things, objects, and systems, rather than with knowledge as either art or science (and see also Westrum, 1991). Although information underlies each of these definitions of the word *technology*, my use of it here accords with everyday parlance. However, I want to broaden this definition a little. Technology is often associated with what is new, modern, and exciting. I want to go beyond thinking of technology in terms of new machines to expand the definition to include such developments as the domestication of animals and of plants.[2] Technology then becomes the purposive ways in which the environment has been structured to benefit humankind. Already there is teleology in the very basis of technology. (I hope the reader can already see the mutuality and the temporal reciprocation such that the chapter may as easily be titled "Technology for Teleology"). A wide interpretation of this

---

[2]It might be of some interest to note that carrots were originally white. Their present appearance comes from selective breeding to engineer the vegetable we see today. Likewise, many animals owe their present appearance and subserve functions that were manufactured over a considerable period of time. I am grateful to Kara Latorella, who pointed out Gary Wilkes's work from the *Mesa Tribune* on an article entitled "Animal behavior put to task for good — and bad — of humans." In this he talked of the use of birds trained to respond on seeing the color yellow being used by air–sea rescue to find downed pilots, the pigeon having much greater visual acuity for yellow than a human observer (and thought to be less affected by the noise and vibration of the rescue helicopter. For a more technical evaluation of this work, see Parasuraman, 1986). Also, another use in which pigeons were trained was to spot defective parts on a production line as a way to relieve human observers of the vigilance burden. Interestingly, this latter project was stopped by protesters who said that reinforcing birds to perform this tedious task was "inhuman." No one protested for the human workers who were then rehired to do the same job.

definition permits the identification of technology that is used by other members of the animal kingdom. I am not unhappy about this inclusionary aspect.

Humans and Their Technology: A Mutual Shaping

What is clear is that technology has fashioned humans as much as humans have fashioned technology. Elsewhere, we (Flach, Hancock, Caird, & Vicente, 1995; Hancock, Flach, Caird, & Vicente, 1995) argued that our contemporary ecology is technology and, as such, it represents the most powerful influence on who we are today. Such is our reliance on technology that we could not survive in the way that we do without that technological support. As Koestler (1978) put it:

> It has to be realized that ever since the first cave dweller wrapped his shivering frame into the hide of a dead animal, man has been, for better or worse, creating for himself an artificial environment and an artificial mode of existence without which he no longer can survive. There is no turning back on housing, clothing, artificial heating, cooked food; nor on spectacles, hearing aids, forceps, artificial limbs, anaesthetics, antiseptics, prophylactics, vaccines, and so forth.

Although at some very basic level use of the environment applies to all living things, the extent to which humans rely on technology is, I think, sufficiently distinct to provide a valid differentiation. The central question becomes, what is the future for a species that relies so heavily on technological support and what happens to that species as the nature of the support itself co-evolves?

Summary

Teleology traditionally refers to ultimate cause or purpose. Here, it refers to human intention for the future. Technology, whether knowledge or artifact, is our major way of changing the world we live in. The emergent focus is on a future for ourselves based on an understanding about the nature and role of technology.

TECHNICAL VERSUS HUMAN CAPABILITIES

> The servant glides by imperceptible approaches into the master; and we have come to such a pass, that even now, man must suffer terribly on ceasing to benefit (from) machines. If all machines were to be annihilated at one moment, so that not a knife nor lever nor rag of clothing nor anything

whatsoever were left to man but his bare body alone that he was born with, and if all knowledge of mechanical laws were taken from him so that he could make no more machines and all machine-made food destroyed so that the race of man should be left as it were naked upon a desert island, we should become extinct in six weeks. A few miserable individuals might linger, but even these in a year or two would become worse than monkeys. Man's very soul is due to the machines; it is a machine-made thing; he thinks as he thinks, and feels as he feels, through the work that machines have wrought upon him, and their existence is quite as much as sine qua non for his, as his is for theirs. (Butler, 1872)

The Direction of Change

I begin the evaluation of our co-evolution through the use of a traditional, dichotomous comparison of human and machine abilities. This is expressed in a simple way in Fig. 22.1. I have already indicated that it is critical to find purpose in technology and to identify purpose at this time. Technology may have been around as long as human beings have cultivated crops, herded cattle, or created their own tools. What makes the present different is the change in the respective roles of the machine and its operator. A traditional view of this change is illustrated here, where the horizontal axis is time, and the vertical axis capability.

The curve labeled (b) represents human capabilities. It is perhaps easiest

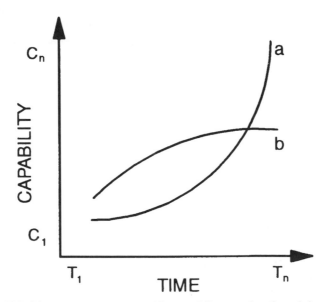

FIG. 22.1. Human versus machine capability as a function of time.

to conceptualize these abilities in terms of physical achievement. As can be seen, human capabilities improve over time, but the rate of that improvement diminishes. These trends can be seen most clearly in world records for forms of locomotion, such as running and swimming. What we know from experience is borne out in these data—human beings get better but by progressively smaller amounts. This holds for individual learning curves as well as collective performance as seen in world records. Eventually, we will reach constraints to physical capabilities. The question is, are intrinsic cognitive functions limited in the same way? We may improve our cognitive capability but, eventually, there is a limit on sensory resolution, memory capacity, decision-making speed, and, not least, motor skills. Defining these limits is of course much more difficult than for structured physical pursuits. (Parenthetically, this may be one reason for the attraction of sports in contemporary society, in which the winner, in most games, is identified unequivocally. Hence the outrage at cheating in major sports, which are seen, often naively, as the one arena of true competition and resolution.) However, unaided by technology, individuals do not posses unbound, endogenous cognitive abilities.

The curve labeled (a) represents technology. There are a myriad of examples of this exponential form of growth (see Card, 1989). For example, the history of the size, speed, and unit cost of computer memory shows this phenomenal technical improvement clearly. Moravec (1988) captured this progression in which he also showed how differing forms of technology have supported continued geometric growth. I should note that this approach, which was developed in conception over a decade ago, was also developed by Vallee (1982), with whose work I have only recently become familiar. Before I discuss the significance of this perspective and the point of intersection, I have to discuss some of the shortcomings and assumptions of this comparative approach.

Fallacies of Generalization

Of course, each curve is a generalization. We know that technology, like resources in whatever form, cannot sustain geometric growth indefinitely (Malthus, 1798; see also Nickerson, 1992). Indeed, for the function cited in computer memory we seem to be rapidly approaching some physical limits to the type of capacity for silicone based systems (Bennett & Landauer, 1985). Futurists are proposing radically different technologies to fracture what seems to be an intrinsic limit to calculational speed. So, although increase in technical capabilities can appear to be geometric, it is only so for some selected portion of their growth. It is much more likely that the curve of technology possesses an inflexion, is shaped more like an ogive, and is, through some intrinsic factor, ultimately self-limiting. With respect to human cognitive capabilities, the individuals who assert that we know all about such limits and what constrains human thought are themselves impoverished of mind and imagination. For every assertion about limits to human intellectual abilities and cognitive capacity, there is another assertion about individual variation and strategic skills that counters it. Thus, specifying an upper boundary to all cognitive functions is naive. However, the recognition of some endogenous limits is, in no small part, the stimulus for technology itself. I consider this integrative perspective later in the chapter. Yet, despite these and other simplifications, the intersection of the curves is of critical importance.

Convergence, Intersection, and Divergence

Although the physical abilities of machines superceded human physical abilities some generations ago, I want to suggest that with respect to cognitive capabilities, it is in our era in which the relationship between the human and the machine has begun to invert. Independent of the objections that I have raised in the previous section, it is within our own generation that certain machine characteristics have outstripped their human equivalents. Pragmatically, what this has meant is that many functions are now performed more quickly and more accurately by machine. Such is the nature of the systems we have, and are building, that many are not controllable without the use of computer assistance. I hasten to add that the supercession of individual human abilities by machine abilities has happened, ability by ability, at different times. As a result, different sorts of systems have had to rely on some degree of machine control at different junctures. This progress from nuclear power stations to fighter aircraft has percolated now to washing machines. It is now the challenge to identify machines that do not have some sort of processor in them.

It is this transference of superiority in specific areas that has seen the birth of imperative automation. By imperative automation, I mean automation we cannot do without. That is, the goal of the system cannot be achieved by manual action alone. We have always exhibited a fascination with automata and have incorporated governors and other forms of automatic or semi-automatic controllers into systems for almost as long as humans have been making machines. But this is discretionary automation. Humans could control the systems and, in many cases, were found to be the cheapest and indeed the preferred way for doing so. However, in our times, we have had to acknowledge that without machine assistance we cannot control some of the things we have built. Many work better if we have machine support, but some work only if we have machine support. Hence the difference is one of acknowledgment that automation is something we cannot do without, rather than something we would prefer to have.

I return to my original question: Why is it important to consider the ascending role of technology now? I conclude that it is important now because our relationship with technology has changed from discretionary use of automation to mandatory use of automation. With that change has come a more subtle move in which machines are no longer simply mindless slaves but have to be considered more in the sense of a partnership, although anthropocentric views of this partnership are likely to be misleading at best. Any time of change is disturbing. However, we live in an age when nonhuman and nonanimal systems have, by circumstance, been granted perhaps the birth of emancipation. We have to ask ourselves questions concerning if, and when, they will demand the right of franchise.

Summary

Human capabilities have progressively been superceded by machine capabilites. In our age, we now build systems that rely on these levels of machine ability that cannot be replicated by any human. Consequently, our relationship with machines has changed. They are no longer unquestioning slaves but are becoming active companions. Are they to become more than this?

IS TECHNOLOGY 'NATURAL'?

One can, of course, argue that the crisis (of technology), too, is "natural," because of man is part of nature. This echoes the views of the earlist Greek philosphers, who saw no difference between matter and consciousness — nature included everything. The British scientist James Lovelock wrote some years ago that "our species with its technology is simply an inevitable part of the natural scene," nothing more than mechanically advanced beavers. In this view, to say we "ended" nature, or even damaged nature, makes no sense, since we are nature, and nothing we can do is "unnatural." This view can be, and is, carried to even greater lengths; Lynn Margulis, for instance, ponders the question of whether robots can be said to be living creatures, since any "invention of human beings is ultimately based on a variety of processes including that of DNA replication from the invention." (McKibben, 1989)

The preceding argument is one in which human and machine are explicitly contrasted as disparate entities; that is, human versus machine abilities. As we progress, I want to argue that this division is unhelpful. To do so, I have to first overcome the assertion that technology and nature are in some way opposed. That is, that technology is not natural.

The Importance of the Question

It might seem, at first, that the question of whether technology is natural is either facile, in the sense that technology being artificial cannot be natural, or pointless, in the sense that the answer makes little difference one way or the other. I suggest that the question is neither facile nor pointless. It is not facile because it forces us to consider what are the boundaries of what we call natural and what artificial means in this context. It is not pointless because our answer biases the very way in which we think about technology and what the ultimate purposes of technology are. However, before I explore this specific case, I would like to found the argument on our propensity to group things together compared with our equally common attempts to differentiate between things.

How Things Are the Same, How Things Are Different

In the behavioral sciences we refer to the characteristic of individual differentiation as idiographic and the patterning of events or things as nomothetic. They form the basis of all statistical procedures, because they are reflections of dispersal and central tendency, respectively. However, the question of grouping or set function goes well beyond the behavioral arena. Indeed, it is a fundamental characteristic of all life and science that we look for similarities and differences in experience. Language represents one explicit recognition of differentiation or separating apart of ideas, objects, and things. Mathematics represents the propensity toward unification. We may start life by distinguishing self from nonself, but it is the richness of language that gives voice to the diversity of the world around us and, parenthetically, it is also language that strikes us dumb with respect to transcendent experiences. In essence, we try to name each of the things we can perceive. Before long, however, we start to try to categorize these things. That is, we seek common characteristics through which we can link individual items together. Thus, differentiation and integration go hand in hand in all perception-action processes.

In all facets of human life, we rejoice in discovering new ways in which things can be unified so that we can extract pattern from (or impose pattern on) experience. For example, we count the recognition of a common force acting on an apple and the moon at one and the same time as one of the great insights of science. Indeed, the concept of number in mathematics is an explicit statement that one object is sufficiently of the same characteristic as another object that they can be put in a common class and recognized as two separate occurences of the same object type—an observation that preceeds the concept of no objects in a set, that being zero. The abstraction of number proceeds from this explicit grouping principle. I argued elsewhere that time is the basis for both this fundamental unification and differentiation and, hence, stands, as Kant implied, as an a priori behavioral and consequently physical construct (Kant, 1787). However, before this journeys too heavily into the metaphysical, I would like to provide a biological example as a precursor to an examination of technology.

Are Humans the Same as Other Animals?

The example concerns the difference between humans and the rest of the animal kingdom. We are aware of Descartes' protestation about the soul as the difference between humans and animals (although I suspect Descartes was not the absolute differentialist he is now painted but was very aware of the traps of his times). It was indeed this barrier that Darwin, without malevolent intent, so thoroughly ruptured. Contemporary scientific debate does not revolve around the contention of common evolution, but one battleground is now established around language. Lurking in the background is the often silent extension into the question of mind and consciousness and the unsaid and now unsayable link to the soul. The arguments center putatively around the nature of the data, but the global agenda of the uniqueness of human creation always hovers in the background.

Why is this? The answer is, I think, understandable. As human beings we have always been, like Cary Grant in *North by Northwest,* the hero of our own story. But our history is a chronicle of our progressive displacement from the center of the universe. From Aristarchus to Copernicus, from Newton to Einstein, the gradual displacement from the physical center has progressed (at times stultified) but never ceased (Koestler, 1959; Nietzsche, 1887). This outfall of science also threatens to displace human beings from the spiritual center of our universe. It is only in the present century that the concatenation of physical relativity and biological unification has served to shatter some of the foundational pillars on which the conventional and comfortable world view was perched.

In 1859, Charles Darwin published the *Origin of Species.* Epic of science though it is, it was a great blow to man. Earlier, man had seen his world displaced from the center of space; he had seen the Empyrean heaven vanish to be replaced by a void filled only with the wandering dust of worlds; he had seen earthly time lengthen until man's duration within it was only a small whisper on the sidereal clock. Finally, now, he was taught that his trail ran backward until in some lost era it faded into the night-world of the beast. Because it is easier to look backward than to look forward, since the past is written in the rocks, this observation, too, was added to the whirlpool. (Eiseley, 1960)

Technology and Natural Laws

In the sense I have conveyed we now have to enquire whether technology and nature are different or whether they are in fact essentially the same. For good or bad, we have come to a situation where strong positive empiricism reigns and technology is the material manifestation of that creed. But is this natural? As I am sure the reader has suspected all along, it all depends on what is considered natural. That is, are we going to use the term in an inclusive or exclusive sense. The inclusive, coarse-grained view is that physical entities obey physical laws. Hence, everything is natural by this definition of nature. But this view is biased by a reificiation of physical laws. To the strict ecologist, to whom these laws and their application is sacrosanct, technology in general and human–machine systems in particular are only extensions of nature. True, they explore exotic regions that cannot be compassed by any living organism alone, but they are bound by the same strictures and constraints and are subject to the pervasive laws. But, in conception, they are founded in human imagination, which is not bound by any such laws. As Koestler (1972) noted, "The contents of conscious experience have no spatio-temporal dimensions; in this respect they re-semble the non-things of quantum physics which also defy definition in terms of space, time, and substance." This unbounding is what makes developments such as virtual reality so intriguing (Hancock, 1992).

Inevitable Technological Failures?

I have purposefully spent some time considering the general level of the question of natural technology to provide a background for the following and more specific example. Consider the following. We have, each of us, noted and commented on the increasing complexity of technical systems. Indeed, one of the major *raison d'etres* for automation is this progressive complexity. I shall not argue what I mean by *complexity* here, because I did that elsewhere by linking complexity to degrees of operational freedom (Hancock & Chignell, 1995; Hancock, Chignell, & Kerr, 1988). I simply assert that technical systems of today are more complex than those of a century ago. Such growing complexity compels comparison with natural ecosystems. Regardless of the link between technology and nature, we can look to nature for models of interacting complex systems with biological players as system components. In so doing, we find that there are intriguing models of systems with mutually adapting agents. In particular, we find that failure through ecosystems propagates readily and that a relationship between the size of failure and the frequency of failure has been posited (see Kauffman, 1993; Raup, 1986; see also Fig. 22.2).

Kauffman suggested a ln/ln relationship in which the log frequency of failure is linear with the log size of failure events. The crux of the argument

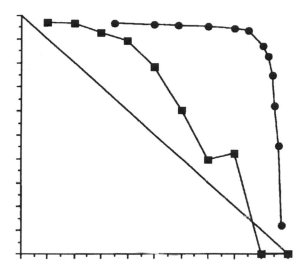

FIG. 22.2.   The illustration shows relationships in log/log space. The axes represent frequency of events and size of events. The linear relationship is a hypothetically perfect one, whereas the two others represent data. The circles show human cohort morbidity, and the squares show extinction events in the Phanerozoic era. (Conception after Kauffman, 1993; extinction events data after Raup, 1986; morbidity data imposed here.)

is that small perturbations are resident in complex systems and that for the vast majority of instances these perturbations are damped out in the system. However, on a periodic basis, described by the ln/ln function, these same perturbations are magnified through the system resulting in correspondingly larger destruction of the elements of the system. In this chapter I cannot explore all the nuances of these important observations, but would strongly recommend reference to Kauffman's (1993) original work. The importance for human factors, of course, is when the entities in the ecosystem are interacting humans and machines. The importance of the general question of the naturalness of technology is now laid bare. If technology is no more than a logical extension of other natural systems, we must expect human–machine systems to be subject to the same effects. The implication is that there will always be failures in such systems and the size and frequency of those failures will be proportional to the complexity of the system. As a result, there will always be intrinsic perturbations of these systems and some of them will always reach such a magnitude as to cause disruption or catastrophic failure as a function of their very nature.

Rejecting Deterministic Failure

It is almost instinctive in us to reject such an assertion. Indeed, there are several objections that immediately present themselves. Surely technology can be natural without having to fail in this manner? Doesn't such an assertion imply that the safety measures that are put in place are bound to be ineffective? As some reflection will show, it comes back to the nature of technology. If we believe that human actions supercede the haphazard exploration of potential systems' spaces by nature, then we are strong advocates for separation and the positive influence of human intention. If we view the combinatorial explosion of system complexity with trepidation and have difficulty distinguishing technology from nature, then we are liable to favor the more fatalistic view as represented in the latter conception. In essence, it means that metaphysical conundrums such as the mystery of mind, the reality of consciousness, and the power of free will are at the heart of how we conceive possible futures for human–machine systems.

Giving Solipsism the Slip

At this juncture, the morass of radical skepticism and fundamental solipsism may beckon the unwary traveler toward the slough of despond. However, the pragmatic solution is simple (see Kurtz, 1994). As I have noted earlier, we all act as though we exercise free will. We discuss consciousness but we are mostly willing to attribute it to others, and we use the concept of mind as something that is a shared characteristic of human experience. Therefore, at a practical level we will continue to believe that our interventions are important, whether they are or not. (Kurtz, 1992). Although the position may be advocated as pragamtism, the proposition certainly remains philosophically doubtful.

Therefore, we will continue to exert our best efforts to support the safety of systems on a pragmatic basis. Of course, as a compromise position, it may well be that safety, being a multidimensional construct, contains some elements that we can influence and others about which we can do nothing. What I advocate here is that, in viewing technology as natural, we have to begin to divorce ourselves from the debilitating concept of blame. In a legalistically redolent society such as the United States, this is very hard. I concede that although there is greed there will always be malfeasance, but it is not this form of blame I am seeking to excise. It is the blame directed toward conscientious, dedicated, professional operators who may make errors, for which some pay with their lives. To apportion blame in such circumstances is to adhere to a physical model in which Aristotelian efficient cause can be distilled for each action. As society begins to reject the concept of a personal deity, dealing fate on a momentary basis, it must also begin to reject this latter physical analog as equally impoverished of mind. If we can begin to conceive pluralistically (and through technologies such as mutual experience in virtual reality such a consciousness may be possible;

see the section entitled "The Supercritical Society" in Hancock & Chignell, 1995), then from this vantage point society only indemnifies itself. It would be as if the cells of an organ took out insurance against individual damage while the whole organ ceased to function. This analogy can be continued through the multiple levels of any holarchy (see Koestler, 1978). In consequence, in a technologically replete world, local kinship becomes a vestigial characteristic. As we mature from the self-centered world of adolesence our reach must be toward a social consciousness in which the myopia of individuality is replaced with the vision of mutuality. Although some secular and religious theorists have advocated similar views, they have, by and large, not been successful in influencing collective behavior. It may be that contemporary technological emancipation will prove the only effective catalyst for such change and growth.

Summary

I claim that technology is transparently natural. But I argue that it can be more. I suggest that we can learn much about complex technological systems by understanding the nature and nuances of complex biological ecosystems. I ask whether catastrophic failure must be an intrinsic characteristic of these linked complex systems, and use this as the acid test for the influence of intention. With respect to disaster it may be that humans see more through tears than they do through telescopes; but they have to be looking in the right temporal direction.

EVOLUTION OF HUMAN AND MACHINE

This is a principal means by which life evolves—exploiting imperfections in copying despite the cost. It is not how we would do it. It does not seem to be how a Deity intent on special creation would do it. The mutations have no plan, no direction behind them; their randomness seems chilling; progress, if any, is agonizingly slow. The process sacrifices all those beings who are now less fit to perform their life tasks because of the new mutation. . . . We want to urge evolution to get where it's going and stop the endless cruelties. But life doesn't know where it's going. It has no long-term plan. There's no end in mind. There's no mind to keep an end in mind. The process is the opposite of teleology. Life is profligate, blind, at this level unconcerned with notions of justice. It can afford to waste multitudes. (Sagan & Druyan, 1992, p. 84)

The past discussion has considered nature and technology almost as though they were static things. However, perhaps the most distinguishing feature of each is their continual change, and consequently the next step is an examination of their respective evolution.

Human Evolution

In considering co-evolution, it is important to consider first the disparities of the separated evolutionary processes. With respect to human evolution we accept a Darwinian-based concept that has been tempered with the continuing discoveries of genetics and molecular biology. In the general run of science, the central concept of the descent of organisms of increasing complexity still holds sway, where a survival-of-the-fittest mechanism has served as the selective process over the eons of geological time. However (it is appropriate to ask here), fit with respect to what? It is usual to emphasize the environment of the organism in this respect. However, this is typically a multi-layered condition. It may be that an organism is perfectly adapted to a specific environment and then an event, such as a volcanic eruption, over the other side of the world changes local conditions. Under these conditions the organism may become extinct. In this respect, survival of the fittest is survival of the survivalists, because excessively specialist adaptations do not pay in the wider order of things (Kauffman, 1993). The best form of adaptation is the ability to adapt, and meta-adaptation is the primary human characteristic as a substrate of a facile brain. We accept that human evolution has taken many millions of years. We also accept that human evolution proceeds at a slow rate such that differential characteristics propagate slowly through the population. Advantages in one context may be disadvantages in another context; hence, with meta-adaptation as the central charateristic it might be expected that human evolution should progress cautiously. In pragmatic terms, evolution proposes and nature disposes. However, what does nature dispose of when technology is the ecology? Under these circumstances, selection of the fittest seems a much less pristine principle.

The problem, of course, is time. It is perhaps the defining human characteristic that we are painfully aware of our own certain death. However, in our brief flight on this planet, we remain almost pathologically infirmed with respect to our temporal vision. Again, as the hero of our own story, it is so difficult for each one of us to conceive of a world without us! However, our appreciation of the passage of geological time remains always an intellectual, not an empathic, exercise. I would like the reader to pause and think on this for a moment.[3] Even with an inkling of the length of

---

[3]The geological year asks the individual to conceive of the time since the origin of the earth (mark this, not the origin of the universe) as a period of one year (a hopelessly embedded temporal structure). Human beings are then said to have been on the planet only during the last

duration, it is clear that our species is a newcomer to the world. In a true sense, each species has some hand in creating other species. All partake in the interplay of environment and resources that sets the frame of selection. The human creation of technology is distinguished only by the apparent intentionality of the act. It is this intentionality that, I think, provides the difference for contemporary minds. I therefore appeal to Darwin for armament to help analyze this view. With respect to intentional selection, Darwin wrote:

> One of the most remarkable features in our domesticated races is that we see in them adaptation, not indeed to the animal's or plant's own good, but to man's use or fancy. Some variations useful to him have probably arisen suddenly, or by one step. . . . But when we compare the dray-horse and the race-horse, the dromedary and camel, the various breeds of sheep fitted either for cultivated land or mountain pasture, with the wool of one breed good for one purpose, and that of another breed for another purpose; when we compare the many breeds of dogs, each good for man in different ways; when we compare the game-cock, so pertinacious in battle, with other breeds so little quarrelsome, with "everlasting layers" which never desire to sit, and with the bantam so small and elegant; when we compare the host of agricultural, culinary, orchard, and flower-garden races of plants, most useful to man at different seasons and of different purposes, or so beautiful in his eyes, we must, I think, look further than to mere variability. We cannot suppose that all breeds were suddenly produced as perfect and as useful as we now see them; indeed, in many cases, we know that this has not been their history. The key is man's power of accumulative selection: nature gives succesive variations; man adds them up in certain directions useful to him. In this sense he may be said to have made for himself usefull breeds.

However, with respect to natural selection, Darwin indicated, "This preservation [in Nature] of favourable individual differences and variations, and the destruction of those which are injurious. I have called Natural Selection, or Survival of the Fittest. Variations neither useful nor injurious would not be affected by natural selection" (Quoted in Sagan & Druyan, 1992).

Some of Darwin's critics could never overcome the absence of intention in evolution. Indeed, without intention it is perfectly reasonable to argue that evolution need not represent progress at all, although this is the typical connotation associated with the word. Not progress, but a series of undirected accidents. Consider, for example, the following:

---

seconds of the last day of the year. In such a picture, the difference between human appearance and technical development is obviously negligible.

The Darwinian process may be described as a chapter of accidents. As such it seems simple, because you do not at first realize all that it involves. But when its whole significance dawns on you, your heart sinks into a heap of sand within you. There is a hideous fatalism about it, a ghastly and damnable reduction of beauty and intelligence, of strength and purpose, of honor and aspiration, to such casually picturesque changes as an avalanche may make in landscape, or a railway accident in a human figure. To call this Natural Selection is a blasphemy, possible to many for whom Nature is nothing but a casual aggregation of inert and dead matter, but eternally impossible to the spirits and souls of the righteous. . . . If this sort of selection could turn an antelope into a giraffe, it could conceivably turn a pond full of amoebas into the French Academy. (G. B. Shaw, quoted in Sagan & Druyan, 1992)

The original arguments against the Darwinian perspective of natural selection as the intrinsic force of evolution were manyfold. For some, at that time as now, the idea of a descent of human beings was anathema to the notion of original creation. If God truly created "man in his own image," then evolution transgressed this edict. In essence, the data took care of this objection, although it is clear that the argument, independent of the data, rolls on today. Perhaps it always will. In addition to theological disputes, more scientific arguments raised against natural selection invoked the blind and accidental nature of selection. It is the case that an individual of any species might represent the fittest of the group and yet, through mere accident or haphazard demise, fail to preferentially reproduce (Raup, 1991). Hence, survival of the fittest, as noted, rapidly devolves to survival of the survivalists. As Waddington (1957) observed:

Survival does not, of course, mean the bodily endurance of a single individual, outliving Methuselah. It implies, in its present-day interpretation, perpetuation as a source for future generations. That individual "survives" best which leaves most offspring. Again, to speak of an animal as "fittest" does not necessarily imply that it is the strongest or most healthy or would win a beauty competition. Essentially, it denotes nothing more than leaving most offspring. The general principle of natural selection, in fact, merely amounts to the statement that the individuals which leave most offspring are those which leave most offspring. It is a tautology.

It is insufficient to argue that any preferential trait has a strong statistical chance of passage because mutation, almost by definition, is a rare and even singular event (but see Miller & Van Loon, 1982). From this view, natural selection is a process by which life explores its myriad possibilities but with no divine intervention or direction. It is this godless and chance nature of evolution that proved upsetting to many who could otherwise accept the observation of development.

What has always been posed as an alternative is the inheritance of learned traits. There is something intrinsically satisfying in this doctrine to those who believe in accountability. Diligence is passed on by the diligent, profligacy by the profligate, skill by the skillful. We still long to see this in operation, hence repeated sports comments about coaches' sons, who by some inheritance, did not have to put long hours in the gymnasium with their father, but somehow inherited the gene for the zone defense. This conception of the inheritance of acquired charateristics is rightly associated with Jean-Baptiste de Lamarck, a strategy that Darwin considered of importance throughout his own lifetime.[4]

Machine Evolution

Investigation revealed that the landing gear and flap handles were similarly shaped and co-located, and that many pilots were raising the landing gear when trying to raise the flaps after landing (Fitts & Jones, 1961). Since then the flap and gear handles have been separated in the cockpit and are even shaped to emulate their functions: in many airplanes, the gear handle is shaped like a wheel and the flap handle is shaped like a wing. Most other diplays and controls that are common to all airplanes have become standardized through trial and error, based on similar errors made over time. But human factors considerations have not been rationally applied to all the devices in the cockpit. (Riley, 1994)

If humans take millenia to evolve and apparently do so by haphazard circumstance, what of technology? More particularly for our present purpose, what of machines? In this realm, Lamarck comes particularly to the forefront. The essence of his laws are that an animal's characteristics and behavior are shaped by adaptation to its natural environment, that special organs grow and diminish according to their use or disuse, and that the adaptive changes that an animal acquires in its lifetime are inherited by its offspring (see Koestler, 1964). Let us consider machines with respect to these principles. Certainly, a machine's physical charateristics and especially its function seem shaped by its immediate environment, especially if we think in terms of contemporary software functions. Its special organs grow and diminsh according to use.

It has been promulgated that evolution proceeds by survival of the fittest. However, let us look at this statement with respect to contemporary society.

---

[4]I have purposely contrasted two extreme views of human evolution here as stereotyped under Darwinian or Lamarckian banners. Neither view in its unalloyed, pristine condition is held in contemporary times, and the questions of evolution have themselves multiplied in complexity (see Kauffman, 1993). My purpose here is to contrast common views of human evolution with machine evolution. I would note that the equivalence I claim is even more valid when the current views of natural evolution are explained for each realm.

At least in the Western world, we are replete with medical facilities. Many of those who are not the fittest frequently survive disadvantage and disease. Others, who are in straightened economic circumstances and do not have simple access to sophisticated medical facilities may frequently not survive, despite early initial advantages. Of course, the problem, as discussed previously, is fit with respect to what? On the machine side, the generalization is that they progress in uniform steps, taking advantage of each previous discovery. However, when we look at technical breakthroughs in more detail, progress is much more serendipitous than they might appear at first blush. Indeed, many steps in machine evolution depend directly on the creative insights of single designers, where design is as much art as it is science. Although a machine may inherit preferred characteristics, such characteristics might become a liability in suceeding generations. Also, some forms of technology can become extinct in the same fashion that natural selection proceeds for animal and plant species.

On the surface, it might therefore appear that machines evolve at their observed rapid pace because of the immediate propagation of preferred characteristics. However, it is important to ask from where the innovations come. If human mutations come from haphazard events such as cosmic rays, machine innovations come from equally haphazard events of human intuition. These events are comparable in all aspects except for their temporal density. Some design innovations in machines are useful for a time and then die out because of supercession of improved abilities (e.g., PCs). Other innovations fail because of economic forces in which rival technologies are paired against each other and one essentially wins (e.g., beta vs. VHS, eight-tracks vs. tapes and CDs). What I want to propose is that the processes of evolution for both human and machine asymptote to common explorational characteristics. But time makes the difference.

Comparative Evolution

It is worth just a moment to summarize and to make explicit the differences in evolution between humans and machines as elaborated previously. The critical difference is cycle time. The average human life is some decades in length, whereas the average machine life is now in the order of years to months. The machine is replaced as soon as a viable alternative is produced. In contrast, we try to save human beings. Some machines are also saveable, for example, secondhand cars. The point is that the landscape of human beings changes slowly compared with machines. Also, as human lifespan is increasing, machine lifespan is diminishing. If this represents the cycle-time differences of a single cycle, we should also recognize that the difference in the respective rates of those cycle time is growing. That is, evolution or change takes places at an increasingly divergent rate. As well as time scale, the respective histories are different in terms of their time.

Summary

At a surface level it appears that humans and machines evolve in a very different manner. However, there is much in common and only time is the essential distinction. They are so divergent in time scale that we see them as more radically different than they are. In reality, these are not separate forms of evolution but go together as we co-evolve.

CONVERGENT EVOLUTION AND CO-EVOLUTION

It makes no sense to talk about cars and power plants and so on as if they were something apart of our lives — they are our lives. (McKibben, 1989)

Convergent Evolution of the Human–Machine Interface

One characteristic in evolutionary landscape is the convergence evolution of entities subjected to the same forces. Before considering co-evolution I first illustrate a case of convergent evolution in the human–machine interface (see also Hancock, 1996). The computer is now the dominant and prefered system that mediates between human and machine. Frequently, of course, the computer itself is the machine of concern. However, for both large-scale complex systems and small appliances, some form of computational medium is pervasive in our society. This ubiquity has distinct effects on human–machine interaction. The generation of a common communication and control medium foster convergent evolution. In essence, it becomes progressively less clear as to which specific system is being controlled. The important differences between systems lie in their distinct response characteristics. However, the computer as intermediary can hide or buffer many of these differences so that what specifically is controlled becomes less of an issue in managing control. Eventually, might the differences between aircraft traffic control, nuclear power stations, and widespread forms of data control become virtually opaque to the operators sitting at their generic control panels viewing their generic displays? Would this be an advisable strategy? Is there any unfathomable rule demanding that there be complex interfaces for complex systems? Indeed, because all systems have to be navigated through some complex phase space of operation, it may be that the metaphor of a boat on an ocean is one that captures many common elements of all systems operation (see Hancock & Chignell, 1995). The task of the interface designer would be to bring the critical variables to the forefront and allow their emergent properties to become the sea lane through which the controller has to pilot his or her craft. This could be done most easily with the four-dimensional wraparound world of virtual reality. Such convergence of evolution is a strong reason for adopting virtual interfaces that take advantage of intrinsic human visuomotor capabilities.

Mutual Co-Evolution

If one basis for a unifying theme is the consideration of technology as facet of the natural environment, we should take the step of recognizing technology as a nascent co-species. Indeed, it is not a species the like of which we have seen before. We have to abandon the perspective that technology is merely an appendage. We have to acknowledge our complete dependence on technology and recognize that we could not be who we are without technological support. We have to free ourselves from the conception that technology is merely just one of the many shaping facets of the environment.

We have denied souls to animals, and still cling hopefully to this supposed difference. Not to do so would be to deny not only our special place in creation, but our very separateness. This separateness is daily and continually sustained in each of us by our existence as a unified individual conscious entity. We are having progressively more trouble sustaining this world view. In an age of virtual reality, of e-mail, of faxes, of teaming, of telecommuting, of collaboration, what is it to be separate anymore? As physical dependence grows, so does a sense of cognitive dependence. Perhaps this is why we collectively clamor for the different, the unique, and the outre in experience, because it supports our vestigial and comfortable view of ourselves as strong independent entities (the myth of the hero dies hard: *Die Hard* or *The Last Action Hero* being good names for movies). We can no longer claim to be simple differentiated entities. Nutritionists have always known that you are what you eat, and organ replacement at least suggests that spare-part grafting is feasible and useful. The cells of the body are all replaced cyclically, and hence what is left as different is an informational pattern with a memory component. But technology is only promising to further the dissolution of that pattern in time and in space. Moravec (1988) desperately grasped at the idea of individual downloading as a preservation of the self, but at best it replicates our present unification.

What I suggest here is that the division is slowly dissolving. The divisions are not simply between ourselves and technology, but between ourselves as individuals. The two antagonistic tendencies of self-assertion and integration have always been locked in some form of battle (see Koestler, 1978). However, technology has joined that fight, not merely as a weapon but as an active combatant. That technology has insinuated itself into society so completely and so unobtrusively to our consciousness attests to its success. I was tempted to quote from Orwell's last paragraph of *Animal Farm* (where the farm animals look from man to pig, and from pig back to man, and see no difference), and to replace pig with machine. However, this still retains the divisive view and fosters the unfortunate arguments about machine replication of human abilities. Co-evolution is much more than this simple recreation of ourselves in a technical surrogate. It is more dynamic, more elusive, and intellectually much more interesting. That co-evolutionary results may initially be crude does not militate against progressively more sophisticated symbiotic interaction.

Summary

The disparity in human versus machine evolution is one, almost exclusively, of cycle time. Convergent evolution is seen in conditions where comparable constraints dictate common process. Co-evolution, the mutual shaping of constituent members of an ecosystem, is dynamic and explorative. The spice of intention adds novelty to human–machine symbiosis. The possibility of emergent machine intention promises a new form of overall mutual evolution that in its embryonic stages, is likely to be seen with respect to human-centered automation.

HUMAN–MACHINE SYMBIOSIS

Either the machine has a meaning to life that we have not yet been able to interpret in a rational manner, or it is itself a manifestation of life and therefore mysterious. (Garrett, 1925)

Mutual Dependence

Could contemporary human beings survive without technology? If we take technology to mean more than machines and to include animal and plant domestication, the answer is no. In these terms, we have shaped the environment that now shapes the individuals and society we are. We have come so far down that evolutionary path that there is no going back. Could contemporary human beings survive without computational machines? Perhaps. But such a society would be radically different from that which we experience in the developed and developing nations of the world. Pragmatically, we will not give up our technology, and practically, its influence spreads daily. We must therefore conclude that there is a human–machine symbiosis. Machines cannot exist without humans, but by the same token, humans (of our kind) do not exist without machines. It may be more comfortable to look on this symbiosis as an elaboration of what it means to belong to the species *Homo Sapiens Sapiens*. For example, no longer are we restricted to storing information and energy endogenously. The immediate and portable availability of each frees individuals in time and space to an extent that is incomparable with the rest of the animal world. Symbiosis, however, is only one stage in the developmental sequence linking humans and machines. Symbiosis results when one entity progresses from total dependence to mutual dependence. The stage after mutual dependence is some degree of nascent independence. For the machine, that is automation of function.

Automation: The Birth of Machine Intention

The critical issue with respect to the topic of the present text is automation. There are, of course, multiple levels of automation. These levels have been discussed extensively by others. But however autonomous a system might appear, there are humans involved. The involvment might be at the conception or design stage or may be at the care and maintenance end, but at some juncture human operators enter the picture. I take this human participation even further. That is, even if we can think of a system that has no maintenance, a system that was designed by a software program so divorced from its programmer that it no longer appears to be based on human action, even then the system will subsume human goals. That is, the motive force for its origin and its function remains essentially human. That is, machines work for human goals, not their own.

Does this always have to be the case? Can we conceive of systems that are not predicated on human intentionality? Indeed we can, because they surround us in nature. Much of the insect world knows nothing of human intentionality and continues its activity relatively unmolested or polluted by our goals. Of course we have to all exist on the same global platform, but insects are rarely of technology in the same way that domesticated animals are (always remembering their adaptation to human products such as pesticides, etc.). Can we then conceive of machines as having intention, at least at a level that, say, an ant has intention? We can certainly conceive of it, but with the result of proliferation of interactions in systems we are now begining to witness it in some of the machines we have created.

I do not advance this as a strong AI position that would protest that machines can think. I appreciate Searle's (1984) argument concerning syntactic versus semantic content, but would suggest that it misses the mark. Each such argument is based on the question of whether machines can think like humans think. I find this a very constrictive approach. Frankly, I would prefer that machines not think like humans think, because we already appear to have thoughtful humans (an empirical statement open to much dispute). Rather, I would hope that if mind is an emergent property of brain function, then machine intention could be an emergent property of machine function. As I have argued with respect to the supercritical society (Hancock & Chignell, 1995), it is indeed an exercise in imagination to understand what characteristics such emergent properties possess. It is to be desparately hoped that they do not merely mimic what already exists.

Design: The Balance of Constraint and Opportunity

May we not fancy that if, in the remotest geological period, some early form of vegetable life has been endowed with the power of reflecting upon the dawning life of animals which was coming into existence alongside of its own, it would have thought itself exceedingly acute if it had surmised that animals would one day become real vegtables? Yet would this be more mistaken than it would be on our part to imagine that because the life of machines is a very different one to our own, there is therefore no higher possible development of life than ours; or that because mechanical life is a very different thing from ours, therefore that it is not life at all?

But I have heard it said, "granted that this is so, and that the vapour-engine has a strength of his own, surely no one will say that it has a will of its own?" Alas! if we look more closely, we shall find that this does not make against the supposition that the vapour-engine is one of the germs of a new phase of life. What is there in this whole world, or in the worlds beyond it, which has a will of its own? The Unknown and Unknowable only! (Butler, 1872)

One of the central questions we face in specifying goals at a societal level is plurality. Individuality is a central pillar of the democratic system. Although pure individuality, like the comparable notion of freedom, is a myth, in the United States at least there is a strong mandate to protect the rights of the individual. This position is set in stark contrast with other societies (e.g., Mao's China), in which the needs of the individual are sublimated with respect to the society. As a consequence, societal aims are frequently expressed as generalities that provide constraint and threat to no one, such as "life, liberty, and the pursuit of happiness." The question to be addressed here is twofold. First, can we continue with such vague societal goals in the face of technology that begs specification? Second, do we want to accept societal goals if they supress our valued individuality?

I think the answers to these questions lie in seeing how technology has addressed human needs, as expressed in the hierarchy of Fig. 22.3 (see also Maslow, 1964). Clearly, technology serves to free society and individuals from the want of the basic physiological needs. (It is, of course, more than

FIG. 22.3.   The classic hierarchy of human needs as proposed by Maslow (1954). Note that this is a descriptive structure, where the individually specified levels bear only a nominal relationship to each other. Putatively, there is an order dependence such that one level is founded on another. The implication is that, for example, friendship is predicated on law. Clearly, these intrinsic relationships need not be dependent. It is clear that technology acts to support the base of the pyramid. That is, technology underlies the availability of food, water, and shelter in contemporary society. Unfortunately, the contemporary function of technology attempts to expand the certainty of basic needs and does not foster vertical transition.

ironic that we have many individuals with more than sufficient monetary resources for several lifetimes who are unable to attain other levels in the noted hierarchy.) However, technology does not protest explicit goals with respect to other levels of the hierarchy. I submit that this is a reflection of the separation of the ideal from the actual, or, more prosaically, the why from the how. My central protestation is that technology, cognizant of the fact or not, shapes the why as much as the how. And in this it has failed. I argue here that the solution lies in the use of technology for education. Indeed, I conclude the present work with such an assertion. However, I am acutely aware that this is inherently a political decision. I do not have room to discourse here on the role of technology in the political process, although it is a matter to which considerable thought need be directed. Rather, I acknowledge political power and assert that any innovations unsupported by such power are liable to have a restricted impact. The quintessential bottom line is that technology must be used to enfranchise, not to enslave.

OUR MUTUAL FUTURE?

Helena: What did you do to him (Radius the Robot)?
Dr: Gall: H'm nothing. There's reaction of the pupils, increase in sensitiveness, and so on. Oh, it wasn't an attack peculiar to the robots.
Helena: What was it then?
Dr. Gall: Heaven alone knows. Stubbornness, fury or revolt—I don't know. And his heart, too.
Helena: How do you mean?
Dr. Gall: It was beating with nervousness like a human heart. Do you know what? I don't believe the rascal is a robot at all now.
Helena: Doctor, has Radius got a soul?
Dr. Gall: I don't know. He's got something nasty. (Capek, 1923)

The Love–Hate Relationship

We have a literal love–hate relationship with the machines we create. On the one hand, we extol their virtues as labor-saving devices, the mechanisms for freeing humans from the bondage of manual work. On the other hand, we decry the loss of employment and human dignity that comes with progressive automation. Our fiction is redolent with visions of technical systems that turn against their human masters. From Samuel Butler's *Erewhon* through Capek's *Rossum's Universal Robots,* to James Cameron's *The Terminator* and *Terminator 2,* we have a collective pathological fear of the machine as master. Yet, at the same time, nothing alters the headlong rush toward ever-more automation as the solution to a growing spectrum of societal problems. One suspects that one's position with respect to technical innovation is very much influenced by the threat that technology poses to yourself and the potential individual gain that is to be had from such development. Scientists can always study it, professors can always pontificate about it, and business people can always seek a profit factor. Hence, I would suspect most who would be inclined to read this would certainly view technology as a valuable enterprise. Should we pose an automatic research machine that can also teach and do away with capitalistic profit there might be some divergence. But even with such a machine, I still suspect that scientists can always study it, certainly professors will always be able to discourse about it, and I would be most surprised if someone could not find a profit in it.

Beyond Anthropomorphism

At present, society shares a comfortable delusion. The delusion is anthropomorphic in nature and relies on biological assumptions. The assumptions are that because machines are not self-replicating in the way that biological systems are self-replicating, then the proliferation of technical systems is controllable. The anthropomorphic delusion is that because machines do not possess intelligence and therefore intention in the same way that humans possess intelligence and intention, then there is an absence of machine intention. Despite our best and desperate efforts at machine intelligence as a surrogate of human intelligence, we still do not have any substantive evidence of machine intelligence, largely because such evidence is constrained to take the form that mimics human intelligence. This is the central fallacy of the Turing test, which requires that evidence of any intelligence is had only through reference to human abilities. This naive, anthropomorphic, and impoverished perspective (but see Akins, 1983; Nagel, 1979) is that which provides a cloak to foster our present warm feeling. To some it is a much colder feeling. In his radical and interesting work, Illich (1973) deplored the contemporary direction of technology for precisely this reason when he noted:

> The re-establishment of an ecological balance depends on the ability of society to counteract the progressive materialization of values. Otherwise man will find himself totally enclosed within his artificial creation, with no exit. Enveloped in a physical, social, and psychological milieu of his own making, he will be a prisoner in a shell of technology, unable to find again the ancient milieu to which he was adapted for hundreds of thousands of years. The ecological balance cannot be re-established unless we recognize again that only persons have ends and that only persons can work toward them. Machines only operate ruthlessly to reduce people to the role of impotent allies in their destructive progress.

In disputing the later contention, it is still important to recognize the validity of the initial premise concerning the reification of material values. Furthermore, given the present treatise on teleology and intention beyond the individual human, his latter points must also be given careful consideration as a condition in which diversification of intention fails.

The societal ambivalence toward machines is, then, never far below the surface. From the Luddites through Capek's robots to Terminator, the theme of machine as enemy runs deep. It is noteworthy that Asimov tried artistically to concoct Laws of Robotics, each of which were sequentially broken within the first few years of robotics research. Are we destined for the dark ruins of 21st-century cities firing hopeful lasers at Skynet's malevolent henchmachines?[5] In its own way, this view is as sterile as the mindless path we progress along in reality. Moravec (1988) postulated that silicon-based intelligence will soon tire of our terrestrial confines and will seek its fulfillment in the celestial spheres, unfettered by the constraints of biological needs. With the exception of the latter, these views are simple linear extrapolations of polarized facets of contemporary development. What is clear is that the uncertainty of future developments depends precisely on nonlinear effects, and prediction under such conditions is problematic to say the least.

Back to Arcadia?

Have I, by these remarks, allied myself with recidivists who seek a return to a happy but mythical golden-age. Am I a committed arcadian, utopian, or even autopian (Hancock & Parasuraman, 1992; More, 1516; Sidney, 1593)? I think not, at this stage. Rather, I seek to put the development of automation and the role of the human, in human-centered automation, in some perspective. Our decisions at the present, nascent stage of machine independence will constrain our possible decisions at subsequent and potentially more volatile stages of human–machine interaction. I do not advocate a giant OFF button, because we have already built systems and the interconnections of those systems that refute the possibility of such a simple strategy. Rather, I advocate that we keep to the forefront of systems design the thought of Francis Bacon, who designed science for the uses of life. In the same way, we must design technology for the uses of human life, because without such a teleology for technology, we are lost indeed.

TWO CULTURES: A TECHNICAL SOCIETY DIVIDED BY TECHNOLOGY

Tools are intrinsic to social relationships. An individual relates himself in action to his society through the use of tools that he actively masters, or by which he is passively acted upon. To the degree that he masters his tools, he can invest the world with his meaning; to the degree that he is mastered by his

---

[5]I should note that this is largely a Western, developed-world preoccupation. Technology is not universally seen in this manner by any stretch of imagination. In addition, differing cultures have widely divergent goals and world views that do not accord with the main themes of our society. Such pluralism must be considered in the aims of global technology (see Moray, 1994).

tools, the shape of the tool determines his own self-image. Convivial tools are those which give each person who uses them the greatest opportunity to enrich the environment with the fruits of his or her vision. Industrial tools deny this possibility to those who use them and they allow their designers to determine the meaning and expectations of others. Most tools today cannot be used in a convivial fashion. (Illich, 1973)

Bifurcations of Society

I have discussed the proposition that our ecology is technology and have advocated a manifest purpose for that technology. I have given considerable space to reflections on the impact of technology in particular as a positive force. However, it is important to give considerations to the downside of technology. I bare this argument on the original observation of C.P. Snow (1964) concerning the two cultures of society. Briefly, Snow observed, "I christened (it) to myself as the "two cultures." For constantly I felt I was moving among two groups — comparable in intelligence, identical in race, not grossly different in social origin, earning about the same incomes, who had almost ceased to communicate at all, who in intellectual, moral and psychological climate had so little in common."

However, the bifurcation I contemplate is one that is much more radical. It does not concern the divergent world views of the educational aristocracy. Rather, it represents the very division of society itself between those empowered by technology and those subjugated by that same technology.

Bifurcations in society are based on the differential control of resources. The obvious manifestation of those resources are physical wealth expressed as goods, lands, or currency. However, control need not be ownership but may be more indirectly expressed. The classic example is the control exercised by medieval clergy who mediated between humans and God through their exclusive access to forms of knowledge. Our development of technology promises to institute a similar form of division. Some will understand the arcane esoterica of technology and will consequently mediate its blessings. Most, however, will not, and will suffer under its benevolent or malevolent oppression. For those who dismiss this as pure fancy, I only ask them to recall the early days of computing when hackers wrote for hackers, or even worse, themselves alone (see also Vallee, 1982). Other mere users, albeit intelligent and well-educated scientists, struggled mightly to understand indecipherable error codes founded themselves on alphanumeric ciphers.

Access Denied

It is comforting to believe that human factors had much to do with improving computer interaction and making communication accessible.

However, the truth lies nearer to the financial drive to sell personal computers in a wider market that mandated more open interaction. Human factors acted in its remedial rather than proactive role. It might now be asumed that computer interaction is open to all. But I require the reader to consider the disadvantages of the illiterate in our society: the older members who have not grown up with technology and who are not facile with its manipulation, the physically and mentally handicapped, the educationally subnormal, and the poor for whom education itself is not seen as a priority. This is not even to mention the majority of the world's population to whom also access is denied. These individuals are not even in the information superhighway slow lane—they have yet to come within a country mile of any entrance ramp and they have, as yet, no vehicle or driver's license.

If I have talked here in general evolutionary terms, then it is these individuals who are threatened with adaptive failure and literal extinction, because they will have failed to adapt to the current ecology, that being technology. Thoughts of this extinction raise critical moral dilemmas in themselves. I further suggest that the bifurcation between technically literate and technically illiterate rapidly grows into rich and poor, priviledged and oppressed. This distinction may divide society in a more profound manner than any of our presently perceived divisions. The only solution to this question is education—full, free education as a right of each member of society and an education that denies any arcane brotherhood disproportionate power by their secretion of knowledge (McConica, 1991).

THE ACTUAL AND THE IDEAL

Dichotomized science claims that it deals only with the actual and the existent and that it has nothing to do with the ideal, that is to say, with the ends, the goals, the purposes of life, i.e., with end-values. (Maslow, 1964)

This work started with a statement of a basic paradox of technology. On the one hand, the explicit aim of technology is to improve the lot of humankind. In actuality, the unintended by-product of contemporary technology seems to be the threat of global destruction, either through acute or chronic influences. Arcadians advocate dismantling technology and living in a world of sylvan beauty, which features peace, harmony, and tranquility. Their hopes are laudable, but their aspirations naive. Peace, harmony, and tranquility are not the hallmarks of a world without technology. Such a world is one of hard physical labor and the ever-present threat of famine at the behest of an uncertain environment or war at the behest of equally unpredicatable neighbors. Humankind has fought for many centuries to rid itself of these uncertainties and the dogma that attends them.

Our ecology is technology. If we are to achieve our individual and collective societal goals it will be through technology. I have tried to argue here that we muct actively guide technical innovation and be wary of the possible alternative facets of machine intention that in no way resembles our own. We must state explicit purpose for the machines we create and these purposes must be expressed at the level of a global society, not that of an individual or even single nation. When expressed at this higher level, we can expose the antagonistic nature of some of the systems we create to an overall good. I further argue that we cannot view ourselves societally, or even individually, as separated from technology. The birth of machine intention poses questions whose answers will become the central force that shape our future. How we interact with machines, the degree of autonomy we permit them, the rate of comparative evolution, and the approach to mutual co-evolution form the manifesto of our potential future. That the ground rules are being set in our time makes the present work on human-centered automation all the more important. I point to our past failure in setting such constraints and ask, "How do we propose to do better in the future?"

Summary

We have long clamored to be involved early in the design process. That is in how an object is designed. However, it is now time to step forward and become involved in the process even earlier; that is, in determining why an object, artifact, or machine is designed in the first place. Human factors is a bastard discipline that sits astride so many divisions. We link art (or design) with science. We deal with the social (human) and the technical (machine). Therefore, it is imperative that in our unique position we must encompass both the Actual (what is) and the Ideal (what should be).

> After all then it comes to this, that the difference between the life of a man and that of a machine is one rather of degree than of kind, though differences in kind are not wanting. An animal has more provision for emergency than a machine. The machine is less versatile; its range of action is narrow; its strength and accuracy in its own sphere are superhuman, but it shows badly in a dilemma; sometimes when its normal action is disturbed, it will lose its head, and go from bad to worse like a lunatic in a raging frenzy; but here, again, we are met by the same consideration as before, namely, that machines are still in their infancy; they are mere skeletons without muscle and flesh. (Samuel Butler, Erewhon, 1872)

CONCLUSION

Superna quaerite: Enquire after higher things.

Early in this work, I suggested that the turn of the millenia was an appropriate juncture for human society to turn from its adolescence to a mature adulthood. This is a comfortable homily in that it sounds most impressive, but in actuality signifies almost nothing. I want to elaborate this statement so that it means something substantive and, I hope, significant.

I claim that for the childhood and adolescence of humankind we have acted as passive victims of an omnipotent environment. Our various names for events that happen to us — *Act of God, fate, Kismet, accident, happenstance, luck* — all connote a conception that life simply happens to us, directed by forces outside our control. I do not claim all natural forces are within human control. I do claim that the passive and victim-laiden attitude is within human control. For much of our existence we have had to label such forces as benevolent or malevolent deities, or a single deity that arbitrates events. Although not wishing to trespass on personal beliefs, I do reject the idea of an individual deity who follows us around to continually control, test, and evaluate. In the absence of a personalized deity, our society still desparately seeks an entity to blame for events that happen. Earlier, I mentioned the "as if" pragmatic approach and postulated that much of society adopts this positive pragmatism. I propose that this form of pragmatism be adopted as a basis for the teleolgy of technology. That is, although we may continue to argue over the existence and role of an omnipotent deity, we assume that mantle on a local scale and become *responsible* for our collective future. This will curtail litigation, because we are indemnifying ourselves.

I started this chapter with, and have made it a theme, the idea that our ecology is technology. I end it in a similar manner by affirming that technology is also fast becoming our contemporary theology. I propose the term *techneology* to cover the concept of intention in technology and its theological referent.

I can affirm that our ecology is technology and that technology is gradually becoming our theology. Although pragmatically useful yet philosophically doubtful, I can only countenance the alternative as voiced by Shakespeare's Macbeth:

> Tomorrow, and tomorrow, and tomorrow,
> Creeps in this petty pace from day to day,
> To the last syllable of recorded time;
> And all our yesterdays have lighted fools
> The way to dusty death. Out, out, brief candle!
> Life's but a walking shadow, a poor player
> That struts and frets his hour upon the stage
> And then is heard no more: it is a tale
> Told by an idiot, full of sound and fury,
> Signifying nothing. (*MacBeth*, Act V, Scene 3, lines 19–28)

ACKNOWLEDGMENTS

I am most grateful for the insightful comments of Raja Parasuraman and Kelly Harwood, who were kind enough to read and comment on an earlier draft of the present work. Their time and effort are much appreciated. I must also express a debt of gratitude to Ron Westrum for his guidance. In directing me to sources that frequently illustrate my naivete, he often humbles but never fails to interest or educate.

REFERENCES

Akins, K. (1983). What is it like to be boring and myopic? In B. Dahlbom (Ed.), *Dennett and his critics* (pp. 124–160). Oxford, England: Blackwell.

Bennett, C. H., & Landauer, R. (1985). The fundamental physical limits of computation. *Scientific American, 253*, 48–56.

Butler, S. (1872). *Erewhon or over the range*. London: Trubner.

Capek, K. (1923). *RUR. [Rossum's Universal Robots]* (Original English Publication). Oxford, England: Oxford University Press.

Card, S. K. (1989). Human factors and artificial intelligence. In P. A. Hancock & M. H. Chignell (Eds.), *Intelligent interfaces: Theory, research, and design* (pp. 27–48). North Holland: Amsterdam.

Eiseley, L. (1960). *The firmament of time*. New York: MacMillan.

Fitts, P., & Jones, R. (1961). Analysis of factors contributing to 460 "pilot error" experiences in operating aircraft controls. In H. Sinaiko (Ed.), *Selected papers on human factors in the design and use of control systems* (00–00). New York: Dover Publications.

Flach, J., Hancock, P. A., Caird, J., & Vicente, K. (Eds.). (1995). *Global perspectives on the ecology of human-machine systems*. Hillsdale, NJ: Lawrence Erlbaum Associates.

Garrett, G. (1925). *Ouroborous or the mechanical extension of mankind*. New York: Dutton.

Hancock, P. A. (1992). On the future of hybrid human-machine systems. In J. A. Wise, V. D. Hopkin, & P. Stager (Eds.), *Verification and validation of complex systems: Human factors* (NATO ASI Series F: Computer and Systems Sciences, pp. 61–85). Berlin: Springer-Verlag.

Hancock, P. A. (1993). What good do we really do? *Ergonomics in Design, 1*, 6–8.

Hancock, P.A. (1996).On convergent technological evolution. *Ergonomics in Design, 4(1)*, 22–29.

Hancock, P. A., & Chignell, M. H. (1995). On human factors. In J. Flach, P. A. Hancock, J. Caird, & K. Vicente (Eds.), *Global approaches to the ecology of human–machine systems* (pp. 14–53). Hillsdale, NJ: Lawrence Erlbaum Associates.

Hancock, P. A., Chignell, M. H., & Kerr, G. (1988). Defining task complexity and task difficulty. *Proceedings of the XXIV International Congress of Psychology*. Sydney, Australia:

Hancock, P. A., Flach, J., Caird, J., & Vicente, K. (Eds.). (1995).*Local applications of the ecology of human-machine systems* Hillsdale, NJ: Lawrence Erlbaum Associates.

Hancock, P. A., & Parasuraman, R. (1992). Human factors and safety issues in intelligent vehicle highway systems (IVHS). *Journal of Safety Research, 23*, 181–198.

Illich, I. (1973). *Tools for conviviality*. New York: Harper and Row.

Kant, I. (1787). *The critique of pure reason*. Riga, Latvia: Hartnocht.

Kauffman, S. A. (1993). *The origins of order: Self-organization and selection in evolution*. Oxford, England: Oxford University Press.

Koestler, A. (1959). *The sleepwalkers: A history of man's changing vision of the universe*. New York: Hutchinson.

Koestler, A. (1964). *The act of creation*. New York: Hutchinson.

Koestler, A. (1972). *The roots of coincidence*. New York: Vintage.

Koestler, A. (1978). *Janus: A summing up*. New York: Vintage.

Kurtz, P. (1992). *The new skepticism: Inquiry and reliable knowledge*. Buffalo, NY: Prometheus.

Kurtz, P. (1994). The new skepticism. *Skeptical Inquirer, 18*, 134–141.

Malthus, T. R. (1798). *An essay on the principle of population, as it affects the future improvement of society with remarks on the speculations of Mr. Godwin, M. Condorcet, and other writers*. London: J. Johnson.

Maslow, A. H. (1964). *Religions, values, and peak experiences*. New York: Penguin.

McConica, J. (1991). *Erasmus*. Oxford, England: Oxford University Press.

McKibben, W. (1989). *The end of nature*. New York: Random House.

Miller, J., & Van Loon, B. (1982). *Darwin for beginners*. Bath: Icon Press.

Moravec, H. (1988). *Mind children: The future of robot and human intelligence*. Cambridge, MA.: Harvard University Press.

Moray, N. (1993). Technosophy and humane factors. In *Ergonomics in Design*. Santa Monica, CA: Human Factors Society.

Moray, N. (1994, August). *Ergonomics and the global problems of the 21st century*. Keynote address given at the International Ergonomics Meeting, Toronto, Canada.

More, T. (1516/1965). *Utopia*. New York: *Washington Square Press*.

Nagel, T. (1979). What is it like to be a bat? In *Mortal questions*. Cambridge, England: Cambridge University Press.

Nickerson, R. S. (1992). *Looking ahead: Human factors challenges in a changing world*. Hillsdale, NJ: Lawrence Erlbaum Associates.

Nietzsche, F. (1887). *Zur genealogie der moral: Eine streitschrift*. Leipzig: C. G. Naumann.

Parasuraman, R. (1986). Vigilance, monitoring, and search. In K. R. Boff, L. Kaufman, & J. P. Thomas (Eds.), *Handbook of perception and human performance* (pp. 43.1–43.39). New York: Wiley.

Raup, D. M. (1986). Biological extinction in earth history. *Science, 231*, 1528.

Raup, D. M. (1991). *Extinctions: Bad genes or bad luck*. New York: W. W. Norton, Inc.

Riley, V. (1994). *Human use of automation*. Unpublished doctoral dissertation, University of Minnesota, Minneapolis.

Russell, P. (1992). *The white hole in time*. San Francisco, CA: Harper.

Sagan, C., & Druyan, A. (1992). *Shadows of forgotten ancestors*. New York: Balantine.

Schneider, W., & Shiffrin, R. M. (1977). Controlled and automatic human information processing I: Detection, search, and visual attention. *Psychological Review, 84*, 1–66.

Searle, J. (1984). *Minds, brains, and science*. Cambridge, MA.: Harvard University Press.

Sedgwick, J. (1993, March). The complexity problem. *The Atlantic Monthly*, pp. 96–104.

Sidney, P. (1593). *The Countess of Pembroke's Arcadia*. London: Ponfonbie.

Snow, C. P. (1964). *The two cultures and a second look*. Cambridge, England: Cambridge University Press.

Tolkkinen, K. (1994, January 27). Machines are making us soft. *Minnesota Daily*.

Vallee, J. (1982). *The network revolution*. Berkeley, CA: And/Or Press.

Waddington, C. H. (1957). *The strategy of the gene*. London: Allen & Urwin.

Westrum, R. (1991). *Technologies and society: The shaping of people and things*. Belmont, CA: Wadsworth.

Author Index

Subject Index

Mode error 9-11, 273-274, 379, 454
Monitoring
  and automation, 101-107, 123-124, 272, 297-
    298
    display separation, 105-107
    manual task load, 101-102
    reliability and consistency, 102-105
    situation awareness, 166-167
  inefficiency, countermeasures against, 107-
    111
    adaptive task allocation, 108-111 *see*
      also Allocation of function
    integrated displays, 107-108
  machine, 95-96
  operational, 93-95

O

Oil and gas pipeline systems, 427-444
Operator reliance on automation, 19-22, 102-
  103, 107, 124-125, *see also*
  Trust@SUBTOPIC 1 =
factors of influence, 22-32

P

Perceptual augmentation, 369-381
Performance measures, *see also* Evaluation of
  automated systems
  air traffic systems, 330-331
  interface evaluation, 79-80
Pilot-automation interaction,coordination, 269-
  271, 274-277
  glass-cockpit training, 269-271
Psychophysiology,
  measures
    advantages and disadvantages, 139-142

Q

Quality control and maintenance, 412-413

R

Representative design, 72-76
Rule-based behavior, 70-71, 284-286

S

Safety
  of automation, 453-454
  in petrochemical and nuclear power plants,
    428-431
Situation awareness, 75, 165-166, 226-228, 325
  impact of automation, 166-172, *see also*
    Monitoring and automation

  active versus passive, 167-168
  benefits, 171-172
  feedback, 168-169
  lack of understanding, 170-171
Skill-based behavior, 70-71, 284-286
Skill degradation, 125
Social effects, 457-459
Stimulus-response compatibility, 127-129
Supervisory control, 92, 183, 221-240, 450-452
  versus active control, 183-184
System observability, 207-208, 329-330, 332

T

Team performance, 50-53, 258-259, 331-332
  communication, 212-213
  individual characteristics, 247-249
  organizational and environmental character-
    istics, 246-247
  task characteristics, 250-253
  team characteristics, 249-250
  team effectiveness model, 244-246
  team processes, 255-257
  work characteristics, 253-255
Technology, 466-467, 471
Teleology, 465-466
Training, 55-56, 246-247
  in aircraft maintenance, 419-420
  in air traffic control, 328-329
  and automation, 124
  in driving, 357-358
  glass cockpit, 269-271
  in maritime, 374-375, 380
  in pipeline control, 436
Trust, *see also* Operator reliance
  20-21, 27-28, 33, 39-40, 95, 124-125, 167,
    247-248, 329-330, 433, 458-459

U

User acceptance, 343-344

V

Vigilance, 97-98, 158, 166-167, 207-208*see*
  also Workload

W

Warning *see also* Alarms
  and caution sytems, *see* Fault management
  collision and avoidance, 344-356
    detection algorithms, 345-347
    driver-in-the-loop, 351-352
    false alarms, 347-348
    information presentation, 351